Proof Theory for Fuzzy Logics

APPLIED LOGIC SERIES

VOLUME 36

SCOPE OF THE SERIES

Logic is applied in an increasingly wide variety of disciplines, from the traditional subjects of philosophy and mathematics to the more recent disciplines of cognitive science, computer science, artificial intelligence, and linguistics, leading to new vigor in this ancient subject. Springer, through its Applied Logic Series, seeks to provide a home for outstanding books and research monographs in applied logic, and in doing so demonstrates the underlying unity and applicability of logic.

For further volumes: http://www.springer.com/series/5632

Proof Theory for Fuzzy Logics

by

George Metcalfe
Vanderbilt University, Nasville, TN, USA

Nicola Olivetti
Paul Cézanne University, Marseille, France

Dov Gabbay
King's College, London, UK

Springer

Dr. George Metcalfe
Vanderbilt University
Department of Mathematics
1326 Stevenson Center
Nashville TN 37240
USA
george.metcalfe@vanderbilt.edu

Dr. Nicola Olivetti
Université Aix-Marseille III
LSIS-UMR CNRS 6168
Campus de Saint Jerôme
Avenue Escadrille Normandie-Niemen
13397 Marseille CX 20
France
nicola.olivetti@univ-cezanne.fr

Prof. Dr. Dov M. Gabbay
King's College London
Department of Computer Science
Strand
London
United Kingdom WC2R 2LS
dov.gabbay@kcl.ac.uk

ISBN: 978-1-4020-9408-8 e-ISBN: 978-1-4020-9409-5

Library of Congress Control Number: 2008938185

Printed on acid-free paper

9 8 7 6 5 4 3 2 1

springer.com

Acknowledgements

A book like this is not written without a great deal of help and encouragement. In particular, we are grateful to our friends and colleagues, Agata Ciabattoni, Petr Cintula, Christian Fermüller, Rostislav Horčík, and Francesco Paoli, for taking the time to read and comment on earlier drafts of the manuscript. Their attention to detail and insightful suggestions are very much appreciated. For thought-provoking discussions and support at various stages in the process, we would also like to thank distinguished researchers in the Mathematical Fuzzy Logic community: Arnon Avron, Francesc Esteva, Lluís Godo, Petr Hájek, Franco Montagna, Daniele Mundici, Hiroakira Ono, and Constantine Tsinakis.

The first author would also like to offer some personal thanks. First, for two years of stimulating conversations and camaraderie, to members of the logic group at the Vienna University of Technology: Matthias Baaz, Rosalie Iemhoff, Norbert Preining, and Sebastiaan Terwijn. And finally, to Nia Stephens, for her much-needed love and understanding as this project neared completion.

Contents

Chapter 1
Introduction

Logics come in many guises: some semantic, some syntactic. In Classical Logic, truth (semantics) may be characterized using lines of 1s in truth tables, validity in Boolean algebras, even winning strategies in dialogue games. At the same time, proofs (syntax) can be defined in Hilbert systems with axioms and a few simple rules, Gentzen systems with simple axioms but many rules, and a wide range of other calculi. These various guises are useful. They exhibit different properties and reveal hitherto unsuspected connections between logics. They assist both deep theoretical investigations and the development of applications. Often there is much more to a logic than a first glance (semantic or syntactic) might suggest.

The aim of this book is to show that certain logics with semantic roots in the real numbers also have natural and useful syntactic characterizations. Typically, these logics arise as the basis for systems dealing with *vagueness*, formalizing reasoning about natural language expression such as "John is tall" or "the water is hot and dirty" in the wider field of Fuzzy Logic.[1] Indeed, they are usually known as *fuzzy logics*. However, logics based on the real numbers turn up in several areas in Logic, Mathematics, and Computer Science.

Example 1.1 (t-norm based logics). One way to build fuzziness into a logic is to make "design choices" at the outset. Take the real unit interval $[0,1]$ as a set of truth values, 0 being falsity and 1 truth, and interpret connectives like "and" and "or" by suitable functions on $[0,1]$. In particular, one strategy popularized by Hájek in [105] is to interpret "and" by a *continuous t-norm*: a binary function that is commutative, associative, increasing, and has 1 as a unit element. Each continuous t-norm then gives rise to a fuzzy logic, e.g.

- *Gödel Logic* G is based on the t-norm $x *_{\mathsf{G}} y = \min(x,y)$, the only t-norm to assign the same truth value to both "A and A" and "A". This key fuzzy logic, also an important "intermediate logic" between Intuitionistic Logic and Classical

[1] Fuzzy Logic, encompassing such diverse topics as fuzzy set theory, fuzzy control, and fuzzy approximation, was introduced by Zadeh in 1965 [223]. Here we investigate only a small subset of what Zadeh has called "Fuzzy Logic in the narrow sense", namely, logical systems for reasoning about vagueness. For more details of the field, we refer to the comprehensive handbook series [72].

Logic, was introduced by Dummett in 1959 [73] as the infinite-valued version of finite-valued logics defined by Gödel in the 1930s [98].

- *Łukasiewicz Logic* Ł, based on the t-norm $x *_{Ł} y = \max(0, x+y-1)$, is the infinite-valued member of a family of many-valued logics introduced in the 1920s by Łukasiewicz [133, 135]. So-called "many-valued" MV-algebras for Ł, defined by Chang in the 1950s [45], are an active field of research in their own right (see e.g. the authoritative monograph [58]). The logic also has a rich and much-studied geometric interpretation based on the elegant 1951 representation theorem of McNaughton [139].
- *Product Logic* P, the third member of this trio of "fundamental t-norm logics", is based on the t-norm $x *_{P} y = x \cdot y$. Introduced by Hájek et al. in 1996 [112], P is a more recent addition to the fuzzy canon, although the implication for this logic appeared already in a 1969 paper of Goguen [100].

Fuzzy logics can also be defined based on classes of t-norms. Most importantly, Hájek's *Basic Logic* BL [105] and Godo and Esteva's *Monoidal t-norm Logic* MTL [77] are the logics of, respectively, all continuous and left-continuous t-norms.

Example 1.2 (Expert systems). Logical reasoning based on the reals is quite common in Expert Systems, MYCIN [198] being a famous early example capable of reasoning under uncertainty. MYCIN diagnoses blood infections using certainty factors taken from the real interval $[-1,1]$ and rules like:

IF the infection is primary-bacteremia
AND the site of the culture is one of the sterile sites
AND the suspected portal of entry is the gastrointestinal tract
THEN there is suggestive evidence (0.7) that infection is bacteroid.

To combine certainty factors MYCIN uses the function:

$$x *_{M} y = \begin{cases} x - y(1-x) & \text{if } \min(x,y) \geq 0 \\ \dfrac{x+y}{1-\min(|x|,|y|)} & \text{if } \min(x,y) < 0 < \max(x,y) \\ x - y(1+x) & \text{if } \max(x,y) \leq 0 \end{cases}$$

This rather complicated-looking function is in fact an isomorphic copy of a *uninorm*: functions defined like t-norms except that the unit element can lie anywhere in $[0,1]$. Unlike t-norms, uninorms allow for "compensatory behaviour" in the sense that new information can have either a negative (decreasing) or a positive (increasing) effect on the combined value. Uninorms were introduced by Yager and Rybalov in [221] and are used to define fuzzy logics generalizing the t-norm approach in [144].

Example 1.3 (Resource-based logics). In resource-based logics how often a formula is used in a proof matters. In some, like Anderson and Belnap's relevance logics [6], they must be used at least once (every formula must be relevant to the proof), in others, like Girard's Linear Logic [97], once exactly (two copies of a formula is not the same as one). One obvious way of modelling resources is to use numbers, e.g.

- In Meyer and Slaney's Abelian Logic A [149] (also one of Casari's logics for modelling comparative reasoning in natural language [43]), conjunction and implication can be interpreted by ordinary addition and subtraction on the real numbers. Truth is then associated with being greater than or equal to 0.
- R-Mingle Logic RM, a member of the Anderson and Belnap family [6], can also have truth values in the reals. In this case, however, one copy of a formula is the same as any number of copies, and conjunction is interpreted by:

$$x *_{\mathsf{RM}} y = \begin{cases} \min(x,y) & \text{if } x \leq -y \\ \max(x,y) & \text{otherwise} \end{cases}$$

so $x *_{\mathsf{RM}} x$, interpreted as "x and x", is just x.

Example 1.4 (Residuated lattices). Real numbers and functions on the reals are good candidates for constructing algebras. Consider the real line $\mathbb{R}$ equipped with the usual order, addition, and subtraction. This is an example of an ordered abelian group. In fact, it is an especially useful example: if an equation holds in this algebra, then it holds in all ordered abelian groups. Such facts find a natural home in the framework of *residuated lattices*, introduced by Ward and Dilworth in the 1930s [217] and intensively investigated (with a more general definition) by Tsinakis and co-workers in [116, 127]. In the commutative case, a residuated lattice is a set L equipped with binary operations $\wedge$, $\vee$, $\odot$, $\to$, and a constant e, written:

$$\langle L, \wedge, \vee, \odot, \to, \mathrm{e} \rangle.$$

where $\langle L, \wedge, \vee \rangle$ is a lattice, $\langle L, \odot, \mathrm{e} \rangle$ is a monoid, and $\to$ is the so-called "residuum" of $\odot$, i.e. $x \odot y \leq z$ iff $x \leq y \to z$ for all $x, y, z \in L$. Interesting candidates for commutative residuated lattices are obtained when L is $[0,1]$ and $\wedge$ and $\vee$ are min and max, respectively. In such cases, the operation $\odot$ is always a uninorm.

These examples should give the reader some idea of the kinds of logics and algebras that we are interested in. As mentioned above, our aim is to show that they have a natural syntactic or "proof-theoretic" characterization. In addition to stipulating when formulas are true using the real numbers, we want algorithmic methods that determine just when this is the case. In this book we make use of two main approaches.

Hilbert systems, traceable back to Frege [84] and popularized by Hilbert [118], generate theorems from a stock of formulas (called axioms) using a small number of rules, often just one, modus ponens: from A and $A \to B$, conclude B. The advantages of this approach are generality – we can think of as many weird and wonderful axioms as we want – and a close kinship with Algebra. On the other hand, when it comes to actually reasoning and working with proofs, Hilbert systems are extremely cumbersome. The problem is that to prove a theorem using rules like modus ponens, it is necessary to guess which formulas should appear in applications of the rule.

Gentzen systems, introduced by Gentzen in the 1930s [93], are much better when it comes to reasoning about proofs. They gain flexibility by dealing with structures, typically *sequents*, that look something like:

$$\Gamma \Rightarrow C$$

where C is a formula and Γ is a structured collection of formulas, usually a set, multiset, or sequence. The sequent arrow $\Rightarrow$ is interpreted as "entails" or reading backwards "follows from". We can then write rules like:

$$\text{If } \Gamma \Rightarrow A \text{ and } \Gamma \Rightarrow B, \text{ then } \Gamma \Rightarrow A \wedge B.$$

Sequent systems, built up from such rules, exist for a wide range of logics. However, fuzzy logics do not fit comfortably into this framework. To get Gentzen systems for these logics we treat sequents "in parallel" using *hypersequents*:

$$S_1 \mid \ldots \mid S_n$$

where $S_1 \ldots S_n$ are sequents and the "|" is read as an "or". Hypersequents were introduced by Avron in 1987 [9] and have been used to define Gentzen systems for many logics not covered by the sequent approach. The extra flexibility is gained by defining rules that act on more than one sequent at the same time.

Most important fuzzy logics can be defined in these two frameworks. For Gentzen systems, many fuzzy logics occur naturally as "hypersequent versions" of ordinary substructural logics. The key property for such systems is the existence of "analytic" proofs: proofs where the formulas occurring are built from the same material (subformulas) as the formula proved. The existence of such proofs – established via the key proof-theoretic technique of *cut elimination* – has nice consequences. In many cases we can deduce decidability and complexity results, obtain interpolation and conservative extension properties, or use the calculus as the basis for automated reasoning methods. Most interesting of all, we can use Gentzen systems to tackle one of the main problems addressed in this book (indeed a central topic in the fuzzy logics literature): so-called "standard completeness", establishing that the semantic and syntactic approaches coincide. For this we make use of another elimination procedure, this time of a special density rule introduced by Takeuti and Titani in [205].

Overview of the Book

The intended audience of this book includes readers unfamiliar with either fuzzy logics or proof theory (possibly both). For the former, we provide an accessible introduction to core techniques and results, semantic as well as syntactic. For the latter, we offer algorithmic presentations of fuzzy logics with applications to traditional problems in the area. A brief overview of the remaining chapters is given below:

Chapter 2 introduces the semantic building blocks of fuzzy logics: ordered sets of truth values equipped with functions for interpreting connectives. We pay particular attention to developing the popular t-norm approach, emphasizing the importance of the "fundamental" Łukasiewicz, Gödel, and Product t-norms. We also generalize the setting to cover residuated uninorms, and define fuzzy logics in the framework of commutative residuated lattices.

Chapter 3 introduces Hilbert systems for fuzzy logics, built by extending a core set of axioms and rules with axioms reflecting key properties. Soundness and completeness results are established with respect to classes of commutative residuated lattices, in particular, classes of linearly ordered, dense, and standard algebras.

Chapter 4 develops Gentzen systems, providing hypersequent calculi for a wide range of fuzzy logics by extending basic systems with additional structural rules. Soundness and completeness for these systems (with the cut rule) is established with respect to the Hilbert systems defined in the previous chapter.

Chapter 5 is the most technically demanding of the book, covering syntactic eliminations of rules from calculi. The central topic of cut elimination (and its variant, cancellation elimination) is developed in detail, as are consequences such as decidability and conservative extension results. We then treat the elimination of the density rule and its applications for proving standard completeness results.

Chapter 6 introduces proof theory for the fundamental fuzzy logics, including sequent and hypersequent calculi for Gödel Logic, Łukasiewicz Logic, and Product Logic. Since the techniques used depend on the logic under investigation, the chapter may (aside from basic notions) be read independently of Chapters 3–5.

Chapter 7 explores uniformity and efficiency issues for the fundamental logics of Chapter 6. Proof systems with uniform logical rules are defined in the framework of relational hypersequents, then refined to give genuine algorithmic presentations following a "logic programming style" goal-directed methodology. These systems are then used to obtain complexity results for the logics.

Chapter 8 treats the addition of first-order quantifiers $\forall$ and $\exists$ to fuzzy logics, both the "standard systems" of Chapter 4 and the tricky case of Łukasiewicz Logic, extending the algebraic, Hilbert system, and Gentzen system presentations of previous chapters. These presentations are then used to obtain Herbrand theorems and Skolemization results for the prenex fragments of fuzzy logics.

Chapter 9 covers a variety of miscellaneous topics, slightly out of the scope of the main text, including modalities and truth-stressers, propositional quantifiers, non-commutative fuzzy logics, finite-valued logics, and comparative logics. The difficult case of Basic Logic is also discussed, along with other open problems in the area.

Finally, let us say what we leave out. There are already a number of fine texts on algebraic aspects of fuzzy and substructural logics [58, 90, 102, 105, 180, 186], and a comprehensive monograph on t-norms and related aggregation operators [129]. Hence we follow the maxim here of including only what is needed, and provide references to the relevant literature at the end of each chapter. With regards to proof theory, we treat just Hilbert and Gentzen systems, while acknowledging that many other proof-theoretic frameworks are available: Natural Deduction, Tableaux, Display Logic, Calculus of Structures, to name just a few. These approaches are too similar to Gentzen systems to offer a really interesting alternative perspective, and – to our eyes at least – are not nearly as natural or convenient. Hypersequents are quite clearly the minimal extension of sequents needed to cover a wide spectrum of fuzzy logics. Perhaps by extending the formalism, further systems could be captured, but with an accompanying loss of clarity and usefulness.

Chapter 2
The Semantic Basis

The focus of this book may be proof theory, but the origins of fuzzy logics are undeniably semantic, rooted in the generalization from the two classical truth values 0 and 1 to the real unit interval $[0,1]$. In this chapter we examine these origins in detail, explaining how basic intuitions about truth values and logical connectives spawn a wide family of interesting and useful logics. Needless to say, not everything of interest is covered here, just enough to support the chapters to come.

2.1 Truth Values

Typically, truth values come in sets: $\{0,1\}$ for Classical Logic, $\{0,\frac{1}{2},1\}$ for three-valued logics, the real unit interval $[0,1]$ for most fuzzy logics, and all kinds of other weird and wonderful collections. We will therefore assume here some familiarity with basic notions of naive set theory. This includes notation for membership $\in$, set-building $\{\ldots:\ldots\}$, the empty set $\emptyset$, subsets $\subseteq$ and $\subset$, union $\cup$, intersection $\cap$, difference $-$, cardinality $|\ldots|$, and ordered n-tuples $\langle x_1,\ldots,x_n\rangle$. Arbitrary sets will be denoted using α and β, where $(\alpha_i)_{i\in I}$ stands for a family of sets indexed by a set I. Direct products of sets will be written as $\alpha\times\beta$ or $\prod_{i\in I}\alpha_i$, and direct powers as α^n (with $\alpha^0=\emptyset$). We also recall that:

- *n-ary relations* on a set α are subsets of α^n, called *unary* if $n=1$ and *binary* if $n=2$. As usual, for a binary relation R, we often write xRy for $\langle x,y\rangle\in R$.
- *functions* from a set α to a set β, written $f:\alpha\to\beta$, are subsets of $\alpha\times\beta$ such that for each $x\in\alpha$, $\langle x,y\rangle\in f$ for exactly one $y\in\beta$, written $f(x)=y$. A function f is *injective* if for every $x,y\in\alpha$, whenever $f(x)=f(y)$, then $x=y$, and *surjective* if for every $y\in\beta$, $f(x)=y$ for some $x\in\alpha$.

We adopt some usual conventions for dealing with functions. If f is a function from the direct power α^n to the set α, then f is called an n-ary function or function with arity n on α. In particular, f is called constant for $n=0$, unary for $n=1$, and binary for $n=2$. In the latter case, $f(a,b)$ may be written in infix notation as afb.

We denote the sets of natural numbers $\{0,1,2,\ldots\}$ by $\mathbb{N}$, integers by $\mathbb{Z}$, rationals by $\mathbb{Q}$, reals by $\mathbb{R}$, extended reals (with extra elements $+\infty$ and $-\infty$) by $\bar{\mathbb{R}}$, and for $\alpha \subseteq \mathbb{R}$, define $\alpha^+ =_{\text{def}} \{x \in \alpha : x > 0\}$ and $\alpha^- =_{\text{def}} \{x \in \alpha : x < 0\}$. We assume familiarity with $\sum$ (sum) and $\prod$ (product) notation, and other basic functions for these sets such as min and max. A set α is called *countable* if there exists an injective function $f : \alpha \to \mathbb{N}$, *countably infinite* if this f is also surjective. We denote *(infinite) sequences*, formally functions $f : \mathbb{N} \to \alpha$ for some set α, by indexed sets $(x_i)_{i \in \mathbb{N}}$.

When dealing with sets of truth values, it is useful to have some built-in notion of *order*: a relation "less true than" that tells us something about the relative size or significance of the values.

Definition 2.1. A binary relation $\leq$ is a *partial order* on a set α iff for all $x, y, z \in \alpha$:

1. $x \leq x$ (reflexivity).
2. If $x \leq y$ and $y \leq x$, then $x = y$ (antisymmetry).
3. If $x \leq y$ and $y \leq z$, then $y \leq z$ (transitivity).

The ordered pair $\langle \alpha, \leq \rangle$ is called a *partially ordered set* (*poset*).

Each partial order $\leq$ for α has an accompanying *strict partial order* defined by:

$$x < y \quad \text{iff} \quad x \leq y \text{ and } x \neq y$$

Using both $\leq$ and $<$, we can define *intervals* for $\langle \alpha, \leq \rangle$:

$$[x,y] =_{\text{def}} \{z \in \alpha : x \leq z \leq y\} \qquad [x,y) =_{\text{def}} \{z \in \alpha : x \leq z < y\}$$
$$(x,y] =_{\text{def}} \{z \in \alpha : x < z \leq y\} \qquad (x,y) =_{\text{def}} \{z \in \alpha : x < z < y\}$$

As usual, we often write $x \geq y$ rather than $y \leq x$, and $x > y$ rather than $y < x$. We also combine pairs in the obvious manner, writing e.g. $x \leq y \leq z$ for $x \leq y$ and $y \leq z$.

Posets are characterized further by considering upper and lower bounds:

Definition 2.2. Let $\langle \alpha, \leq \rangle$ be a poset and $\beta \subseteq \alpha$:

- $a \in \alpha$ is an *upper bound* for β if $x \leq a$ for all $x \in \beta$.
- $a \in \alpha$ is a *lower bound* for β if $a \leq x$ for all $x \in \beta$.
- $a \in \alpha$ is the *supremum* of β ($a = \sup \beta$) if a is an upper bound of β and $a \leq b$ for every upper bound b of β.
- $a \in \alpha$ is the *infimum* of β ($a = \inf \beta$) if a is a lower bound of β and $b \leq a$ for every lower bound b of β.

Definition 2.3. A poset $\langle \alpha, \leq \rangle$ is:

- a *lattice* if every $\{x, y\} \subseteq \alpha$ has a supremum $x \vee y$ and infimum $x \wedge y$.
- *bounded* if the *bounds* $\sup \alpha$ and $\inf \alpha$ exist in α.
- *complete* if for any $\beta \subseteq \alpha$, both $\sup \beta$ and $\inf \beta$ exist in α.
- *linearly ordered* (and called a *chain*) if $x \leq y$ or $y \leq x$ for all $x, y \in \alpha$.

- *dense* if whenever $x < y$ for some $x, y \in \alpha$, then $x < z < y$ for some $z \in \alpha$.
- *well-ordered* if there is no sequence $(x_i)_{i \in \mathbb{N}}$ in α with $x_{i+1} < x_i$ for all $i \in \mathbb{N}$.

Example 2.4. The "classical" structure $\langle \{0,1\}, \leq \rangle$ (where $\leq$ is the usual ordering on the reals) is the best known poset of truth values, but there are other popular choices:

- Dividing the interval [0 1] into finer and finer distinctions gives truth values suitable for finite-valued logics:

$$\{0, \frac{1}{n-1}, \ldots, \frac{n-2}{n-1}, 1\} \quad \text{for } n = 2, 3, \ldots$$

  Each of these posets is a linearly and well-ordered complete bounded (but not dense) lattice.

- As limits of this dividing process, we obtain infinite sets of truth values, the real unit interval $[0,1]$ or, remaining countable, the rational unit interval $[0,1] \cap \mathbb{Q}$; both are linearly but not well-ordered dense bounded lattices, and $[0,1]$ is also complete.
- Doing away with bounds, we might also consider the sets of reals $\mathbb{R}$, rationals $\mathbb{Q}$, integers $\mathbb{Z}$, or natural numbers $\mathbb{N}$: all linearly ordered, where both $\mathbb{R}$ and $\mathbb{Q}$ are dense and only $\mathbb{N}$ is well-ordered. None of these posets are complete, but $\mathbb{R}$, $\mathbb{Z}$, and $\mathbb{N}$ may be "completed" by adding top and bottom elements $+\infty$ and $-\infty$ as required to obtain $\bar{\mathbb{R}}$, $\bar{\mathbb{Z}}$, and $\bar{\mathbb{N}}$, respectively.
- Any linearly ordered set with a smaller cardinality than $\mathbb{R}$ can be "normalized" to an order isomorphic (see below) poset of real numbers. We can even restrict our attention just to subsets of $[0,1]$. However, for many logics, the truth values are not linearly ordered. For example, in the "First Degree Entailment" Logic FDE, there exist values for being both truth and false b, or neither n. The resulting order can be represented using the "four corners of truth" Hasse diagram:

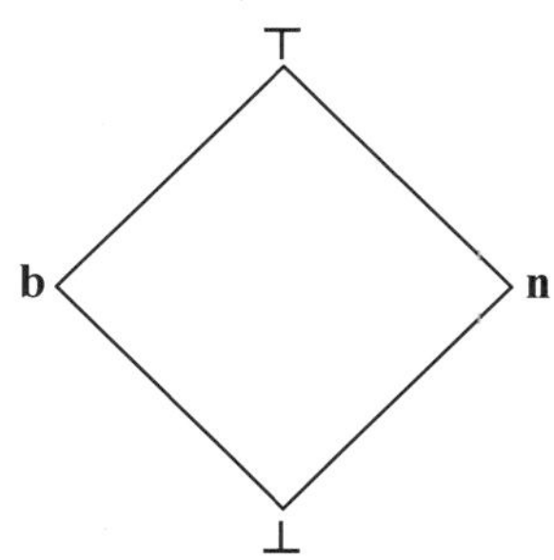

  where a value connected upwards by lines to another value is "less true than" that value. Clearly this poset is a complete bounded and well-ordered lattice, but it is not linearly or densely ordered.

Further useful examples of posets are obtained by defining the lexicographic or "dictionary" order on direct products of posets.

Definition 2.5. Let $\langle \alpha_1, \leq_1 \rangle, \ldots, \langle \alpha_n, \leq_n \rangle$ be posets. The *lexicographic order* $\leq_d$ is defined on the direct product $\alpha_1 \times \ldots \times \alpha_n$ by:

$$(a_1, \ldots, a_n) \leq_d (b_1, \ldots, b_n) \text{ iff } a_i = b_i \text{ for } i = 1 \ldots k \text{ and } a_{k+1} <_{k+1} b_{k+1} \text{ or } k = n.$$

Moreover, if $\langle \alpha_1, \leq_1 \rangle, \ldots, \langle \alpha_n, \leq_n \rangle$ are well-ordered, then so is $\langle \alpha_1 \times \ldots \times \alpha_n, \leq_d \rangle$.

Of course in Mathematics, two posets with ostensibly different elements can often be "matched up" and treated as equivalent.

Definition 2.6. An *order isomorphism* from the poset $\langle \alpha_1, \leq_1 \rangle$ to the poset $\langle \alpha_2, \leq_2 \rangle$ is a surjective function $h : \alpha_1 \longrightarrow \alpha_2$ such that for all $x, y \in \alpha_1$:

$$x \leq_1 y \quad \text{iff} \quad h(x) \leq_2 h(y)$$

In this case, the two posets are said to be *order isomorphic*.

Example 2.7. Consider the function $h : (0,1) \rightarrow \mathbb{R}$ defined by:

$$h(x) = \log(\frac{x}{1-x})$$

It is easy to check that h is an order isomorphism from the poset $(0,1)$ to the poset $\mathbb{R}$ (with the usual orders). Moreover, assuming $\log 0 = -\infty$ and $\log \infty = \infty$, the above function extends also to an order isomorphism from $[0,1]$ to $\bar{\mathbb{R}}$.

This means that we can quite often treat the intervals $[0,1]$, $[0,1)$, $(0,1]$, and $(0,1)$ as canonical sets of truth values, without worrying that other choices of a and b for $[a,b]$, $[a,b)$, $(a,b]$, $[a,b]$ would give different results.

2.2 Ands and Ors

Given a set of truth values for a logic, the next issue is naturally how to interpret usual logical connectives like "and", "or", "if ... then", and "not". In Fuzzy Logic, such connectives are typically interpreted "truth-functionally", i.e. by functions on the set of truth values.[1] In particular, appropriate functions for "ands" (conjunctions) and "ors" (disjunctions) are *aggregation operators* that combine values into one "representative" value. Such functions are often constructed for a particular task or application by hand. Here we take a different approach. We make "design choices" for logics, specifying basic properties that conjunctions and disjunctions should possess.

[1] This is not always the case in Logic; for example, the *probability* of "A and B" is not a function of the probability of A and the probability of B.

2.2.1 Basic Properties

We begin with a selection of basic properties of functions on posets.

Definition 2.8. For a poset $\langle \alpha, \leq \rangle$, a binary function $* : \alpha^2 \to \alpha$ is:

- *associative* iff $(x * y) * z = x * (y * z)$ for all $x, y, z \in \alpha$.
- *commutative* iff $x * y = y * x$ for all $x, y \in \alpha$.
- *unital* iff $e * x = x * e = x$ for some $e \in \alpha$ for all $x \in \alpha$.
- *idempotent* iff $x * x = x$ for all $x \in \alpha$.

An n-ary function $h : \alpha^n \to \alpha$ is:

- *increasing* (*or decreasing*) in argument i if whenever $y \leq z$ (or $z \leq y$):

$$h(x_1, \ldots, x_{i-1}, y, x_{i+1}, \ldots, x_n) \leq h(x_1, \ldots, x_{i-1}, z, x_{i+1}, \ldots, x_n)$$

- *strictly increasing* (*or decreasing*) in argument i if whenever $y < z$ (or $z < y$):

$$h(x_1, \ldots, x_{i-1}, y, x_{i+1}, \ldots, x_n) < h(x_1, \ldots, x_{i-1}, z, x_{i+1}, \ldots, x_n)$$

Assuming some basic knowledge of limits, we also adopt the following definitions:

Definition 2.9. A binary function $* : [0,1]^2 \to [0,1]$ is:

- *continuous* iff for all $x, y \in \alpha$, given a sequence $(x_i)_{i \in \mathbb{N}}$ in α such that $x = \lim_{i \to \infty} x_i$, then also $\lim_{i \to \infty}(x_i * y) = x * y$.
- *left-continuous* iff for all $x, y \in \alpha$, given a sequence $(x_i)_{i \in \mathbb{N}}$ in α such that $x > x_i$ for all $i \in \mathbb{N}$ and $x = \lim_{i \to \infty} x_i$, then also $\lim_{i \to \infty}(x_i * y) = x * y$.

Example 2.10. One of the best known aggregation operators for real numbers is the *arithmetic mean*. The binary version of this function on $[0,1]$ is:

$$x *_{\mathsf{A}} y = \frac{x+y}{2}$$

It is easy to see that $*_{\mathsf{A}}$ is commutative, strictly increasing in both arguments, continuous, and idempotent. However, $*_{\mathsf{A}}$ is not associative:

$$x *_{\mathsf{A}} (y *_{\mathsf{A}} z) = x/2 + y/4 + z/4 \quad \text{but} \quad (x *_{\mathsf{A}} y) *_{\mathsf{A}} z = x/4 + y/4 + z/2$$

There is also no unit element for $*_{\mathsf{A}}$: if $x *_{\mathsf{A}} y = x$, then $y = x$.

Commutativity tells us that the order of arguments does not matter, and associativity that bracketing is unimportant. If $*$ is associative, then we often drop brackets altogether and for $n \in \mathbb{N}^+$, write $x_1 * \ldots * x_n$ for $x_1 * (x_2 * (\ldots (x_{n-1} * x_n) \ldots)$ and let:

$$x_*^{(n)} =_{\text{def}} \overbrace{x * \ldots * x}^{n}$$

Commutativity and associativity are, arguably, essential features of "and-ness" and "or-ness", while the fact that the function is increasing in both arguments supports the intuition that making A or B more true cannot make "A and B" or "A or B" any less true.[2] It is also convenient that such functions be unital, i.e. a particular element, thought of perhaps as "the least true truth value", should play the role of an identity for "and" or "or". Other properties are perhaps not so essential. For example, idempotence might fail for a fuzzy logic. The truth value of "A and A" could be different to the truth value of A: a repeated statement could be more or less true than the statement alone. Also, while assuming continuity is convenient and plausible for functions interpreting "and", left-continuity suffices (at least technically) for the approach taken here.

2.2.2 *t*-Norms

We will stick with the design choices of commutative, associative, increasing, unital functions to interpret "ands" and "ors" throughout this book (or at least until the very last chapter). Typically, we will also take our basic poset of truth values to be the real unit interval $[0,1]$ equipped with the usual ordering $\leq$. If we now add one more quite plausible assumption for interpreting "ands" – that the unit element is 1 – we arrive at the following well known and much studied class of functions.

Definition 2.11. A *t-norm* is a function $* : [0,1]^2 \to [0,1]$ satisfying:[3]

1. $x * y = y * x$ (commutativity).
2. $(x * y) * z = x * (y * z)$ (associativity).
3. $x \leq y$ implies $x * z \leq y * z$ (monotonicity).
4. $1 * x = x$ (identity).

One easy consequence of this definition is that 0 is always an "annihilating element". That is, $0 * x = x * 0 = 0$ for all $x \in [0,1]$ for any t-norm $*$. We can use conditions 1–4 to show this as follows:

$$\begin{aligned} 0 * x &= x * 0 && \text{(commutativity)} \\ &\leq 1 * 0 && \text{(monotonicity)} \\ &= 0 && \text{(identity)} \end{aligned}$$

There are uncountably many t-norms, many of which can be classified into families possessing special properties or representations, such as the so-called Frank or Hamacher t-norms. Here we concentrate on classifying families of t-norms using the plausible properties of conjunctions identified in Definition 2.8.

We start with some fundamental examples of continuous t-norms.

[2] Of course non-commutative and non-associative logics can be interesting (see Chapter 9 for a discussion of the former) but not necessarily for applications in Fuzzy Logic.

[3] The prefix notation $T(x,y)$ is often used for t-norms. Here, to emphasize the interpretation of logical connectives via t-norms, we prefer the infix notation $x * y$.

Definition 2.12 (Fundamental t-norms).

$$\textit{Łukasiewicz t-norm: } x *_{Ł} y =_{\text{def}} \max(0, x+y-1)$$
$$\textit{Gödel t-norm: } x *_{G} y =_{\text{def}} \min(x, y)$$
$$\textit{Product t-norm: } x *_{P} y =_{\text{def}} x \cdot y$$

Continuity is a nice property for interpreting "and"; it means that the function is not over-sensitive to slight changes in its arguments. The fundamental t-norms play a special role in this respect. It turns out that *any* continuous t-norm is "locally isomorphic" to one of these three. We will make this precise below. First, we make the following observation.

Proposition 2.13. $*_G$ *is the only idempotent t-norm.*

Proof. Let $*$ be an idempotent t-norm. If $x \leq y$, then:

$$\begin{aligned} x &= x * x \quad \text{(idempotence)} \\ &\leq x * y \quad \text{(monotonicity)} \\ &\leq x * 1 \quad \text{(monotonicity)} \\ &= x \qquad \text{(identity)} \end{aligned}$$

We have sandwiched $x * y$ between x on the left and right, so $x * y = x$. Similarly, if $y \leq x$, then $x * y = y$. So $x * y = \min(x, y)$. □

More generally, any binary function $*$ can have "idempotents": elements a such that $a * a = a$. The endpoints 0 and 1 are idempotents for any t-norm $*$ since always $1 * 1 = 1$ and $0 * 0 = 0$. If $*$ is continuous, then idempotents separate intervals of $[0, 1]$ where $*$ acts like either the Łukasiewicz or the product t-norm.

Definition 2.14. A continuous t-norm $*$ is *Archimedean* iff it has no idempotent elements except 0 and 1; i.e. $x * x = x$ implies $x = 0$ or $x = 1$.[4]

We now describe a method for constructing t-norms using other t-norms, the idea being to show that all continuous t-norms can be constructed in this way from continuous Archimedean t-norms.

Definition 2.15. Let $([a_i, b_i])_{i \in I}$ be a family of intervals satisfying:

1. $0 \leq a_i < b_i \leq 1$ for all $i \in I$.
2. $(a_i, b_i) \cap (a_j, b_j) = \emptyset$ if $i \neq j$ for all $i, j \in I$.

Then the *ordinal sum* $\sum_{i \in I}([a_i, b_i], *_i)$ of a family of t-norms $(*_i)_{i \in I}$ is the function $* : [0, 1]^2 \to [0, 1]$ defined by:

$$x * y = \begin{cases} a_k + (b_k - a_k)\left(\dfrac{x - a_k}{b_k - a_k} *_k \dfrac{y - a_k}{b_k - a_k}\right) & \text{if } x, y \in [a_k, b_k] \\ \min(x, y) & \text{otherwise} \end{cases}$$

[4] An equivalent definition (for continuous t-norms) is to say that a continuous t-norm $*$ is Archimedean iff for all $x, y \in (0, 1)$, there exists some $n \in \mathbb{N}^+$ such that $x_*^{(n)} < y$.

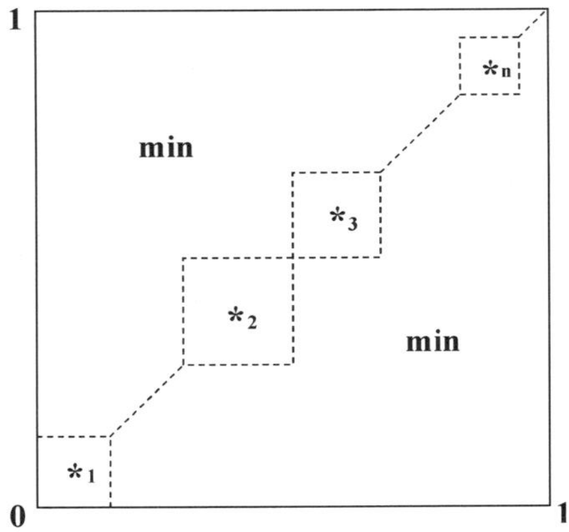

Fig. 2.1 Construction of an ordinal sum

A visual representation of this construction is given in Fig. 2.1. If I is empty, then the ordinal sum is just the minimum function, i.e. the Gödel t-norm. If I is a singleton indexing the whole real unit interval, then the ordinal sum consists of one t-norm.

For continuous t-norms we have the following special situation:

Theorem 2.16. *Each continuous t-norm $*$ is the ordinal sum of a family of continuous Archimedean t-norms.*

Proof. Consider the set $E_* = \{x \in [0,1] : x * x = x\}$ of idempotents of $*$. Suppose that y is the supremum of a set of idempotents of $*$; i.e. $y = \sup \alpha$ where $x * x$ for all $x \in \alpha$. By continuity, $y * y = \sup \alpha * \sup \alpha = \sup\{x * x : x \in \alpha\} = \sup \alpha = y$. So y is also idempotent. The same holds if y is the infimum of a set of idempotents of $*$. Hence E_* is a union of bounded intervals and for some index set I:

$$[0,1] - E_* = \bigcup_{i \in I} (a_i, b_i)$$

where $0 \leq a_i < b_i \leq 1$ for $i \in I$ and $(a_i, b_i) \cap (a_j, b_j) = \emptyset$ for $i \neq j$.

It easy to check that the restriction of $*$ to $[a_i, b_i]$ is mapped to a continuous Archimedean t-norm by $f_i(x) = (x - a_i)/(b_i - a_i)$ for each $i \in I$. Moreover, suppose that $x \in (a_i, b_i)$ and $y \in (a_j, b_j)$ where $i \neq j$ and $x < y$. Then there must be an idempotent z of $*$ such that $x \leq z \leq y$. By continuity, since $z * 0 = 0 \leq x \leq z = z * 1$, there exists $u \in [0,1]$ such that $z * u = x$. But then $x \geq x * y \geq x * z = (u * z) * z = u * (z * z) = u * z = x$. So $x * y = x = \min(x, y)$ as required. □

Let us take a closer look now at continuous Archimedean t-norms:

Theorem 2.17. *The following are equivalent:*

(1) $*$ *is a continuous Archimedean t-norm.*

(2) $*$ *has an* additive generator*; that is, there exists a strictly decreasing continuous function* $t : [0,1] \to [0,\infty]$ *such that* $t(1) = 0$*, and for all* $x, y \in [0,1]$*:*

$$x * y = t^{(-1)}(t(x) + t(y))$$

where $t^{(-1)}(x) = t^{-1}(\min(x,\ t(0)))$.

Proof. It is straightforward to show that (2) implies (1). The continuity, commutativity, and monotonicity of $*$ are immediate and $x * 1 = t^{(-1)}(t(x) + t(1)) = t^{(-1)}(t(x) + 0) = t^{(-1)}(t(x)) = x$ so the identity condition is satisfied. For associativity, we use the fact that $t(x * y) = \min(t(x) + t(y), t(0))$ to derive:

$$\begin{aligned}(x * y) * z &= t^{(-1)}(t(x * y) + t(z)) \\ &= t^{(-1)}(t(x) + t(y) + t(z)) \\ &= t^{(-1)}(t(x) + t(y * z)) \\ &= x * (y * z)\end{aligned}$$

Finally, $*$ is Archimedean. If $x * x = x$, then $t(x) = \min(t(x) + t(x), t(0))$, and so $t(x) = t(0)$, $t(x) = 0$, or $t(x) = \infty$. But t is strictly decreasing, so $x = 1$ or $x = 0$.

For the other direction, recall the definition of $x_*^{(n)}$ for $n \in \mathbb{N}$. We extend this to positive rational numbers as follows. For $n \in \mathbb{N}^+$ and $m \in \mathbb{N}$ let:

$$x_*^{(1/n)} =_{\text{def}} \sup\{y \in [0,1] : y_*^{(n)} < x\} \qquad \text{and} \qquad x_*^{(m/n)} =_{\text{def}} (x_*^{(1/n)})_*^{(m)}$$

Notice that $x_*^{(m/n)} = x_*^{(km/kn)}$ for any $k \in \mathbb{N}^+$, so this operation is well-defined (i.e. if $r = s$, then $x_*^r = x_*^s$). Now observe that if $x_*^{(n)} = x_*^{(n+1)}$ for some $n \in \mathbb{N}$, then by a simple induction:

$$x_*^{(n)} = x_*^{(2n)} = (x_*^{(n)})_*^2$$

So, since $*$ is Archimedean, $x_*^{(n)} \in \{0,1\}$; i.e. $x_*^{(n)} > x_*^{(n+1)}$ whenever $x \in (0,1)$.

We can now define the required additive generator. First, fix an arbitrary $a \in (0,1)$ and define $h : \mathbb{Q} \cap [0,\infty) \to [0,1]$ by:

$$h(r) = a_*^{(r)}$$

Since $*$ is continuous and by the Archimedean property $\lim_{n\to\infty} x_*^{(1/n)} = 1$, we have that h is also continuous. Moreover for any $x \in [0,1]$, $m, p \in \mathbb{N}$, and $n, q \in \mathbb{N}^+$:

$$\begin{aligned}x_*^{((m/n)+(p/q))} &= x_*^{((mq+np)/nq)} \\ &= (x_*^{(1/nq)})_*^{(mq+np)} \\ &= (x_*^{(1/nq)})_*^{(mq)} * (x_*^{(1/nq)})_*^{(np)} \\ &= x_*^{(m/n)} * x_*^{(p/q)}\end{aligned}$$

So for all $r, s \in \mathbb{Q} \cap [0,\infty)$:

$$h(r+s) = a_*^{(r+s)} = a_*^{(r)} * a_*^{(s)} \leq a_*^{(r)} = h(r)$$

I.e. h is decreasing. Moreover, h is strictly decreasing for any $(m/n), (p/q) \in \mathbb{Q} \cap [0, \infty)$ with $h(m/n) > 0$, since:

$$h((m/n) + (p/q)) \leq h((mq+1)/(nq)) = (a_*^{(1/nq)})_*^{(mq+1)} < (a_*^{(1/nq)})_*^{(mq)} = h(m/n)$$

Using the fact that h is decreasing and continuous on $\mathbb{Q} \cap [0, \infty)$, we can define $\bar{h}(x) : [0, \infty] \to [0, 1]$ by:

$$\bar{h}(x) = \inf\{h(r) : r \in \mathbb{Q} \cap [0, x]\}$$

Again, $\bar{h}$ is continuous and decreasing, strictly decreasing for $\bar{h}(x) > 0$, and $\bar{h}(x + y) = \bar{h}(x) * \bar{h}(y)$ for all $x, y \in [0, \infty]$. Define $t : [0, 1] \to [0, \infty]$ by:

$$t(x) = \sup\{y \in [0, \infty] : \bar{h}(y) > x\} \qquad \text{(where } \sup \emptyset = 0\text{)}$$

Then it is straightforward to check that t satisfies the required conditions of (2). □

As key examples, note that an additive generator for the Łukasiewicz t-norm $*_{\text{Ł}}$ is $t_{\text{Ł}}(x) = 1 - x$, while for the product t-norm $*_{\text{P}}$, we have $t_{\text{P}}(x) = -\log x$. These two t-norms form the prototypes for any continuous Archimedean t-norm, and hence (by Theorem 2.16) the building blocks for any continuous t-norm.

Definition 2.18. Let $*$ be a t-norm. Then $a \in [0, 1]$ is:

- a *nilpotent* of $*$ iff there exists $n \in \mathbb{N}^+$ such that $a_*^{(n)} = 0$.
- a *zero divisor* of $*$ iff $a * b = 0$ for some $b \in (0, 1)$.

An Archimedean t-norm is *strict* if its only nilpotent is 0, and *nilpotent* otherwise.

Lemma 2.19. *Each continuous Archimedean t-norm $*$ is:*

either *strict and $\langle [0,1], * \rangle$ is order isomorphic to $\langle [0,1], *_{\text{P}} \rangle$;*
or *nilpotent and $\langle [0,1], * \rangle$ is order isomorphic to $\langle [0,1], *_{\text{Ł}} \rangle$.*

Proof. Consider any continuous Archimedean t-norm $*$. By the previous theorem, $*$ has an additive generator t. We have two cases:

(a) Suppose that $t(0) = \infty$. If $a_*^{(n)} = 0$ for some $n \in \mathbb{N}^+$, then $t(a_*^{(n)}) = t(0) = \infty$. But $t(a_*^{(n)}) = nt(a)$ so $t(a) = \infty$. Since t is strictly decreasing, $a = 0$, so $*$ is strict. Moreover, $f(x) = e^{-t(x)}$ is an order isomorphism from $\langle [0,1], * \rangle$ to $\langle [0,1], *_{\text{P}} \rangle$.

(b) Suppose that $t(0) \neq \infty$. Since $t(1) = 0$, by continuity there exists $c \in (0, 1]$ such that $t(c) = t(0)/2$. But then $c * c = t^{(-1)}(t(c) + t(c)) = t^{(-1)}(t(0)) = 0$, so $*$ is nilpotent. Moreover, $g(x) = 1 - (t(x)/(t(0))$ is the required order isomorphism from $\langle [0,1], * \rangle$ to $\langle [0,1], *_{\text{Ł}} \rangle$. □

Finally, we can put these facts together and obtain the following characterization.

Theorem 2.20. *Each continuous t-norm is the ordinal sum of a family of t-norms order isomorphic to either the Łukasiewicz or product t-norm.*

Partly as a result of this elegant representation theorem, continuous t-norms – in particular, the fundamental t-norms – are the most commonly used aggregation operators in Fuzzy Logic. However, important non-continuous t-norms also occur regularly in the literature. In particular, the *nilpotent minimum* t-norm:

$$x *_{\mathsf{N}} y =_{\text{def}} \begin{cases} \min(x,y) & \text{if } x+y>1 \\ 0 & \text{otherwise} \end{cases}$$

is left-continuous but not continuous, while the *drastic product* t-norm:

$$x *_{\mathsf{D}} y =_{\text{def}} \begin{cases} 0 & \text{if } x,y \in [0,1) \\ \min(x,y) & \text{otherwise} \end{cases}$$

is not even left-continuous. In fact $*_{\mathsf{D}}$ is the least t-norm according to the partial ordering defined on the set of t-norms by:

$$*_1 \leq_T *_2 \quad \text{iff} \quad x *_1 y \leq x *_2 y \ \text{ for all } x,y \in [0,1]$$

The Gödel t-norm $x *_{\mathsf{G}} y = \min(x,y)$ is greatest with respect to this ordering. That is, $*_{\mathsf{D}} \leq_T * \leq_T *_{\mathsf{G}}$ for all t-norms $*$.

2.2.3 t-Conorms

Many properties that are natural for interpreting "ands" are also very natural for interpreting "ors". Again, it is reasonable to assume that such functions are commutative, associative, and increasing. The point of divergence comes from the fact that 0 is now a more suitable identity element.

Definition 2.21. A *t-conorm* is a function $* : [0,1]^2 \to [0,1]$ satisfying:

1. $x * y = y * x$ (commutativity).
2. $(x * y) * z = x * (y * z)$ (associativity).
3. $x \leq y$ implies $x * z \leq y * z$ (monotonicity).
4. $0 * x = x$ (identity).

Just as for t-norms, 0 is an "annihilator", so 1 plays this role for t-conorms; i.e. $1 * x = x * 1 = 1$ for all $x \in [0,1]$. In fact, there is a strong correspondence between t-norms and t-conorms. Each t-norm can be used to define a dual t-conorm and vice versa. More precisely, a function $\circ : [0,1]^2 \to [0,1]$ is a t-conorm iff there exists a t-norm $*$ such that for all $x,y \in [0,1]$:

$$x * y = 1 - ((1-x) \circ (1-y))$$

In this case, $*$ is called the *dual t-norm* of $\circ$, and $\circ$ the *dual t-conorm* of $*$.

Example 2.22. The dual t-conorms of the fundamental t-norms are:

$$\textit{Bounded sum t-conorm: } x \circ_{\text{Ł}} y =_{\text{def}} \min(1, x+y)$$
$$\textit{Maximum t-conorm: } x \circ_{\text{G}} y =_{\text{def}} \max(x, y)$$
$$\textit{Probabilistic sum t-conorm: } x \circ_{\text{P}} y =_{\text{def}} x+y-x\cdot y$$

All the representation theorems for t-norms of the previous section have dual versions for t-conorms. However, since here we treat conjunction connectives as primitive and disjunction connectives as defined, we leave the derivation of such results as exercises for the interested reader.

2.2.4 Uninorms

As we have just seen, "ands" and "ors" interpreted by t-norms and t-conorms have overlapping properties: commutativity, associativity, and monotonicity. The difference lies solely with the identity element: 1 for "and" and 0 for "or". Let us see now what happens if we allow this element to be *any* number in $[0,1]$.

Definition 2.23. A *uninorm* is a function $* : [0,1]^2 \to [0,1]$ satisfying:

1. $x * y = y * x$ (commutativity).
2. $(x*y)*z = x*(y*z)$ (associativity).
3. $x \leq y$ implies $x*z \leq y*z$ (monotonicity).
4. $\mathrm{e}_* * x = x$ (identity).

For $\mathrm{e}_* = 1$ or $\mathrm{e}_* = 0$, we get t-norms or t-conorms, but for $\mathrm{e}_* \in (0,1)$, we get something new: functions with "compensatory behaviour". While for a t-norm $*$, always $x*y \leq x$, and for a t-conorm $\circ$, always $x \circ y \geq x$, for uninorms, $x*y$ can be less than or greater than x. In this case, the identity element e_* can be interpreted as the score or truth value given to a "neutral statement", and truth is naturally associated with values greater than or equal to e_*, i.e. the set $[\mathrm{e}_*, 1]$.

Let us take a closer look at the behaviour of uninorms, beginning with the values taken at the extremal points 0 and 1. It is easy to see that $0*0=0$ and $1*1=1$ for all uninorms. However, the final "classical" value $0*1 = 1*0$ is not fixed. It can be 1 or 0, being either *conjunctive* like a t-norm, or *disjunctive* like a t-conorm.

Proposition 2.24. *For any uninorm $*$, one of these two conditions holds:*

(1) $x*0 = 0*x = 0$ *for all* $x \in [0,1]$ *and* $*$ *is called* conjunctive.
(2) $x*1 = 1*x = 1$ *for all* $x \in [0,1]$ *and* $*$ *is called* disjunctive.

Proof. Let $*$ be a uninorm. We show first that for any $x, y \in [0,1]$, if $x \leq 0*1 \leq y$, then $x*y = 0*1$. For $x \leq 0*1 \leq y$ we have by associativity and monotonicity that $0*1 = (0*0)*1 = 0*(0*1) \leq x*y \leq (0*1)*1 = 0*(1*1) = 0*1$, so $x*y = 0*1$ as required. Now suppose that $\mathrm{e}_* \leq 0*1$. By the previous claim, $1 = \mathrm{e}_* * 1 = 0*1$.

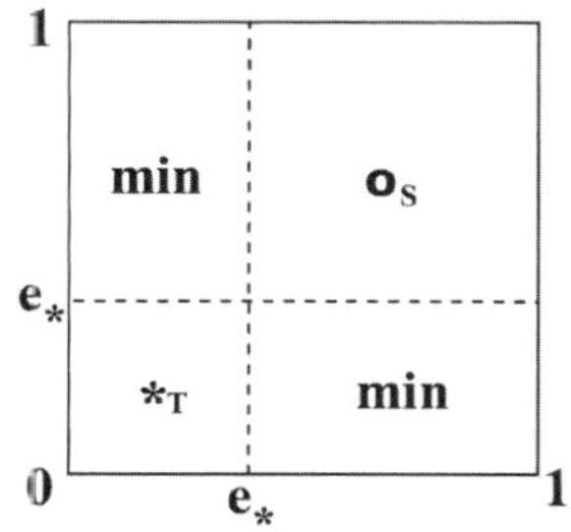

Fig. 2.2 The structure of a uninorm with identity $e_* \in (0,1)$

So by monotonicity, $x*1 = 1*x = 1$ for all $x \in [0,1]$. The case where $0*1 < e_*$ and $*$ is disjunctive is symmetrical. □

Hence conjunctive and disjunctive uninorms coincide with classical conjunction and disjunction respectively on the set $\{0,1\}$. More generally, as shown in Fig. 2.2, a uninorm with identity $e_* \in (0,1)$ exhibits a "block-like" structure on the unit square, where the lower corner $[0,e_*]^2$ is isomorphic to a t-norm, and the upper corner $[e_*,1]^2$ is isomorphic to a t-conorm.

Proposition 2.25. *If $*$ is a uninorm with identity $e_* \in (0,1)$, then:*

(i) $x *_{\mathrm{T}} y =_{\mathrm{def}} \dfrac{(e_* x) * (e_* y)}{e_*}$ *is a t-norm.*

(ii) $x \circ_{\mathrm{S}} y =_{\mathrm{def}} \dfrac{(e_* + (1-e_*)x) * (e_* + (1-e_*)y) - e_*}{1-e_*}$ *is a t-conorm.*

(iii) $$x * y = \begin{cases} e_*(\dfrac{x}{e_*} *_{\mathrm{T}} \dfrac{y}{e_*}) & \text{if } x,y \in [0,e_*] \\ (e_* + (1-e_*))(\dfrac{x-e_*}{1-e_*} \circ_{\mathrm{S}} \dfrac{y-e_*}{1-e_*}) & \text{if } x,y \in [e_*,1] \end{cases}$$

(iv) $\min(x,y) \le x*y \le \max(x,y)$ *for all* $(x,y) \in [0,1]^2 - ([0,e_*]^2 \cup [e_*,1]^2)$.

Proof. We just check (i), leaving (ii), (iii), and (iv) as exercises. First notice that $*_{\mathrm{T}}$ is clearly commutative and increasing by the corresponding conditions for $*$, so we can just check identity and associativity as follows:

$$\begin{aligned} x *_{\mathrm{T}} 1 &= \frac{(e_* x) * (e_* 1)}{e_*} \\ &= \frac{(e_* x) * e_*}{e_*} \\ &= \frac{e_* x}{e_*} \\ &= x \end{aligned} \qquad \begin{aligned} (x *_{\mathrm{T}} y) *_{\mathrm{T}} z &= \frac{e_* \dfrac{(e_* x) * (e_* y)}{e_*} * (e_* z)}{e_*} \\ &= \frac{((e_* x) * (e_* y)) * (e_* z)}{e_*} \\ &= \frac{(e_* x) * ((e_* y) * (e_* z))}{e_*} \\ &= x *_{\mathrm{T}} (y *_{\mathrm{T}} z) \end{aligned}$$

□

We can investigate various classes of uninorms, just as we did with t-norms. Note, however, that assuming continuity gives nothing new in this case.

Proposition 2.26. *If $*$ is a continuous uninorm, then $*$ is a t-norm or a t-conorm.*

Proof Suppose that $*$ is a continuous conjunctive uninorm. Since $0*1=0$ and $1*1=1$, by continuity, there exists $x \in [0,1]$ such that $e_* = x*1$. But then $1 = e_* * 1 = (x*1)*1 = x*(1*1) = x*1 = e_*$ so $*$ is a t-norm. Similar reasoning shows that a continuous disjunctive uninorm is a t-conorm. □

However, there do exist new uninorms continuous on the open interval $(0,1)$.

Example 2.27. The following "cross-ratio" uninorm is conjunctive and continuous everywhere except at $(1,1)$:

$$x *_{\mathrm{CR}} y = \begin{cases} \dfrac{xy}{xy+(1-x)(1-y)} & \text{if } \{x,y\} \neq \{0,1\} \\ 0 & \text{otherwise} \end{cases}$$

Isomorphic versions of this t-norm have been used to combine degrees of belief in expert systems such as MYCIN and PROSPECTOR (see the historical remarks at the end of the chapter for references).

Conjunctive uninorms that are strictly increasing on $(0,1)$ and continuous on the half-open interval $[0,1)$ can be classified in a nice way with the previous example as a prototype. The proof of the following theorem, omitted here, is very similar to the proof of Theorem 2.17 for Archimedean continuous t-norms.

Theorem 2.28. *For any uninorm $*$ with $e_* \in (0,1)$, the following are equivalent:*

(1) *$*$ is strictly increasing on $(0,1)$ and continuous on $[0,1)$.*
(2) *$*$ has an additive generator; that is, there exists a strictly increasing continuous function $h : [0,1] \to \bar{\mathbb{R}}$ such that $h(0) = -\infty$, $h(e_*) = 0$, $h(1) = +\infty$, and:*

$$x * y = h^{-1}(h(x)+h(y)) \quad \textit{for all } (x,y) \in [0,1]^2 - \{(0,1),(1,0)\}$$

Clearly, all conjunctive (or disjunctive) uninorms characterized by this theorem (called *representable* uninorms) are order isomorphic. So we can take the cross-ratio uninorm as representative of this class in the same way that e.g. the product t-norm represents the class of strict Archimedean t-norms.

What about idempotent uninorms? For t-norms there is just one example: the Gödel t-norm min. For uninorms, however, we can identify a new class of functions:

Theorem 2.29. *For a uninorm $*$ with $e_* \in (0,1)$, the following are equivalent:*

(1) *$*$ is left-continuous and idempotent.*
(2) *For $f(x) = \sup\{y \in [0,1] : x*y \leq e_*\}$:*

$$x * y = \begin{cases} \min(x,y) & \textit{if } y \leq f(x) \\ \max(x,y) & \textit{otherwise} \end{cases}$$

Proof. It is easy to check that (1) follows from (2). For the other direction, note that $f(x) = \sup\{y \in [0,1] : x * y \leq e_*\}$ exists for $x \in [0,1]$ by the left-continuity of $*$, and:

$$y \leq f(x) \quad \text{iff} \quad x * y \leq e_* \quad \text{iff} \quad x \leq f(y)$$

Suppose without loss of generality that $x \leq y$. If $y \leq f(x)$, then $x = x * x \leq x * y = x * (x * y) \leq x * e_* = x$; i.e. $x * y = x$. Otherwise $y > f(x)$, so $x * y \geq e_*$ and we have $y = y * y \geq x * y = (x * y) * y \geq e_* * y = y$; i.e. $x * y = y$. □

Example 2.30. Taking $f(x) = 1 - x$ in the previous theorem, we get:

$$x *_S y = \begin{cases} \min(x,y) & \text{if } x + y \leq 1 \\ \max(x,y) & \text{otherwise} \end{cases}$$

This uninorm is order isomorphic to functions used in certain "Sugihara" algebras over $\overline{\mathbb{R}}$ for the relevance logic RM.

2.3 Nots and Ifs

"Ands" and "ors" are important, but arguably the core connective of a logic is implication "if...then". Below, we take a detailed look at how to interpret such a connective, examining various options and exploring connections with uninorms. Let us begin, however, by considering the useful related notion of a negation "not".

Definition 2.31. A *negation* is a function $\mathrm{n} : [0,1] \to [0,1]$ satisfying:

1. $x \leq y$ implies $\mathrm{n}(y) \leq \mathrm{n}(x)$ (antitonicity).
2. $\mathrm{n}(0) = 1$ and $\mathrm{n}(1) = 0$ (boundary conditions).

Definition 2.32. A negation n is:

- *strict* if it is strictly decreasing and continuous.
- an *involution* if $\mathrm{n}(\mathrm{n}(x)) = x$ for all $x \in [0,1]$.
- *strong* if it is a strict involution, and *weak* otherwise.

Example 2.33. The most widely used function for interpreting "not" in Fuzzy Logic is the strong negation:

$$\mathrm{n}_{Ł}(x) =_{\mathrm{def}} 1 - x$$

There are also popular negations which are strict but not strong such as $\mathrm{n}_2(x) =_{\mathrm{def}} 1 - x^2$, or more generally, $\mathrm{n}_k(x) =_{\mathrm{def}} 1 - x^k$ for $k \geq 2$. Important negations that are weak (and not strict) include:

$$\mathrm{n}_G(x) =_{\mathrm{def}} \begin{cases} 1 & \text{if } x = 0 \\ 0 & \text{otherwise} \end{cases} \qquad \mathrm{n}_D(x) =_{\mathrm{def}} \begin{cases} 1 & \text{if } x < 1 \\ 0 & \text{if } x = 1 \end{cases}$$

In fact, n_G and n_D are, respectively, the least and greatest negations; i.e. $n_G(x) \leq n(x) \leq n_D(x)$ for any negation n and all $x \in [0,1]$.

Let us now consider some desirable properties for implications:

Definition 2.34. A binary function $\rightarrow: [0,1]^2 \rightarrow [0,1]$ satisfies:

- *exchange* if $x \rightarrow (y \rightarrow z) = y \rightarrow (x \rightarrow z)$ for all $x,y,z \in [0,1]$.
- *left antitonicity* if $x \leq y$ implies $y \rightarrow z \leq x \rightarrow z$ for all $x,y,z \in [0,1]$.
- *right isotonicity* if $x \leq y$ implies $z \rightarrow x \leq z \rightarrow y$ for all $x,y,z \in [0,1]$.
- *e-degree ranking* for $e \in [0,1]$, if $e \leq x \rightarrow y$ iff $x \leq y$ for all $x,y \in [0,1]$.
- the *left boundary condition* if $0 \rightarrow x = 1$ for all $x \in [0,1]$.
- the *right boundary condition* if $x \rightarrow 1 = 1$ for all $x \in [0,1]$.
- the *normality condition* if $1 \rightarrow 0 = 0$.
- *left neutrality* if $1 \rightarrow x = x$ for all $x \in [0,1]$.
- *contraposition* for a negation n if $x \rightarrow y = n(y) \rightarrow n(x)$ for all $x,y \in [0,1]$.
- *generalized modus ponens* for a uninorm $*$ if $x * (x \rightarrow y) \leq y$ for all $x,y \in [0,1]$.

Example 2.35. Consider the function (obtained e.g. as $n_Ł(x) \circ_G y$):

$$x \rightarrow_{KD} y =_{def} \max(1-x, y)$$

This function, the *Kleene-Dienes implication*, generalizes the characterization $x \rightarrow y = \neg x \vee y$ of classical implication, and is used frequently in Fuzzy Logic. The exchange, left antitonicity and right isotonicity, left and right boundary, normality, and left-neutrality conditions, and contraposition with respect to $n_Ł$ are all satisfied by $\rightarrow_{KD}$, but the e-degree ranking property fails for any $e \in [0,1]$. More seriously, there is no uninorm $*$ such that the generalized modus ponens principle holds. For this last reason it is often claimed that $\rightarrow_{KD}$ is not a true implication.

The previous example touches on an important point. An implication should ideally "tie in" somehow with the conjunction of the logic. At an intuitive level: for a conjunction $*$ and implication $\rightarrow$, the generalized modus ponens property $x * (x \rightarrow y) \leq y$ should hold, and $x \rightarrow y$ should be maximal (as true as possible) subject to this restriction. This motivates the following definition:

Definition 2.36. A commutative function $*: \alpha^2 \rightarrow \alpha$ for a poset $\langle \alpha, \leq \rangle$ is *residuated* iff there exists a function $\rightarrow_*: \alpha^2 \rightarrow \alpha$ called the *residuum* of $*$, such that:

$$x * y \leq z \quad \text{iff} \quad x \leq y \rightarrow_* z \quad \text{for all } x,y,z \in \alpha$$

When such a function exists, easily:

$$x \rightarrow_* y = \max\{z \in \alpha : z * x \leq y\}$$

Moreover, residuated uninorms can be characterized as follows:

Proposition 2.37. *A uninorm is residuated iff it is left-continuous and conjunctive.*

Proof. For the left-to-right direction, note first that $0 \leq 1 \rightarrow_* 0$. Hence, by residuation, $0 * 1 \leq 0$; i.e. $*$ is conjunctive. Now for $x, y \in [0,1]$, let $(x_i)_{i \in \mathbb{N}}$ be a sequence such that $x = \sup_{i \in \mathbb{N}} x_i$. Define $w = \sup_{i \in \mathbb{N}}(y * x_i)$. Clearly $w \leq y * x$. Also, for each $i \in \mathbb{N}$, $y * x_i \leq w$. So by residuation, $x_i \leq y \rightarrow_* w$. It follows that $x \leq y \rightarrow_* w$, and so by residuation again, $y * x \leq w$. Hence $*$ is left-continuous in its second argument, and so by commutativity, also in the first.

For the right-to-left direction, suppose that $*$ is left-continuous and conjunctive. Let $y \rightarrow_* z =_{\text{def}} \sup\{w \in [0,1] : w * y \leq z\}$, noting that this exists since $0 = 0 * y \leq z$, and the set is bounded from above by 1. If $x * y \leq z$, then clearly $x \leq y \rightarrow_* z$. If $x \leq y \rightarrow_* z$, then $x * y \leq \sup\{w \in [0,1] : w * y \leq z\} * y$. By left-continuity, $x * y \leq \sup\{w * y : w * y \leq z\} \leq z$ as required. □

Residua for the fundamental t-norms are easily calculated:

$$\textit{Łukasiewicz implication: } x \rightarrow_{\text{Ł}} y =_{\text{def}} \min(1, 1 - x + y)$$

$$\textit{Gödel implication: } x \rightarrow_{\text{G}} y =_{\text{def}} \begin{cases} 1 & \text{if } x \leq y \\ y & \text{otherwise} \end{cases}$$

$$\textit{Product implication: } x \rightarrow_{\text{P}} y =_{\text{def}} \begin{cases} 1 & \text{if } x \leq y \\ y/x & \text{otherwise} \end{cases}$$

Moreover, it is straightforward to verify that residua of uninorms (or in some cases, just certain examples) satisfy many of the key implication properties.

Proposition 2.38. *Let $*$ be a residuated uninorm:*

(a) $\rightarrow_*$ *obeys the exchange, left antitonicity and right isotonicity, left and right boundary, and normality conditions, and the e_*-degree ranking and generalized modus ponens properties.*

(b) $\rightarrow_*$ *obeys the left neutrality condition iff $*$ is a t-norm.*

(c) $\rightarrow_{\text{CR}}$ *and* $\rightarrow_{\text{Ł}}$ *obey contraposition for the negation* $\mathrm{n}_{\text{Ł}}(x) = 1 - x$.

Residuation is a natural way of obtaining implications from uninorms, but it is not the only way. We could proceed instead as in Example 2.35 and generalize the classical reading of "A implies B" as "not A or B". That is, take an involutive negation such as $\neg x =_{\text{def}} 1 - x$ and t-conorm (or disjunctive uninorm) $\circ$, and define $x \rightarrow y =_{\text{def}} \neg x \circ y$. Functions obtained in this way are called S-implications, while the residuum based functions studied in this book are called R-implications.

One widely followed practice for defining a negation makes use of both an implication and the least element (usually 0 in the case of fuzzy logics) or, more generally, an arbitrary element f. That is, given a residuated uninorm $*$ and $\mathrm{f} \in [0,1]$, let:

$$\neg_*^{\mathrm{f}} x =_{\text{def}} x \rightarrow_* \mathrm{f}$$

It is easy to show that $\neg_*^{\mathrm{f}}$ satisfies antitonicity and $\neg_*^{\mathrm{f}} 0 = 1$. Also, if $\mathrm{f} = 0$, then $\neg_*^{\mathrm{f}} 1 = 0$ and $\neg_*^{\mathrm{f}}$ is a negation. In particular, we can obtain the Łukasiewicz negation as $n_Ł(x) = x \to_Ł 0$ and the Gödel (product) negation encountered in Example 2.33 as $n_G(x) = x \to_G 0 = x \to_P 0$.

A nice feature of continuous t-norms is that together with their residua they can always be used to define the functions min and max.

Proposition 2.39. *For any continuous t-norm* $*$*:*

(a) $x \leq y$ *iff* $x \to_* y = 1$.

(b) $x * (x \to_* y) = \min(x,y)$.

(c) $\min((x \to_* y) \to_* y, (y \to_* x) \to_* x) = \max(x,y)$.

Proof. (a) $x \leq y$ iff $x * 1 \leq y$ iff $1 \leq x \to_* y$. (b) If $x \leq y$, then $x * (x \to_* y) = x * 1 = x$. Suppose that $y \leq x$. Then by the continuity of $*$, since $x * 0 = 0 \leq y \leq x = x * 1$, there exists z such that $x * z = y$. So clearly also $x * (x \to_* y) = y$. (c) If $x \leq y$, then $x \to_* y = 1$ and since $y \leq (y \to_* x) \to_* x$, we get $\min((x \to_* y) \to_* y, (y \to_* x) \to_* x) = y$. The case of $y \leq x$ is symmetrical. □

Finally, note that a bi-implication "iff" connective can be interpreted using min and implication as follows:

$$x \Leftrightarrow_* y =_{\text{def}} \min((x \to_* y), (y \to_* x))$$

It is easy to see that $x = y$ iff $e_* \leq x \Leftrightarrow_* y$ for any residuated uninorm $*$. Hence $\Leftrightarrow_*$ gives us a way of expressing (strict) equality for fuzzy logics.

2.4 Ordered Algebraic Structures

Let us take a moment to recap. So far we have been looking at different partially ordered sets and a wide range of functions acting on these sets. By making definite choices – of a poset and a selection of functions – we get examples of *ordered algebraic structures*, or more simply, *algebras*.

2.4.1 Basic Notions

First, we need a way to identify the arities of functions occurring in our algebras.

Definition 2.40. A *type* is a set $\nu = (n_i)_{i \in I}$ of natural numbers indexed by a set I.

Algebras are then characterized as structures of a particular type.

Definition 2.41. An *algebra* of type $\nu = (n_i)_{i \in I}$ is an ordered pair:

$$\mathbf{A} = \langle L_\mathbf{A}, F_\mathbf{A} \rangle$$

where $L_\mathbf{A}$ is a non-empty set called the *universe* of $\mathbf{A}$ and $F_\mathbf{A} = (f_i^\mathbf{A})_{i \in I}$ is a set of functions $f_i^\mathbf{A} : L_\mathbf{A}^{n_i} \to L_\mathbf{A}$ called the *fundamental operations* of $\mathbf{A}$.

Following convention, when the type is finite, we write $\mathbf{A} = \langle L_{\mathbf{A}}, \{f_1^{\mathbf{A}}, \ldots, f_m^{\mathbf{A}}\}\rangle$ without set brackets as $\mathbf{A} = \langle L_{\mathbf{A}}, f_1^{\mathbf{A}}, \ldots, f_m^{\mathbf{A}}\rangle$.

Example 2.42. Taking the min function (the Gödel t-norm) for conjunction, max for disjunction, and $\mathrm{n_Ł}(x) = 1 - x$ for negation, we have an algebra of type $(2,2,1)$:

$$\langle [0,1], \min, \max, \mathrm{n_Ł} \rangle$$

This is the algebra most commonly encountered in Fuzzy Logic. However, as discussed in Example 2.35, the implication definable here as $x \rightarrow_{\mathsf{KD}} y = \max(1-x, y)$ is not very satisfactory.

Often we treat classes of algebras of the same type, that is, algebras having the same number of fundamental operations with the same arity. Consider for example the class of algebras of type $(2,2,1)$:

$$\langle [0,1], *, \circ, \mathrm{n} \rangle$$

where $*$ is a t-norm, $\circ$ is a t-conorm, and n is a strong negation. The algebra $\langle [0,1], \min, \max, \mathrm{n_Ł} \rangle$ of Example 2.42 is just one special member of this class.

Example 2.43. Lattices, introduced in Definition 2.3 as special kinds of posets, can be defined equivalently as algebras of type $(2,2)$:

$$\langle L, \wedge, \vee \rangle$$

such that $\langle L, \leq \rangle$ is a lattice with ordering $x \leq y$ iff $x \wedge y = x$, where $x \wedge y = \inf\{x,y\}$ and $x \vee y = \sup\{x,y\}$. Similarly, a bounded lattice is an algebra $\langle L, \wedge, \vee, \bot, \top \rangle$ such that $\langle L, \wedge, \vee \rangle$ is a lattice, $\bot = \inf L$, and $\top = \sup L$.

Another simple but important class of algebras of interest in Fuzzy Logic is the class of monoids:

Definition 2.44. A *monoid* is an algebra of type $(2,0)$:

$$\langle L, \odot, \mathrm{e} \rangle$$

such that $\odot$ is an associative binary function on L with unit element e.

We have already encountered many examples of monoids and lattices. In particular, $\langle [0,1], *, \mathrm{e}_* \rangle$ is a monoid for any uninorm $*$, while the crucial lattice for Fuzzy Logic is $\langle [0,1], \min, \max \rangle$.

However, there is another useful way of describing monoids. We can define the class of all algebras $\langle L, \odot, \mathrm{e} \rangle$ of type $(2,0)$ that satisfy the following equations:

$$(x \odot y) \odot z = x \odot (y \odot z) \qquad x \odot \mathrm{e} = x \qquad \mathrm{e} \odot x = x$$

In this case, we can say that monoids form an equational class or variety.

Definition 2.45. A *variety* is the class of all algebras of a given type that satisfy a particular set of equations.[5]

Example 2.46. The class of lattices, already presented in two ways, can be defined a third time as the class of algebras $\langle L_{\mathbf{A}}, \wedge, \vee \rangle$ of type $(2,2)$ satisfying the equations:

$$\begin{array}{rlrl} x \wedge x &= x & x \vee x &= x \\ x \wedge y &= y \wedge x & x \vee y &= y \vee x \\ x \wedge (y \wedge z) &= (x \wedge y) \wedge z & x \vee (y \vee z) &= (x \vee y) \vee z \\ x \wedge (x \vee y) &= x & x \vee (x \wedge y) &= x \end{array}$$

Bounded lattices are algebras $\langle L_{\mathbf{A}}, \wedge, \vee, \bot, \top \rangle$ of type $(2,2,0,0)$ satisfying these equations and also $x \wedge \bot = \bot$ and $x \vee \top = \top$.

Often we are interested in the equations satisfied by a particular algebra or class of algebras. In this case, it is helpful to consider the variety consisting of *all* algebras satisfying these equations.

Definition 2.47. If $\mathcal{K}$ is a class of algebras of the same type, then $\mathcal{V}(\mathcal{K})$, the *variety generated by* $\mathcal{K}$, denotes the smallest variety containing all the algebras in $\mathcal{K}$.

Finally, let us introduce some definitions for relating algebras.

Definition 2.48. Let $\mathbf{A} = \langle L_{\mathbf{A}}, (f_i^{\mathbf{A}})_{i \in I} \rangle$ and $\mathbf{B} = \langle L_{\mathbf{B}}, (f_i^{\mathbf{B}})_{i \in I} \rangle$ be algebras of the same type $\nu = (n_i)_{i \in I}$. A function $\phi : L_{\mathbf{A}} \to L_{\mathbf{B}}$ is a *homomorphism* from $\mathbf{A}$ to $\mathbf{B}$ if for all $i \in I$ and $a_1, \ldots, a_{n_i} \in L_{\mathbf{A}}$:

$$\phi(f_i^{\mathbf{A}}(a_1, \ldots, a_{n_i})) = f_i^{\mathbf{B}}(\phi(a_1), \ldots, \phi(a_{n_i}))$$

If ϕ is injective, then ϕ is called an *embedding* of $\mathbf{A}$ into $\mathbf{B}$, and if ϕ is also surjective, then ϕ is an *isomorphism* from $\mathbf{A}$ onto $\mathbf{B}$ and the algebras are called *isomorphic*.

Example 2.49. Consider the algebras of type $(2,2,0)$:

$$\langle (0,1], *_{\mathrm{P}}, \to_{\mathrm{P}}, 1 \rangle \quad \text{and} \quad \langle \mathbb{R}^- \cup \{0\}, +, \to_+, 0 \rangle$$

where $*_{\mathrm{P}}$ and $\to_{\mathrm{P}}$ are the product t-norm and its residuum, respectively, $+$ is usual addition, and $x \to_+ y = \min(0, y - x)$. Define $\phi : (0,1] \to \mathbb{R}^- \cup \{0\}$ by $\phi(x) = \log x$. Then $\phi(x *_{\mathrm{P}} y) = \log(xy) = \log x + \log y = \phi(x) + \phi(y)$ and $\phi(x \to_{\mathrm{P}} y) = \log(\min(1, y/x)) = \min(\log 1, \log(y/x)) = \min(0, \log y - \log x) = \phi(x) \to_+ \phi(y)$. Since ϕ is both injective and surjective, it is an isomorphism and the two algebras are isomorphic. This means that in practice we can often switch between the two when performing calculations or establishing properties.

---

[5] By a famous theorem of Birkhoff, a variety is, equivalently, a class of algebras of the same type that is closed under homomorphic images, subalgebras, and direct products.

2.4.2 Commutative Residuated Lattices

In theory, we can accommodate all kinds of algebras for building fuzzy logics. In practice, however, it is more convenient and interesting to settle on fixed selections of operations. In particular, the class of *residuated lattices* – combining the notions of a lattice of truth values with a monoid operation and its residuum – provides a versatile framework for dealing with a wide range of non-classical logics, including fuzzy logics. Here we will restrict our attention to the commutative case and (as is usual for substructural logics) add an extra constant to the type to model negation.

Definition 2.50. A *pointed commutative residuated lattice (pcrl)* is an algebra:

$$\mathbf{A} = \langle L_{\mathbf{A}}, \wedge, \vee, \odot, \rightarrow, \mathrm{e}, \mathrm{f} \rangle$$

with binary operations $\wedge$, $\vee$, $\odot$, $\rightarrow$, and constants e and f such that:

1. $\langle L_{\mathbf{A}}, \wedge, \vee \rangle$ is a lattice.
2. $\langle L_{\mathbf{A}}, \odot, \mathrm{e} \rangle$ is a commutative monoid.
3. $x \odot y \leq z$ iff $x \leq y \rightarrow z$ for $x, y, z \in L_{\mathbf{A}}$.

We also define operations:

$$\neg x =_{\text{def}} x \rightarrow \mathrm{f} \qquad x \oplus y =_{\text{def}} \neg x \rightarrow y$$
$$0.x =_{\text{def}} \mathrm{f} \qquad (n+1).x =_{\text{def}} x \oplus n.x$$
$$x^0 =_{\text{def}} \mathrm{e} \qquad x^{n+1} =_{\text{def}} x \odot x^n$$

For convenience, a *commutative residuated lattice (crl)* may be regarded as a pcrl where $\mathrm{f} = \mathrm{e}$, and the type can be shortened accordingly.

Example 2.51. Examples of (pointed) commutative residuated lattices occur in many branches of Mathematics, perhaps most pertinently:

$$\mathbf{Z} = \langle \mathbb{Z}, \min, \max, +, -, 0 \rangle$$
$$\mathbf{Q} = \langle \mathbb{Q}, \min, \max, +, -, 0 \rangle$$
$$\mathbf{R} = \langle \mathbb{R}, \min, \max, +, -, 0 \rangle$$

where $-$ is the binary subtraction function, and residuation in such cases amounts to the fact that $x + y \leq z$ iff $x \leq z - y$.

Pointed commutative residuated lattices contain most of the functions that we need for defining logics: $\odot$ and $\rightarrow$ for conjunction and implication, $\wedge$ and $\vee$ for weak order defining conjunction and disjunction, e for truth, and f for falsity. However, for many fuzzy logics, the truth value set is bounded (most commonly by 0 and 1) and it is helpful to also represent these in the algebra:

Definition 2.52. A *bounded pcrl (bpcrl)* is an algebra:

$$\mathbf{A} = \langle L_{\mathbf{A}}, \wedge, \vee, \odot, \rightarrow, \mathrm{e}, \mathrm{f}, \bot, \top \rangle$$

such that $\langle L_{\mathbf{A}}, \wedge, \vee, \odot, \rightarrow, \mathrm{e}, \mathrm{f}\rangle$ is a pcrl and $\langle L_{\mathbf{A}}, \wedge, \vee, \bot, \top\rangle$ is a bounded lattice.

From a fuzzy perspective, the most important examples of bpcrls are based on residuated uninorms.

Definition 2.53. For any residuated uninorm $*$ and $\mathrm{f} \in [0,1]$, we define the bpcrl:

$$\mathbf{A}(*, \mathrm{f}) =_{\mathrm{def}} \langle [0,1], \min, \max, *, \rightarrow_*, \mathrm{e}_*, \mathrm{f}, 0, 1\rangle$$

Note that the "falsity" value f is maintained here for uniformity. It only really plays a part in the logic when the negation $\neg x =_{\mathrm{def}} x \rightarrow \mathrm{f}$ is an involution. Indeed for t-norm based logics $\mathrm{f} = 0$, and for the logic based on the cross-ratio uninorm, $\mathrm{f} = \mathrm{e}_*$.

Let us denote the classes of pcrls and bpcrls by $\mathcal{CRL}^+$ and $\mathcal{BCRL}^+$, respectively. We then observe the following significant fact:

Proposition 2.54. *$\mathcal{CRL}^+$ and $\mathcal{BCRL}^+$ are varieties.*

Proof. The equations for a monoid have already been mentioned above, and those for (bounded) lattices are given in Example 2.46. The following equations then guarantee the residuation property:

$$\begin{array}{ll} x \odot (y \vee z) = (x \odot y) \vee (x \odot z) & \quad (x \odot (x \rightarrow y)) \vee y = y \\ x \rightarrow (y \wedge z) = (x \rightarrow y) \wedge (x \rightarrow z) & \quad (x \rightarrow (x \odot y)) \wedge y = y \end{array}$$

Let $\mathbf{A}$ be an algebra satisfying these equations, and suppose first that $x \leq y \rightarrow z$; i.e. $x \vee (y \rightarrow z) = y \rightarrow z$. Then:

$$y \odot (y \rightarrow z) = y \odot (x \vee (y \rightarrow z)) = (y \odot x) \vee (y \odot (y \rightarrow z))$$

But $(y \odot (y \rightarrow z)) \vee z = z$, so $z = (y \odot x) \vee (y \odot (y \rightarrow z)) \vee z = (y \odot x) \vee z$; i.e. $y \odot x \leq z$.

For the other direction, if $x \odot y \leq z$, i.e. $(x \odot y) \wedge z = x \odot y$, then:

$$(y \rightarrow (x \odot y)) \wedge (y \rightarrow z) = y \rightarrow ((x \odot y) \wedge z) = y \rightarrow (x \odot y)$$

But then $x \wedge (y \rightarrow z) = x \wedge (y \rightarrow (x \odot y)) \wedge (y \rightarrow z) = x \wedge (y \rightarrow (x \odot y)) = x$; i.e. $x \leq y \rightarrow z$ as required. □

(Bounded) (pointed) commutative residuated lattices provide a common framework for many important classes of algebras. The only caveat is that we must be rather loose with the type. Algebras are said to be *term equivalent* if the fundamental operations of one are definable using the fundamental operations of the other.

- *Boolean algebras*, the algebras of Classical Logic, are term equivalent to the variety $\mathcal{BA}$ of pcrls satisfying $x \odot y = x \wedge y$ and $(x \rightarrow \mathrm{f}) \rightarrow \mathrm{f} = x$.
- *Heyting algebras*, the algebras of Intuitionistic Logic, are term equivalent to the variety $\mathcal{HA}$ of pcrls satisfying $x \odot y = x \wedge y$ and $\mathrm{f} = \mathrm{f} \wedge x$.
- *MV-algebras*, the algebras of Łukasiewicz Logic, are term equivalent to the variety $\mathcal{MV}$ of pcrls satisfying $x \vee y = (x \rightarrow y) \rightarrow y$ and $\mathrm{f} = \mathrm{f} \wedge x$.

Table 2.1 Properties for commutative residuated lattices

Name	Condition
integral	$\mathrm{e} \leq (x \to \mathrm{e}) \wedge (\mathrm{f} \to x)$
involutive	$x = \neg\neg x$
idempotent	$x = x \odot x$
square-increasing	$x \leq x \odot x$
square-decreasing	$x \odot x \leq x$
n-contractive	$x^{n-1} \leq x^n$
splitting	$\mathrm{e} \leq x \vee \neg x$
strict	$\mathrm{e} \leq \neg(x \wedge \neg x)$
distributive	$x \wedge (y \vee z) = (x \wedge y) \vee (x \wedge z)$
prelinear	$\mathrm{e} \leq (x \to y) \vee (y \to y)$

- *Lattice-ordered abelian groups (abelian ℓ-groups)* are term equivalent to the variety $\mathcal{ALG}$ of pcrls satisfying $x \odot (x \to \mathrm{e}) = \mathrm{e}$ and $\mathrm{f} = \mathrm{e}$.

Other important varieties are obtained by mixing and matching the conditions listed in Table 2.1, recalling that an inequality $t \leq s$ for these algebras can always be rewritten as an equality $t \wedge s = t$. For example, $\mathcal{ICRL}$ and $\mathcal{DCRL}$ are the varieties of integral and distributive pcrls, respectively.

Classes of algebras $\mathbf{A}(*,\mathrm{f})$ where $*$ is a residuated uninorm and $\mathrm{f} \in [0,1]$ do not form varieties, however. There is no set of equations that are satisfied only by non-empty collections of these algebras. Instead, we focus on varieties generated by these classes. For example, let $\mathcal{K} = \{\mathbf{A}(*,0) : * \text{ is a } t\text{-norm}\}$. We will see later that $\mathcal{V}(\mathcal{K})$, the variety generated by $\mathcal{K}$, is exactly the class of integral prelinear bpcrls.

2.4.3 The Dedekind-MacNeille Completion

In proving generation results for varieties – specifically, generation by algebras based on the real numbers – we will make key use of *Dedekind-MacNeille completions*, based on Dedekind's construction of the real numbers from the rationals. Key features of the construction are the fact that the rationals can be embedded into the reals, and that every subset of the reals has both a supremum and an infimum.

The construction was generalized by MacNeille to arbitrary posets as follows.

Definition 2.55. For a poset $P = \langle \alpha, \leq \rangle$, let:

$$\mathsf{DM}(P) =_{\mathrm{def}} \{\beta \subseteq \alpha : (\beta^u)^l = \beta\}$$

where $\beta^u =_{\mathrm{def}} \{x \in \alpha : y \leq x \text{ for all } y \in \beta\}$ and $\beta^l =_{\mathrm{def}} \{x \in \alpha : x \leq y \text{ for all } y \in \beta\}$.

For any (even infinite) subset $D \subseteq \mathsf{DM}(P)$, the infimum and supremum of D according to the ordering $\subseteq$ are the intersection $\cap D$ and union $\cup D$, respectively. Moreover, it is easy to check that $((\cap D)^u)^l = \cap D$ and $((\cup D)^u)^l = \cup D$. Hence:

Lemma 2.56. $\langle \mathsf{DM}(P), \subseteq \rangle$ *is a complete lattice.*

Just as Dedekind extended the operations of the rationals to the reals, so the operations of an arbitrary pcrl or bpcrl **A** can be extended from $L_{\mathbf{A}}$ to $\mathsf{DM}(L_{\mathbf{A}})$ to obtain a new algebra.

Definition 2.57. For any pcrl or bpcrl **A**, $\mathsf{DM}(\mathbf{A})$ is an algebra of the same type with universe $\mathsf{DM}(L_{\mathbf{A}})$ and fundamental operations defined (where appropriate) by:

$$\begin{aligned} \alpha \wedge_{\mathsf{DM}} \beta &=_{\text{def}} \alpha \cap \beta & \mathrm{e}_{\mathsf{DM}} &=_{\text{def}} \{\mathrm{e}\}^l \\ \alpha \vee_{\mathsf{DM}} \beta &=_{\text{def}} ((\alpha \cup \beta)^u)^l & \mathrm{f}_{\mathsf{DM}} &=_{\text{def}} \{\mathrm{f}\}^l \\ \alpha \odot_{\mathsf{DM}} \beta &=_{\text{def}} (\{x \odot y : x \in \alpha, y \in \beta\}^u)^l & \bot_{\mathsf{DM}} &=_{\text{def}} \{\bot\} \\ \alpha \to_{\mathsf{DM}} \beta &=_{\text{def}} \{x \in L_{\mathbf{A}} : x \odot y \in \beta \text{ for all } y \in \alpha\} & \top_{\mathsf{DM}} &=_{\text{def}} L_{\mathbf{A}} \end{aligned}$$

Theorem 2.58. *For any (bounded) pcrl* **A**:

(a) $\mathsf{DM}(\mathbf{A})$ *is a (bounded) pcrl.*

(b) *If* **A** *is integral, involutive, idempotent, square decreasing or increasing, n-contractive, splitting, strict, linearly or densely ordered, then so is* $\mathsf{DM}(\mathbf{A})$.

(c) $\Phi(x) = \{x\}^l$ *is an embedding of* **A** *into* $\mathsf{DM}(\mathbf{A})$.

Proof. It is easy to see that $\mathsf{DM}(\mathbf{A})$ satisfies the (bounded) lattice and monoid properties. For residuation, observe that:

$$\begin{aligned} \alpha \odot_{\mathsf{DM}} \beta \subseteq \gamma \quad &\text{iff} \quad x \odot y \in \gamma \text{ for all } x \in \alpha \text{ and } y \in \beta \\ &\text{iff} \quad x \in \beta \to_{\mathsf{DM}} \gamma \text{ for all } x \in \alpha \\ &\text{iff} \quad \alpha \subseteq \beta \to_{\mathsf{DM}} \gamma \end{aligned}$$

It is straightforward also to verify that if **A** is integral, involutive, idempotent, square decreasing or increasing, n-contractive, splitting, strict, linearly or densely ordered, then the same holds for $\mathsf{DM}(\mathbf{A})$. Finally for (c), Φ is clearly a homomorphism (preserving infinite joins and meets) and injective since if $x \neq y$, then either $x \not\leq y$ or $y \not\leq x$, so $\{x\}^l \neq \{y\}^l$. □

The Dedekind-MacNeille completion will be useful in the next chapter when we use it to show that certain logics are characterized by algebras **A** where $L_{\mathbf{A}} = [0, 1]$. Intuitively, it allows us to step from an algebra that is dense and linearly ordered to one that is isomorphic to an algebra on the real numbers. Note, however, that not all our desired properties are preserved by the Dedekind-MacNeille completion. Most importantly, divisibility – key for the logic of continuous t-norms – may be lost in this construction.

2.5 Languages and Logics

So far in this chapter we have seen a great many algebras suitable for interpreting fuzzy logics. What we have not yet seen is a general definition of these logics.

Table 2.2 Common connectives

Connectives	Arity	Meaning
$\wedge, \odot$	2	conjunction: "... and ..."
$\vee, \oplus$	2	disjunction: "... or ..."
$\to$	2	implication: "if... then..."
$\leftrightarrow$	2	bi-implication: "... if and only if ...' '
$\neg$	1	negation: "not ..."
e, $\top$	0	truth
f, $\bot$	0	falsity

There is a good reason for this of course. Logics based on the real numbers have been introduced with different motivations and in a variety of contexts. To place all of these logics in a single general framework would be difficult, perhaps even impossible. Nevertheless, as we will see below, there do exist standard uniform presentations which cover the most important systems of Fuzzy Logic, and have useful and interesting connections with proof theory.

Our starting point is – as usual – a (formal) language, an essential ingredient of any logic, whatever the presentation. Languages provide the basic materials for making (vague) statements like "John is tall", "I am hungry and thirsty", "If the clothes are dirty, then the water is hot", etc. In propositional logics, we deal with such statements (propositions) using variables to stand for arbitrary basic propositions, and logical connectives to combine propositions into more complex ones:

Definition 2.59. A *(propositional) language* of type $\nu = (n_i)_{i\in I}$ is a set $\mathcal{L} = (\star_i)_{i\in I}$ where $\star_i$ is an n_i-ary function symbol called an *$\mathcal{L}$-connective* of arity n_i.

In this book we will need more connectives than is usually supplied for Classical Logic. A selection of these is displayed in Table 2.2, together with their arities and some clues as to their expected behaviour. Note that for simplicity, we use here the same symbols for connectives as for operations of (bounded) pcrls, relying on context to distinguish between the different uses.

Formulas for a language $\mathcal{L}$ (and their subformulas), denoted $A, B, C\ldots$, are built up out of a set of variables, denoted $p, q, r, \ldots$, and connectives as follows:

Definition 2.60. Let $\mathcal{L}$ be a language and X a set of variables. Then the set $\mathrm{Fm}_{\mathcal{L}}(\mathsf{X})$ of *$\mathcal{L}$-formulas over* X is the smallest set such that:

(1) $\mathsf{X} \subseteq \mathrm{Fm}_{\mathcal{L}}(\mathsf{X})$.
(2) $\star(A_1,\ldots,A_n) \in \mathrm{Fm}_{\mathcal{L}}(\mathsf{X})$ for each n-ary $\star \in \mathcal{L}$ and $A_1,\ldots,A_n \in \mathrm{Fm}_{\mathcal{L}}(\mathsf{X})$.

The *subformulas* of a formula are defined inductively by:

(1) A is a subformula of A for all $A \in \mathrm{Fm}_{\mathcal{L}}(\mathsf{X})$.
(2) each subformula of A_i for $i = 1\ldots n$ is a subformula of $\star(A_1,\ldots,A_n)$ for all n-ary $\star \in \mathcal{L}$ and $\{A_1,\ldots,A_n\} \subseteq \mathrm{Fm}_{\mathcal{L}}(\mathsf{X})$.

The *complexity* of a formula is defined inductively by:

Table 2.3 Languages appearing in this book

Label	Connectives
$\mathcal{L}_B$	$\wedge, \vee, \odot, \to, \mathrm{f}, \mathrm{e}, \bot, \top$
$\mathcal{L}_F$	$\wedge, \vee, \odot, \to, \mathrm{f}, \mathrm{e}$
$\mathcal{L}_G$	$\wedge, \vee, \to, \bot, \top$
$\mathcal{L}_I$	$\wedge, \to, \bot, \top$
$\mathcal{L}_T$	$\odot, \to, \bot$
$\mathcal{L}_C$	$\odot, \to$
$\mathcal{L}_A$	$\wedge, \to$
$\mathcal{L}_L$	$\to, \bot$
$\mathcal{L}_\to$	$\to$

(1) $\mathrm{cp}(p) = 1$ for $p \in \mathsf{X}$.
(2) $\mathrm{cp}(\star(A_1,\ldots,A_n)) = 1 + \sum_{i=1}^{n} \mathrm{cp}(A_i)$ for n-ary $\star \in \mathcal{L}$ and $\{A_1,\ldots,A_n\} \subseteq \mathrm{Fm}_{\mathcal{L}}(\mathsf{X})$.

We call formulas with complexity 1, *atoms* or *atomic* formulas.

Definition 2.61. A set of formulas $T \subseteq \mathrm{Fm}_{\mathcal{L}}(\mathsf{X})$ is called an $\mathrm{Fm}_{\mathcal{L}}(\mathsf{X})$-*theory*.

Connectives may also be defined as *abbreviations* of other connectives; in particular:

$$\neg A =_{\mathrm{def}} A \to \mathrm{f} \qquad A \oplus B =_{\mathrm{def}} \neg A \to B$$
$$0.A =_{\mathrm{def}} \mathrm{f} \qquad (n+1).A =_{\mathrm{def}} A \oplus n.A$$
$$A^0 =_{\mathrm{def}} \mathrm{e} \qquad A^{n+1} =_{\mathrm{def}} A \odot A^n$$

Here the defined connectives, which mirror exactly the defined operations for (bounded) pcrls, should be thought of just as syntactic conveniences, not present in the language itself.

For each binary connective $\star$ we swap freely between prefix $\star(x,y)$ and infix notation $x \star y$. We disregard brackets where readability is not at stake, and assume that $\neg$ binds more tightly than other connectives, e.g. reading $\neg p \wedge q$ as $\neg(p) \wedge q$ rather than $\neg(p \wedge q)$. For convenience, when $\star \in \{\wedge, \vee, \odot, \oplus\}$, we sometimes abuse notation and write $\star\{A_1,\ldots,A_n\}$ for $A_1 \star (A_2 \star \ldots (A_{n-1} \star A_n)\ldots)$. We also define (when these make sense for the language at hand):

$$\bigvee \emptyset =_{\mathrm{def}} \bot \qquad \bigwedge \emptyset =_{\mathrm{def}} \top \qquad \odot \emptyset =_{\mathrm{def}} \mathrm{e} \qquad \oplus \emptyset =_{\mathrm{def}} \mathrm{f}$$

A selection of the languages used in this book is presented in Table 2.3.

A logic can be presented semantically as a language together with a class of algebras of the same type. That is, for every connective of the language there is an associated operation of the algebra with the same arity. Indeed, as already mentioned, we abuse notation here by using the same symbols for both connectives and their algebraic counterparts. We begin then by defining mappings from formulas into algebras of the same type, fixing for now, a countably infinite set of variables X, and writing $\mathrm{Fm}_{\mathcal{L}}$ rather than $\mathrm{Fm}_{\mathcal{L}}(\mathsf{X})$. For convenience, we will also assume that the language $\mathcal{L}$ of our formulas and theories matches the language of bpcrls or pcrls, $\mathcal{L}_B$ or $\mathcal{L}_F$, as appropriate.

Definition 2.62. Let $\mathcal{L} = (\star_i)_{i\in I}$ be a language and $\mathbf{A} = \langle L_{\mathbf{A}}, (f_i)_{i\in I}\rangle$ an algebra of the same type $\nu = (n_i)_{i\in I}$. An **A**-*valuation* for $\mathcal{L}$ is a function $v : \mathrm{Fm}_{\mathcal{L}} \to L_{\mathbf{A}}$ such that for all $i \in I$:

$$v(\star_i(A_1,\ldots,A_{n_i})) = f_i(v(A_1),\ldots,v(A_{n_i}))$$

Definition 2.63. Let **A** be a (bounded) pcrl:

- A formula $A \in \mathrm{Fm}_{\mathcal{L}}$ is **A**-*valid* if $v(A) \geq \mathrm{e}$ for all **A**-valuations v.
- An **A**-valuation v is an **A**-*model* of an $\mathrm{Fm}_{\mathcal{L}}$-theory T if $v(A) \geq \mathrm{e}$ for all $A \in T$.
- We write $T \models_{\mathbf{A}} A$ if every **A**-model of T is an **A**-model of $\{A\}$.

For a class of (bounded) pcrls $\mathcal{K}$, we write $T \models_{\mathcal{K}} A$ if $T \models_{\mathbf{A}} A$ for all $\mathbf{A} \in \mathcal{K}$.

Logics can then be presented semantically via a class of algebras. The valid formulas or tautologies of the logic are those formulas that are valid in every algebra of the class. In particular, *substructural logics* are often characterized as logics based on classes of (bounded) (pointed) (commutative) residuated lattices. However, for fuzzy logics, something more is required. There are many options, but in this book we will mostly encounter logics based on bpcrls where the universe is $[0,1]$, the lattice operations are min and max, and $\odot$, $\to$, and e are a uninorm, its residuum, and unit, respectively. That is, we treat logics based on algebras of the form $\mathbf{A}(*,\mathrm{f})$ where $*$ is a uninorm and f an arbitrary member of $[0,1]$.

Definition 2.64. A *logic* L based on a class of (bounded) pcrls $\mathcal{K}$ is the set:

$$\{(T,A) : T \models_{\mathcal{K}} A\}$$

We write $T \models_{\mathrm{L}} A$ for $(T,A) \in \mathrm{L}$ and say that A is L-*valid* if $\models_{\mathrm{L}} A$.

Logics based on single algebras include *Łukasiewicz Logic* Ł, *Gödel Logic* G, and *Product Logic* P, the fundamental fuzzy logics, defined via the algebras $\mathbf{A}(*_{Ł},0)$, $\mathbf{A}(*_{\mathsf{G}},0)$, and $\mathbf{A}(*_{\mathsf{P}},0)$, respectively. Logics based on classes of algebras include *Basic Logic* BL and *Monoidal t-norm Logic* MTL, defined via the algebras $\mathbf{A}(*,0)$ where $*$ is a continuous t-norm and a residuated t-norm, respectively. Other logics studied in the literature are obtained by requiring that the uninorm be idempotent or n-contractive, or that the negation be involutive. Table 2.4 provides a reference, shortening residuated to res., involutive to inv., idempotent to idem., and contractive to cont.

2.6 Historical Remarks

The study of t-norms – "t" for triangular – has a long and distinguished history, originating in the context of statistical metric spaces with the 1942 paper of

Table 2.4 Fuzzy logics

Logic	Name	Class of algebras
UL	Uninorm Logic	$\{\mathbf{A}(*,\mathrm{f}) : *$ a res. uninorm, $\mathrm{f} \in [0,1]\}$
IUL	Involutive UL	$\{\mathbf{A}(*,\mathrm{f}) : *$ an inv. res. uninorm, $\mathrm{f} \in [0,1]\}$
UML	Uninorm Mingle Logic	$\{\mathbf{A}(*,\mathrm{f}) : *$ an idem. res. uninorm, $\mathrm{f} \in [0,1]\}$
IUML	Involutive UML	$\{\mathbf{A}(*,\mathrm{e}) : *$ an idem. inv. res. uninorm$\}$
MTL	Monoidal *t*-norm Logic	$\{\mathbf{A}(*,0) : *$ a res. *t*-norm$\}$
IMTL	Involutive MTL	$\{\mathbf{A}(*,0) : *$ an inv. *n*-cont. res. *t*-norm$\}$
$\mathrm{MTL_n}$	*n*-Contractive MTL	$\{\mathbf{A}(*,0) : *$ an *n*-cont. res. *t*-norm$\}$
$\mathrm{IMTL_n}$	*n*-Contractive IMTL	$\{\mathbf{A}(*,0) : *$ an inv. *n*-cont. res. *t*-norm$\}$
SMTL	Strict MTL	$\{\mathbf{A}(*,0) : *$ a res. *t*-norm; $\neg_*^0$ is $\mathrm{n_G}\}$
PMTL	Product MTL	$\{\mathbf{A}(*,0) : *$ a restricted cancellative res. *t*-norm$\}$
BL	Basic Logic	$\{\mathbf{A}(*,0) : *$ a continuous *t*-norm$\}$
SBL	Strict BL	$\{\mathbf{A}(*,0) : *$ a continuous *t*-norm; $\neg_*^0$ is $\mathrm{n_G}\}$
Ł	Łukasiewicz Logic	$\mathbf{A}(*_{\text{Ł}},0)$
G	Gödel Logic	$\mathbf{A}(*_{\mathrm{G}},0)$
P	Product Logic	$\mathbf{A}(*_{\mathrm{P}},0)$
A	Abelian Logic	$\mathbf{R}$
CHL	Cancellative Hoop Logic	$\langle (0,1], *_{\mathrm{P}}, \to_{\mathrm{P}}, 1\rangle$
CRL	Cross-Ratio Logic	$\mathbf{A}(*_{\mathrm{CRL}}, \frac{1}{2})$

Menger [140] (see also [196] for more details). The key ordinal sum representation of continuous *t*-norms described above was established by Mostert and Shields in 1957 [157]. This result and a wealth of further material on *t*-norms and related operators may be found in the 2000 monograph of Klement, Mesiar, and Pap [129]. The particular use of *t*-norms in Fuzzy Logic extends back to Zadeh's 1965 paper [223], and was further developed by Goguen in 1969 [100] and Pavelka in 1979 [181] among many others. The systematic approach followed in this chapter was expounded by Hájek for continuous *t*-norms in his 1998 monograph [105] and extended to left-continuous *t*-norms by Godo and Esteva in 2001 [77]. Other useful references for the *t*-norm based methodology are the books of Gottwald [102], Novák et al. [167], and Turunen [211].

Uninorm aggregation operators were introduced explicitly by Yager and Rybalov in 1996 [221], but had already appeared in the 1970s and 1980s in papers by Silvert [199] (the cross-ratio uninorm), Czogala and Drewniak [65] (representable uninorms), and Dombi [71] (idempotent uninorms), and as combining functions for expert systems such as MYCIN [198] (see also [113]). The structural properties of representable and idempotent uninorms and their residua described in this chapter were obtained (using earlier results) by De Baets, Fodor, Rybalov, and Yager in [68, 69, 83]. The extension of Hájek's *t*-norm approach to uninorm based logics was developed by Metcalfe and Montagna in [144]. The particular case of the cross-ratio uninorm was treated with other "$[0,1)$-continuous uninorm logics" by Gabbay and Metcalfe in [87].

For a general introduction to partially ordered sets, lattices, and the fundamentals of Universal Algebra we refer to the textbooks [40, 67]. The more specific study of residuated lattices originated in the first half of the 20th century with the theory of ring ideals, in particular, with Ward and Dilworth's 1939 paper [217]. Such structures have appeared repeatedly – often with conflicting definitions – in the Algebra and Logic literature. Particularly important for fuzzy logics are the 1985 paper of Ono and Komori [177] which treats algebras for "contraction-free" substructural logics, and the 1995 paper of Höhle [119] which puts forwards integral commutative residuated lattices as suitable algebras for investigating fuzziness. Also significant is the 1999 monograph of Cignoli, D'Ottaviano, and Mundici [58] which provides an intensive study of Chang's MV-algebras, the algebraic semantics of Łukasiewicz logic. The definitions of (pointed) (bounded) (commutative) residuated lattices used in this chapter follow the uniform and comprehensive approach described by Tsinakis and co-workers in the early 2000s [116, 127, 210]. These structures and their connections with substructural logics are also investigated in detail in the 2007 book of Galatos, Jipsen, Kowalski, and Ono [90].

Finally, the Dedekind-MacNeille completion described above is based on MacNeille's 1937 poset generalization of Dedekind's construction of the reals [136]. Such completions have proved useful in many different contexts for non-classical logics, including – as we will see in future chapters – establishing certain completeness results, and even for proving cut admissibility (see e.g. [175]).

Chapter 3
Hilbert Systems

In the last chapter, we considered semantic aspects of fuzzy logics: what it means for a formula to be valid in some algebra. Here we turn our attention to syntax and what it means for formulas (or other structures) to be derivable in a particular proof system. The challenge is of course to show that the two coincide – derivable formulas are valid in suitable algebras and vice versa.

Hilbert systems are perhaps the best known of the vast panorama of proof systems existing in the literature. They consist of a (usually small) set of rules that generate distinguished formulas (theorems) from an initial set of formulas (axioms). Such systems provide a flexible framework for presenting a wide spectrum of logics. Usually, they are easily matched with appropriate classes of algebras. On the other hand, Hilbert systems are not always so convenient for proof search or investigating algorithmic aspects of logics. For these tasks we will need the Gentzen system framework described in Chapter 4.

3.1 Structures and Systems

Several proof frameworks appear in this book, so we will make the core definitions here as general as possible, beginning with the elements of the derivations themselves. For Hilbert systems these are just formulas, but in later chapters we will make use of sequents (ordered pairs of multisets of formulas), hypersequents (multisets of sequents), and even more complicated constructions. In general, for a proof system we need only assume some distinguished "set of structures" $\mathcal{S}$. Inferences and rules for the system are then constructed from members of this set as follows:

Definition 3.1. An *inference* for a set of structures $\mathcal{S}$ is an ordered pair consisting of a structure $W \in \mathcal{S}$ called the *conclusion*, and a finite set (possibly empty) of structures $W_1, \ldots, W_n \in \mathcal{S}$ called the *premises*, written as either $W_1, \ldots, W_n \;/\; W$ or:

$$\frac{W_1 \;\ldots\; W_n}{W}$$

An *(inference) rule* (r) for $\mathcal{S}$ is a set of inferences for $\mathcal{S}$, called *instances* of (r), and a *proof system (calculus)* $\mathbb{C}$ consists of a set of structures $\mathcal{S}$ and a set of rules for $\mathcal{S}$.

The structures for an arbitrary[1] Hilbert system HL (the "H" is for Hilbert) are formulas $\mathrm{Fm}_{\mathcal{L}}(\mathsf{X})$ for some language $\mathcal{L}$ and a countably infinite set of variables X. As in the last chapter, let us assume for now that X is fixed, and just speak of formulas in $\mathrm{Fm}_{\mathcal{L}}$. Typically, rules for Hilbert systems are presented via *schema* – formulas with variables replaced by formula meta-variables $A, B, C \ldots$ – where substituting actual formulas for the meta-variables gives instances of the rule. Schema with no premises are called *axiom schema* and their instances are called *axioms*. Hilbert systems themselves are often called axiomatizations: they "axiomatize" some logic.

Example 3.2. For $\mathcal{L} = \{\neg, \wedge, \vee, \to\}$, the inference for $\mathrm{Fm}_{\mathcal{L}}$:

$$\frac{\neg(p \wedge r) \vee (p \to q) \quad p \wedge r}{p \to q}$$

is an instance of the "Ackermann gamma rule" for $\mathrm{Fm}_{\mathcal{L}}$ defined by the schema:

$$\frac{\neg A \vee B \quad A}{B}$$

Example 3.3. One famous Hilbert system for Classical Logic – for convenience, call it HCPC – consists of $\mathrm{Fm}_{\mathcal{L}}$ for $\mathcal{L} = \{\to, \neg\}$ with axiom schema:

(CL1) $A \to (B \to A)$
(CL2) $(A \to (B \to C)) \to ((A \to B) \to (A \to C))$
(CL3) $(\neg A \to \neg B) \to (B \to A)$

and the "modus ponens" rule:

$$\frac{A \quad A \to B}{B} \ (\mathrm{MP})$$

Intuitively, a structure W "follows from" a set of structures $\mathcal{S}$ in a proof system $\mathbb{C}$ if, starting with members of $\mathcal{S}$ and rule instances of $\mathbb{C}$ with no premises, we can use rule instances of $\mathbb{C}$ to arrive at W. More precisely, a derivation is a certain kind of "labelled finite tree" (in the set-theoretic sense), defined as follows:

Definition 3.4. A *finite tree* is a finite poset $\langle \alpha, \leq \rangle$ with a distinguished element x_0 called the *root* such that:

1. $x_0 \leq x$ for all $x \in \alpha$.
2. $\langle \{y \in \alpha : y \leq x\}, \leq \rangle$ is linearly ordered for all $x \in \alpha$.

The members of α are called *nodes* and:

---

[1] We will use the roman font L, HL, GL, etc. throughout this book to denote an arbitrary logic or system, reserving sans serif MTL, HUL, GP, etc. for particular cases.

- each node x such that $\{y \in \alpha : x < y\} = \emptyset$ is called a *leaf*.
- for each leaf x, the set $\{y \in \alpha : y \leq x\}$ is called a *branch*.
- a node x is a *child* of a *parent* node y if $y < x$ and $\{z \in \alpha : y < z < x\} = \emptyset$.
- the *height* of the tree is $\sup\{|\{y \in \alpha : y \leq x\}| : x \in \alpha\}$.

A *finite tree labelled by* β consists of a finite tree $\langle \alpha, \leq \rangle$ and a function $f : \alpha \to \beta$.

Definition 3.5. Let $\mathbb{C}$ be a calculus for a set of structures $\mathcal{S}$. A $\mathbb{C}$-*derivation* d of $W \in \mathcal{S}$ from a finite set $\mathcal{T} \subseteq \mathcal{S}$ is a finite tree of height $\text{ht}(d)$ labelled by $\mathcal{S}$ such that:

1. W labels the root and is called the *end-structure* of d.
2. For each node x labelled W_0, either $W_0 \in \mathcal{T}$ or the child nodes of x are labelled $W_1, \ldots, W_n$ and $W_1, \ldots, W_n \;/\; W_0$ is an instance of a rule of $\mathbb{C}$.

$W \in \mathcal{S}$ is $\mathbb{C}$-*derivable from* $\mathcal{T} \subseteq \mathcal{S}$ if there is a $\mathbb{C}$-derivation d of W from a finite set $\mathcal{T}^F \subseteq \mathcal{T}$, written $d; \mathcal{T} \vdash_{\mathbb{C}} W$ or simply $\mathcal{T} \vdash_{\mathbb{C}} W$.

Hilbert system derivations are often displayed in linear format. In this case, an HL-derivation of a formula A from a set of formulas (i.e. a theory) T is written as a sequence of formulas $A_1, \ldots, A_n$ such that $A = A_n$ and for each $i = 1 \ldots n$, A_i is either an axiom of HL, a member of T, or follows from previous formulas in the sequence using a rule of HL. Formulas derivable from the empty theory $\emptyset$ (often written as an empty space) are called *theorems* of HL.

Example 3.6. Derivations of the "identity" theorems $A \to A$ in the calculus HCPC of Example 3.3 can be written in linear format as follows:

1. $A \to ((A \to A) \to A)$	(CL1)
2. $(A \to ((A \to A) \to A)) \to ((A \to (A \to A)) \to (A \to A))$	(CL2)
3. $(A \to (A \to A)) \to (A \to A)$	(MP) 1,2
4. $A \to (A \to A)$	(CL1)
5. $A \to A$	(MP) 3,4

Notice that even derivations of simple theorems like this can be tricky to find.

Before we consider Hilbert systems in greater detail, let us first point out some useful properties of proof systems that follow immediately from the definitions. In particular, a structure derivable from a set of structures $\mathcal{T}$ is derivable from any superset of $\mathcal{T}$ (expansion) and some finite subset of $\mathcal{T}$ (compactness).

Proposition 3.7. *If* $\mathcal{T} \vdash_{\mathbb{C}} W$ *for some calculus* $\mathbb{C}$ *based on* $\mathcal{S}$, *then:*

(a) $\mathcal{T}^+ \vdash_{\mathbb{C}} W$ *whenever* $\mathcal{T} \subseteq \mathcal{T}^+ \subseteq \mathcal{S}$.
(b) $\mathcal{T}^F \vdash_{\mathbb{C}} W$ *for some finite subset* $\mathcal{T}^F$ *of* $\mathcal{T}$.

We also define some general properties of rules.

Definition 3.8. For a calculus $\mathbb{C}$ based on $\mathcal{S}$, a rule (r) for $\mathcal{S}$ is:

- $\mathbb{C}$-*derivable* if $W_1, \ldots, W_n \vdash_{\mathbb{C}} W$ for each instance $W_1, \ldots, W_n \;/\; W$ of (r).

- $\mathbb{C}$-*admissible* if for each instance $W_1, \ldots, W_n \,/\, W$ of (r), whenever $\vdash_{\mathbb{C}} W_i$ for $i = 1 \ldots n$, then $\vdash_{\mathbb{C}} W$.
- $\mathbb{C}$-*invertible* if for each instance $W_1, \ldots, W_n \,/\, W$ of (r), whenever $\vdash_{\mathbb{C}} W$, then $\vdash_{\mathbb{C}} W_i$ for $i = 1 \ldots n$.

For an algebra **A** or logic L (assuming that $\models_{\mathbf{A}} W$ or $\models_{\mathrm{L}} W$ is defined for $W \in \mathcal{S}$), a rule (r) for $\mathcal{S}$ is:

- **A**-*sound* (or L-*sound*) if for each instance $W_1, \ldots, W_n \,/\, W$ of (r), whenever $\models_{\mathbf{A}} W_i$ (or $\models_{\mathrm{L}} W_i$) for $i = 1 \ldots n$, then $\models_{\mathbf{A}} W$ (or $\models_{\mathrm{L}} W$).
- **A**-*invertible* (or L-*invertible*) if for each instance $W_1, \ldots, W_n \,/\, W$ of (r), whenever $\models_{\mathbf{A}} W$ (or $\models_{\mathrm{L}} W$), then $\models_{\mathbf{A}} W_i$ (or $\models_{\mathrm{L}} W_i$) for $i = 1 \ldots n$.

Finally, let us define the general notion of an extension of a calculus.

Definition 3.9. Let $\mathbb{C}_i$ be a calculus based on $\mathcal{S}_i$ with rules R_i for $i = 1, 2$. If $\mathcal{S}_1 \subseteq \mathcal{S}_2$ and $R_1 \subseteq R_2$, then $\mathbb{C}_2$ is called an *extension* of $\mathbb{C}_1$.

$\mathbb{C}_2$ is usually obtained by adding schematic rules $(\mathrm{r}_1), \ldots, (\mathrm{r}_n)$ to $\mathbb{C}_1$, and in this case we write $\mathbb{C}_2 = \mathbb{C}_1 + (\mathrm{r}_1) + \cdots + (\mathrm{r}_n)$. Also, we note the following useful substitution property for Hilbert systems with schematic rules, established by an easy induction on the height of a derivation. Let us write $A[p/B]$ for the result of substituting all occurrences of a variable p in a formula A with a formula B.

Lemma 3.10. *For any Hilbert system* HL *with schematic rules: if* $\vdash_{\mathrm{HL}} A$*, then* $\vdash_{\mathrm{HL}} A[p/B]$.

3.2 Core Axioms and Rules

The goal of this chapter is to define Hilbert systems systematically for a range of fuzzy logics, the challenge being to capture semantic aspects such as linearity and continuity using axioms. Our first step is to define elementary Hilbert systems that characterize validity for general classes of (bounded) pcrls. We then tackle particular logics as extensions of these systems.

The "implicational core" consists of just three axiom schema and one rule:

Definition 3.11 (Axioms for Implication).

(B)	$(A \to B) \to ((B \to C) \to (A \to C))$	(transitivity)
(C)	$(A \to (B \to C)) \to (B \to (A \to C))$	(permutation)
(I)	$A \to A$	(reflexivity)

Definition 3.12 (Modus Ponens).

$$\frac{A \quad A \to B}{B}\ (\mathrm{MP})$$

Example 3.13. The following derivation uses transitivity, permutation, and modus ponens to establish a useful "suffixing" law for implicational formulas:

1. $(C \to A) \to ((A \to B) \to (C \to B))$ (B)
2. $((C \to A) \to ((A \to B) \to (C \to B))) \to ((A \to B) \to ((C \to A) \to (C \to B)))$ (C)
3. $(A \to B) \to ((C \to A) \to (C \to B))$ (MP)

(B), (C), (I), and (MP) provides an axiomatization BCI for the implicational fragment of Linear Logic, one of the most famous substructural logics. To axiomatize the so-called "multiplicative" fragments, we add axioms connecting $\odot$ and $\to$ as conjunction and implication, and fix e as the unit for $\odot$.

Definition 3.14 (Axioms for Multiplicative Conjunction).

$$(\odot 1)\ A \to (B \to (A \odot B))$$
$$(\odot 2)\ (A \to (B \to C)) \to ((A \odot B) \to C)$$

Definition 3.15 (Axioms for the Multiplicative Unit).

$$(\mathrm{e}1)\ \mathrm{e}$$
$$(\mathrm{e}2)\ A \to (\mathrm{e} \to A)$$

Example 3.16. Certain properties of $\odot$ are inherited directly from $\to$. For example, we can prove the commutativity of $\odot$ as follows:

1. $A \to (B \to (A \odot B))$ $(\odot 1)$
2. $(A \to (B \to (A \odot B))) \to (B \to (A \to (A \odot B)))$ (C)
3. $(B \to (A \to (A \odot B)))$ (MP) 1, 2
4. $(B \to (A \to (A \odot B))) \to ((B \odot A) \to (A \odot B))$ $(\odot 2)$
5. $(B \odot A) \to (A \odot B)$ (MP) 3, 4

To characterize the "additive" connectives $\wedge$ and $\vee$, we need more axioms and also a further rule:

Definition 3.17 (Axioms for Additive Conjunction).

$$(\wedge 1)\ (A \wedge B) \to A$$
$$(\wedge 2)\ (A \wedge B) \to B$$
$$(\wedge 3)\ ((A \to B) \wedge (A \to C)) \to (A \to (B \wedge C))$$

Definition 3.18 (Adjunction Rule).

$$\frac{A \quad B}{A \wedge B}\ (\mathrm{ADJ})$$

Definition 3.19 (Axioms for Additive Disjunction).

$$(\vee 1)\ A \to (A \vee B)$$
$$(\vee 2)\ B \to (A \vee B)$$
$$(\vee 3)\ ((A \to C) \wedge (B \to C)) \to ((A \vee B) \to C)$$

Sometimes the schema $A \to (B \to (A \wedge B))$ is used in axiomatizations instead of (ADJ). However, in such cases, "weakening" formulas $A \to (B \to A)$ are derivable, which may not always be a desirable feature of the logic under consideration.

Example 3.20. A nice example using (ADJ) is provided by the following derivation of the commutativity law for $\wedge$:

1.	$(A \wedge B) \to A$	$(\wedge 1)$
2.	$(A \wedge B) \to B$	$(\wedge 2)$
3.	$((A \wedge B) \to B) \wedge ((A \wedge B) \to A)$	(ADJ) 1,2
4.	$(((A \wedge B) \to B) \wedge ((A \wedge B) \to A)) \to ((A \wedge B) \to (B \wedge A))$	$(\wedge 3)$
5.	$(A \wedge B) \to (B \wedge A)$	(MP) 3,4

Many of the logics we consider also make use of the additive constants $\bot$ and $\top$: respectively, the "bottom falsity" and "top truth". For fuzzy logics, these represent the endpoints 0 and 1 of the real unit interval. Intuitively, axioms for these constants tell us that "$\bot$ proves everything" and "$\top$ is proved by everything".

Definition 3.21 (Axioms for the Additive Constants).

$$(\bot)\ \bot \to A$$
$$(\top)\ A \to \top$$

In this setting, $\top$ and $\bot$ are different from e and f (the latter has no core axioms to characterize it). In fact, as we will see, identifying e with $\top$ and f with $\bot$ is one route to logics with weakening.

3.3 Axiomatic Extensions

By combining all the axioms and rules above, we obtain HMAILL: a Hilbert system for Multiplicative Additive Intuitionistic Linear Logic in the language $\mathcal{L}_{\mathrm{B}} = \{\wedge, \vee, \odot, \to, \mathrm{f}, \mathrm{e}, \bot, \top\}$, recalled for the reader's convenience in its entirety in Fig. 3.1. Its additive-constant-free cousin HMAILL$^-$ in the language $\mathcal{L}_{\mathrm{F}} = \{\wedge, \vee, \odot, \to, \mathrm{f}, \mathrm{e}\}$ will be the most elementary Hilbert system treated in this book. Axiomatizations for all the other logics, fuzzy or otherwise, that we encounter are obtained by adding further axiom schema to this system. In particular, we will make use of the following conventions:

Definition 3.22. An HL-*extension* for a Hilbert system HL based on $\mathrm{Fm}_{\mathcal{L}_{\mathrm{F}}}$ or $\mathrm{Fm}_{\mathcal{L}_{\mathrm{B}}}$ consists of HL extended with axiom schema based on $\mathrm{Fm}_{\mathcal{L}_{\mathrm{B}}}$. Also, for any HMAILL-extension HL, the Hilbert system HL$^-$ is HL with $(\bot)$ and $(\top)$ removed.

$$
\begin{array}{rl}
(\mathrm{B}) & (A \to B) \to ((B \to C) \to (A \to C)) \\
(\mathrm{C}) & (A \to (B \to C)) \to (B \to (A \to C)) \\
(\mathrm{I}) & A \to A \\
(\odot 1) & A \to (B \to (A \odot B)) \\
(\odot 2) & (A \to (B \to C)) \to ((A \odot B) \to C) \\
(\mathrm{e}1) & \mathrm{e} \\
(\mathrm{e}2) & A \to (\mathrm{e} \to A) \\
(\wedge 1) & (A \wedge B) \to A \\
(\wedge 2) & (A \wedge B) \to B \\
(\wedge 3) & ((A \to B) \wedge (A \to C)) \to (A \to (B \wedge C)) \\
(\vee 1) & A \to (A \vee B) \\
(\vee 2) & B \to (A \vee B) \\
(\vee 3) & ((A \to C) \wedge (B \to C)) \to ((A \vee B) \to C) \\
(\bot) & \bot \to A \\
(\top) & A \to \top
\end{array}
$$

$$
\frac{A \quad A \to B}{B}\ (\mathrm{MP}) \qquad \frac{A \quad B}{A \wedge B}\ (\mathrm{ADJ})
$$

Fig. 3.1 The Hilbert system HMAILL

To perform derivations in Hilbert systems, it will be helpful to have some shortcuts:

Lemma 3.23. *For any* HMAILL^-*-extension* HL*:*

(i) *If* $T \vdash_{\mathrm{HL}} A$ *and* $T \vdash_{\mathrm{HL}} A \to B$*, then* $T \vdash_{\mathrm{HL}} B$.
(ii) *If* $T \vdash_{\mathrm{HL}} A$ *and* $T \vdash_{\mathrm{HL}} B$*, then* $T \vdash_{\mathrm{HL}} A \wedge B$ *and* $T \vdash_{\mathrm{HL}} A \odot B$.
(iii) *If* $T \vdash_{\mathrm{HL}} A \to B$ *and* $T \vdash_{\mathrm{HL}} B \to C$*, then* $T \vdash_{\mathrm{HL}} A \to C$.
(iv) *If* $T \vdash_{\mathrm{HL}} A \to B$ *and* $T \vdash_{\mathrm{HL}} A \to C$*, then* $T \vdash_{\mathrm{HL}} A \to (B \wedge C)$.
(v) *If* $T \vdash_{\mathrm{HL}} A \to C$ *and* $T \vdash_{\mathrm{HL}} B \to C$*, then* $T \vdash_{\mathrm{HL}} (A \vee B) \to C$.
(vi) $T \vdash_{\mathrm{HL}} A \to (B \to C)$ *iff* $T \vdash_{\mathrm{HL}} B \to (A \to C)$.

Proof. (i) Suppose that $d_1; T \vdash_{\mathrm{HL}} A$ and $d_2; T \vdash_{\mathrm{HL}} A \to B$. Then $d; T \vdash_{\mathrm{HL}} B$ where d is a tree with root labelled B having child nodes labelled A and $A \to B$ themselves roots of trees d_1 and d_2. (ii) Similarly to (i), if $d_1; T \vdash_{\mathrm{HL}} A$ and $d_2; T \vdash_{\mathrm{HL}} B$, then $d; T \vdash_{\mathrm{HL}} A \wedge B$ where d is the tree obtained by attaching the roots of d_1 and d_2 as child nodes of a root labelled $A \wedge B$. Also, since $T \vdash_{\mathrm{HL}} A \to (B \to (A \odot B))$, by two applications of (i), $T \vdash_{\mathrm{HL}} A \odot B$. Each case of (iii)–(vi) then follows easily using (i), (ii), and a particular axiom of HMAILL^-. □

We also collect some useful theorems.

Lemma 3.24. *The following are derivable in any* HMAILL^-*-extension:*

(i) $(A \wedge B) \to (B \wedge A)$
(ii) $((A \wedge B) \wedge C) \to (A \wedge (B \wedge C))$
(iii) $A \to (A \wedge A)$
(iv) $(A \vee B) \to (B \vee A)$
(v) $((A \vee B) \vee C) \to (A \vee (B \vee C))$

(vi) $(A \vee A) \to A$
(vii) $A \to (A \wedge (A \vee B))$
(viii) $(A \vee (A \wedge B)) \to A$
(ix) $(A \odot B) \to (B \odot A)$
(x) $((A \odot B) \odot C) \to (A \odot (B \odot C))$
(xi) $(A \odot (A \to B)) \to B$
(xii) $A \leftrightarrow (A \odot \mathrm{e})$
(xiii) $(A \leftrightarrow B) \to ((A \odot C) \leftrightarrow (B \odot C))$
(xiv) $(A \leftrightarrow B) \to ((A \to C) \leftrightarrow (B \to C))$
(xv) $(A \leftrightarrow B) \to ((C \to A) \leftrightarrow (C \to B))$
(xvi) $(A \leftrightarrow B) \to ((A \wedge C) \leftrightarrow (B \wedge C))$
(xvii) $(A \leftrightarrow B) \to ((A \vee C) \leftrightarrow (B \vee C))$

Proof. We just give a derivation for (xi) and leave the rest as exercises:

1. $(A \to B) \to (A \to B)$	(I)
2. $A \to ((A \to B) \to B)$	Lemma 3.23 (vi) 1
3. $(A \to ((A \to B) \to B)) \to ((A \odot (A \to B)) \to B)$	($\odot$2)
4. $(A \odot (A \to B)) \to B$	(MP) 2, 3

□

As we will see soon enough, derivability in *any* HMAILL$^-$-extension corresponds to validity in a suitable class of (bounded) pcrls. This generality notwithstanding, we first describe some key axioms that capture important properties for substructural and fuzzy logics. For the reader's convenience, these axioms and the Hilbert systems they help to axiomatize are collected in Tables 3.1 and 3.2, respectively.

3.3.1 Truth, Falsity, Negation

The axiom schema and rules of HMAILL do not mention the constant f explicitly. Nevertheless, the defined connective $\neg A =_{\mathrm{def}} A \to \mathrm{f}$ already possesses many properties of a negation. Namely:

- *contraposition*: $(A \to B) \to (\neg B \to \neg A)$ is an instance of (B).
- *non-contradiction*: $\neg(A \odot \neg A)$ is an instance of $(A \odot (A \to B)) \to B$.
- *the weak DeMorgan law*: $\neg(A \vee B) \leftrightarrow (\neg A \wedge \neg B)$ is HMAILL$^-$-derivable. One direction, $(\neg A \wedge \neg B) \to \neg(A \vee B)$, is an instance of the axiom ($\vee$3), while the other direction is established as follows:

1. $A \to (A \vee B)$	$(\vee 1)$
2. $(A \to (A \vee B)) \to (\neg(A \vee B) \to \neg A)$	(B)
3. $\neg(A \vee B) \to \neg A$	(MP) 1,2
4. $B \to (A \vee B)$	$(\vee 2)$
5. $(B \to (A \vee B)) \to (\neg(A \vee B) \to \neg B)$	(B)
6. $\neg(A \vee B) \to \neg B$	(MP) 4,5
7. $\neg(A \vee B) \to (\neg A \wedge \neg B)$	Lemma 3.23 (iv) 3,6

For the *strong* De Morgan law, one direction $(\neg A \vee \neg B) \to \neg(A \wedge B)$ is HMAILL^--derivable (an easy exercise), but the other is not.

- *involution*: $A \to \neg\neg A$ is an instance of the HMAILL^--theorem $A \to ((A \to B) \to B)$. However, the other direction $\neg\neg A \to A$ is not HMAILL^--derivable.

To obtain stronger properties for negation, more axioms are needed. In particular, an involutive negation can be characterized by the following schema:

Definition 3.25 (Involution Axioms).

$$(\mathrm{INV}) \quad \neg\neg A \to A$$

Adding (INV) to HMAILL gives an axiomatization HMALL for Multiplicative Additive Linear Logic. Several important fuzzy logics are axiomatized as HMALL-extensions, including Łukasiewicz Logic and Involutive Monoidal t-norm Logic. Indeed, by adding the involution axioms, we obtain an "involutive version" of any HMAILL^--extension.

Example 3.26. Adding (INV) to a logic gives tighter connections between connectives. E.g. in any HMALL^--extension we can derive:

$$(A \to B) \leftrightarrow \neg(A \odot \neg B)$$

(INV) also allows us to complete the missing part of De Morgan's laws. Using the weak DeMorgan law, $\vdash_{\mathsf{HMAILL}^-} \neg(\neg A \vee \neg B) \to (\neg\neg A \wedge \neg\neg B)$. So by contraposition and Lemma 3.23 (iii), $\vdash_{\mathsf{HMAILL}^-} \neg(\neg\neg A \wedge \neg\neg B) \to \neg\neg(\neg A \vee \neg B)$. But then easily $\vdash_{\mathsf{HMALL}^-} \neg(A \wedge B) \to \neg(\neg\neg A \wedge \neg\neg B)$ and $\vdash_{\mathsf{HMALL}^-} \neg\neg(\neg A \vee \neg B) \to (\neg A \vee \neg B)$ using (INV). So using Lemma 3.23 (iii) twice, $\vdash_{\mathsf{HMAILL}^-} \neg(A \wedge B) \to (\neg A \vee \neg B)$.

Negation may be characterized further by fixing relationships between truth constants, e.g. stipulating whether e is more or less true than f.

Definition 3.27 (Axioms for e and f).

$$(\mathrm{e})\ \mathrm{f} \to \mathrm{e}$$
$$(\mathrm{f})\ \mathrm{e} \to \mathrm{f}$$

These rather innocuous looking axioms can have interesting consequences. For example, in the presence of (e), the following rule is admissible for any HMAILL^--extension HL:

$$\frac{A \quad B}{A \oplus B}$$

If $\vdash_{\mathrm{HL}} A$ and $\vdash_{\mathrm{HL}} B$, then also $\vdash_{\mathrm{HL}} \mathrm{e} \to B$. So by (e) and Lemma 3.23 (iii), $\vdash_{\mathrm{HL}} \mathrm{f} \to B$. But $\vdash_{\mathrm{HL}} A \to ((A \to \mathrm{f}) \to \mathrm{f})$. So by Lemma 3.23 (i), $\vdash_{\mathrm{HL}} (A \to \mathrm{f}) \to \mathrm{f}$ and by Lemma 3.23 (iii), $\vdash_{\mathrm{HL}} (A \to \mathrm{f}) \to B$.

3.3.2 Distributivity and Prelinearity

To axiomatize fuzzy logics, in particular, to ensure that the logics are characterized by *linearly ordered* algebras, we require two rather special axiom schema.

Definition 3.28 (Distributivity Axioms).

$$\text{(DIS)} \quad (A \wedge (B \vee C)) \to ((A \wedge B) \vee (A \wedge C))$$

Definition 3.29 (Prelinearity Axioms).

$$\text{(PRL)} \quad (A \to B) \vee (B \to A)$$

These are the fundamental axiom schema of the fuzzy logics in this book. Adding them to HMAILL and HMALL gives Hilbert systems HUL and HIUL for the elementary fuzzy logics Uninorm Logic and Involutive Uninorm Logic, respectively.

Note that prelinearity alone is not in general enough to secure linearity: distributivity is also required. A good example showing this (non-distributive RM) will be given in the next chapter. Nevertheless, it is possible to combine the two in a nice way in the single axiom schema:

$$\text{(LIN)} \quad ((A \to B) \wedge \mathrm{e}) \vee ((B \to A) \wedge \mathrm{e})$$

(LIN) can be used to derive (PRL) as follows:

1. $((A \to B) \wedge \mathrm{e}) \vee ((B \to A) \wedge \mathrm{e})$	(LIN)
2. $((A \to B) \wedge \mathrm{e}) \to (A \to B)$	($\wedge$1)
3. $((B \to A) \wedge \mathrm{e}) \to (B \to A)$	($\wedge$1)
4. $(A \to B) \to ((A \to B) \vee (B \to A))$	($\vee$1)
5. $(B \to A) \to ((A \to B) \vee (B \to A))$	($\vee$2)
6. $((A \to B) \wedge \mathrm{e}) \to ((A \to B) \vee (B \to A))$	Lemma 3.23 (iii) 2,4
7. $((B \to A) \wedge \mathrm{e}) \to ((A \to B) \vee (B \to A))$	Lemma 3.23 (iii) 3,5
8. $(((A \to B) \wedge \mathrm{e}) \vee ((B \to A) \wedge \mathrm{e})) \to ((A \to B) \vee (B \to A))$	Lemma 3.23 (v) 6,7
9. $(A \to B) \vee (B \to A)$	Lemma 3.23 (i) 1,8

We can also prove (DIS) using (LIN) but the derivation is more complicated and is left here as an exercise for the interested reader.

Finally, this is a good place to establish some properties of axiomatizations with distributivity that will be helpful later on for proving completeness results.

Lemma 3.30. *Let* HL *be any* HMAILL$^-$*-extension plus* (DIS)*:*

(a) *If* $\vdash_{\mathrm{HL}} (A \wedge B) \to C$, *then* $\vdash_{\mathrm{HL}} ((A \vee D) \wedge (B \vee D)) \to (C \vee D)$.

(b) *If* $\vdash_{\mathrm{HL}} (A \odot B) \to C$, *then* $\vdash_{\mathrm{HL}} (((A \vee D) \wedge \mathrm{e}) \odot ((B \vee D) \wedge \mathrm{e})) \to (C \vee D)$.

Proof. (a) Suppose that $\vdash_{\mathrm{HL}} (A \wedge B) \to C$. Then using $(\vee 1) - (\vee 3)$ and Lemma 3.23 (iii) and (v):

$$\vdash_{\mathrm{HL}} ((A \wedge B) \vee D) \to (C \vee D)$$

So using (DIS) and Lemma 3.23 (i), $\vdash_{\mathrm{HL}} ((A \vee D) \wedge (B \vee D)) \to (C \vee D)$ as required.

(b) Using the core axioms for the connectives and Lemma 3.23 (various parts):

$$\vdash_{\mathrm{HL}} (A \wedge \mathrm{e}) \to ((B \wedge \mathrm{e}) \to ((A \odot B) \vee D)) \quad \text{and} \quad \vdash_{\mathrm{HL}} (D \wedge \mathrm{e}) \to ((B \wedge \mathrm{e}) \to ((A \odot B) \vee D))$$

Hence by Lemma 3.23 (v) and (vi):

$$\vdash_{\mathrm{HL}} ((A \wedge \mathrm{e}) \vee (D \wedge \mathrm{e})) \to ((B \wedge \mathrm{e}) \to ((A \odot B) \vee D))$$

But then using (DIS) and Lemma 3.23 (i):

$$\vdash_{\mathrm{HL}} ((A \vee D) \wedge \mathrm{e}) \to ((B \wedge \mathrm{e}) \to ((A \odot B) \vee D))$$

Also using the core axioms for the connectives and Lemma 3.23 (various parts):

$$\vdash_{\mathrm{HL}} ((A \vee D) \wedge \mathrm{e}) \to ((D \wedge \mathrm{e}) \to ((A \odot B) \vee D))$$

Hence using Lemma 3.23 (v) and (vi):

$$\vdash_{\mathrm{HL}} ((A \vee D) \wedge \mathrm{e}) \to (((B \wedge \mathrm{e}) \vee (D \wedge \mathrm{e})) \to ((A \odot B) \vee D))$$

So now using (DIS), $(\odot 2)$, and Lemma 3.23 (various parts):

$$\vdash_{\mathrm{HL}} (((A \vee D) \wedge \mathrm{e}) \odot ((B \vee D) \wedge \mathrm{e})) \to ((A \odot B) \vee D)$$

But finally, if $\vdash_{\mathrm{HL}} (A \odot B) \to C$, then $\vdash_{\mathrm{HL}} ((A \odot B) \vee D) \to (C \vee D)$, using $(\vee 1) - (\vee 3)$ and Lemma 3.23 (iii) and (v). So by Lemma 3.23 (iii), $\vdash_{\mathrm{HL}} (((A \vee D) \wedge \mathrm{e}) \odot ((B \vee D) \wedge \mathrm{e})) \to (C \vee D)$ as required. □

3.3.3 Weakening

Weakening axioms are key for axiomatizing t-norm based logics. They capture the fact that the unit element e is also the top element $\top$ of the algebra (in particular, for uninorms $\mathrm{e}_* = 1$), and consequently, always $x \odot y \leq x$. Here we make use of the fact that f and e are in the language of HMAILL$^-$ to give the following characterization:

Definition 3.31 (Weakening Axioms).

$$(\mathrm{W})\ (A \to \mathrm{e}) \wedge (\mathrm{f} \to A)$$

If $\bot$ and $\top$ are in the language, then using (W), these constants collapse to f and e, respectively. That is, $\bot \to \mathrm{f}$ and $\mathrm{f} \to \bot$ follow from ($\bot$) and (W), and $\mathrm{e} \to \top$ and $\top \to \mathrm{e}$ from ($\top$) and (W). In this case, (W) can even be replaced with the single axiom $(\top \to \mathrm{e}) \wedge (\mathrm{f} \to \bot)$.

Example 3.32. The weakening axiom schema more commonly found in the literature, in particular, when e or f are not present in the language, is $A \to (B \to A)$. These formulas are derivable using (W) as follows:

1. $A \to (\mathrm{e} \to A)$	(e2)
2. $\mathrm{e} \to (A \to A)$	Lemma 3.23 (vi) 1
3. $(B \to \mathrm{e}) \wedge (\mathrm{f} \to B)$	(W)
4. $((B \to \mathrm{e}) \wedge (\mathrm{f} \to B)) \to (B \to \mathrm{e})$	($\wedge$1)
5. $B \to \mathrm{e}$	(MP) 3,4
6. $B \to (A \to A)$	Lemma 3.23 (iii) 2,5
7. $A \to (B \to A)$	Lemma 3.23 (vi) 6

However, note that while formulas $A \to \mathrm{e}$ are derivable using the schema $A \to (B \to A)$, this is not true for formulas $\mathrm{f} \to A$. So (W) is a little stronger.

Extending HMAILL and HMALL with the weakening axioms gives axiomatizations HML and HAMALL for Monoidal Logic and Affine Multiplicative Additive Linear Logic, respectively. Adding prelinearity and distributivity (or weakening axioms to HUL and HIUL) gives HMTL and HIMTL for Monoidal t-norm Logic and Involutive Monoidal t-norm Logic, respectively.

Example 3.33. For fuzzy logics with weakening, the axiom (PRL) is sufficient to prove (DIS). Also, we can make use of an implicational version of (PRL):

$$((A \to B) \to C) \to (((B \to A) \to C) \to C)$$

Let F be an instance of this schema, and HL an HMTL^--extension. Easily using (W), $\vdash_{\mathrm{HL}} (A \to B) \to F$ and $\vdash_{\mathrm{HL}} (B \to A) \to F$. So by Lemma 3.23 (v), $\vdash_{\mathrm{HL}} ((A \to B) \vee (B \to A)) \to F$. But then using (PRL) and Lemma 3.23 (i), $\vdash_{\mathrm{HL}} F$. On the other hand, replace C in the schema by $(A \to B) \vee (B \to A)$, and observe that $(A \to B) \to ((A \to B) \vee (B \to A))$ and $(B \to A) \to ((A \to B) \vee (B \to A))$ are HMAILL^--derivable. Then by Lemma 3.23 (i), $(A \to B) \vee (B \to A)$ is derivable in any extension of HMAILL^- with the schema.

3.3.4 Contraction and Mingle

Contraction and mingle axioms are also important for axiomatizing substructural logics. They tell us when a repeated formula $A \odot A$ is more or less true than A alone, and together ensure that the monoid operation (or uninorm) is *idempotent*.

Definition 3.34 (Contraction and Mingle Axioms).

$$(\mathrm{C}_2)\ A \to (A \odot A)$$
$$(\mathrm{M})\ (A \odot A) \to A$$

Indeed, sometimes these axioms are combined to give:

$$(\mathrm{ID})\quad A \leftrightarrow (A \odot A)$$

Extending HUL or HIUL with the contraction and mingle axioms (or just (ID)) gives Hilbert systems HUML and HRM for Uninorm Mingle Logic and the relevance logic RM, respectively. An axiomatization HIUML for Involutive Uninorm Mingle Logic is obtained by adding (f) to HRM. In the case of logics with weakening, (M) is already derivable. Hence an axiomatization for Gödel Logic HG is HMTL plus (C_2), while adding (C_2) to HIMTL (or indeed HAMALL) gives an axiomatization HCL for Classical Logic.

Example 3.35. Sometimes an implicational axiom schema is used instead of (C_2), namely, $(A \to (A \to B)) \to (A \to B)$. It is easy to see that each can be derived using the other. For one direction:

1. $A \to (A \odot A)$	(C_2)
2. $(A \odot A) \to (((A \odot A) \to B) \to B)$	Lemma 3.24 (xi)
3. $A \to (((A \odot A) \to B) \to B)$	Lemma 3.23 (iii) 1, 2
4. $((A \odot A) \to B) \to (A \to B)$	Lemma 3.23 (vi) 3
5. $(A \to (A \to B)) \to ((A \odot A) \to B)$	($\odot$2)
6. $(A \to (A \to B)) \to (A \to B)$	Lemma 3.23 (iii) 4, 5

For the other direction, just replace B with $A \odot A$ in $(A \to (A \to B)) \to (A \to B)$ and note that $A \to (A \to (A \odot A))$ is an instance of ($\odot$1).

We also mention a common generalization of the contraction axioms:

Definition 3.36 (N-Contraction Axioms).

$$(\mathrm{C_n})\ A^{n-1} \to A^n$$

Axiomatizations $\mathrm{HMTL_n}$ and $\mathrm{HIMTL_n}$ for "n-contractive" MTL and IMTL where $n \geq 2$ are obtained by extending HMTL and HIMTL with ($\mathrm{C_n}$), noting that for $n = 2$, this gives Hilbert systems for Gödel Logic and Classical Logic, respectively.

3.3.5 *Divisibility*

Choosing axioms to characterize properties is not always as easy as the above cases might suggest. In particular, it is not entirely obvious how to characterize the property of continuity for t-norms. Following intensive investigation, however, an elegant solution has emerged making use of the related notion of *divisibility*.

Definition 3.37 (Divisibility Axioms).

$$\text{(DIV)} \quad (A \wedge B) \to (A \odot (A \to B))$$

In logics with weakening, $(A \odot (A \to B)) \to (A \wedge B)$ is derivable. Hence in this case, divisibility corresponds to characterizing $A \wedge B$ as $A \odot (A \to B)$. Also, (DIV) could be replaced with:

$$(A \odot (A \to B)) \to (B \odot (B \to A))$$

Our axiomatization HBL for Basic Logic (the logic of continuous t-norms) is obtained by extending HMTL with (DIV). However, $\wedge$ and $\vee$ are definable in terms of the other connectives for this logic. Hence a more common axiomatization based on formulas of the reduced language $\mathcal{L}_{\mathrm{T}} = \{\odot, \to, \bot\}$ (with $A \wedge B =_{\mathrm{def}} A \odot (A \to B)$ and $A \vee B =_{\mathrm{def}} ((A \to B) \to B) \wedge ((B \to A) \to A))$ consists of the following axiom schema with (MP):

$$\begin{array}{rl}
\text{(A1)} & (A \to B) \to ((B \to C) \to (A \to C)) \\
\text{(A2)} & (A \odot B) \to A \\
\text{(A3)} & (A \odot B) \to (B \odot A) \\
\text{(A4)} & (A \odot (A \to B)) \to (B \odot (B \to A)) \\
\text{(A5a)} & (A \to (B \to C)) \to ((A \odot B) \to C) \\
\text{(A5b)} & ((A \odot B) \to C) \to (A \to (B \to C)) \\
\text{(A6)} & ((A \to B) \to C) \to (((B \to A) \to C) \to C) \\
\text{(A7)} & \bot \to A
\end{array}$$

Simple extensions of HBL include two of the most famous and important fuzzy logics. An axiomatization HŁ for Łukasiewicz Logic is obtained by extending HBL with (INV), while Gödel Logic can be axiomatized as HBL extended with (C_2).

3.3.6 Excluded Middle and Non-Contradiction

Many familiar theorems of Classical Logic fail in logics lacking weakening or contraction. However, there may be good reasons to add them to a Hilbert system as extra axioms. In particular, consider the well-known classical principles:

Definition 3.38 (Excluded Middle Axioms).

$$\text{(EM)} \quad A \vee \neg A$$

Definition 3.39 (Non-Contradiction Axioms).

$$\text{(NC)} \quad \neg(A \wedge \neg A)$$

Adding (NC) to HMTL and HBL gives Hilbert systems HSMTL and HSBL, respectively, for the "strict negation" logics SMTL and SBL. In these logics, the t-norm based negation $x \rightarrow_* 0$ is 1 if $x = 0$ and 0 otherwise. It is easy to see that this is the case iff the (NC) axioms are valid in the algebra $\mathbf{A}(*, 0)$.

The excluded middle axiom schema (EM) may seem odd, given that we are aiming for fuzzy logics. Indeed, adding this to any HMAILL$^-$-extension with weakening gives either triviality or Classical Logic. For logics without weakening, however, (EM) can be used to split truth values into exactly two parts: the true (e.g. $\geq e_*$) and the false (e.g. $< e_*$).

3.3.7 Cancellation

Other interesting axioms deal with the cancellation property "if $A \odot C$ implies $B \odot C$, then A implies B". In particular, to axiomatize Product Logic and Cross Ratio Logic, a form of cancellation is needed that holds except when the cancelled formula takes the top or bottom truth value.

Definition 3.40 (Cancellation Axioms).

$$\text{(CAN)}\ (A \rightarrow (A \odot B)) \rightarrow B$$
$$\text{(RCAN)}\ (\top \rightarrow A) \vee (A \rightarrow \bot) \vee ((A \rightarrow (A \odot B)) \rightarrow B)$$

An axiomatization HP for Product Logic is obtained by extending HBL with (RCAN), while HCHL for Cancellative Hoop Logic is obtained by extending HBL$^-$ with (CAN), (e), and (f). For logics with weakening, the disjunct $\top \rightarrow A$ can be removed from (RCAN). Alternatively, (RCAN) can be replaced by (NC) together with:

$$(\Pi)\ \neg\neg A \rightarrow ((A \rightarrow (A \odot B)) \rightarrow B)$$

To see that (RCAN) helps us to derive the (NC) axioms, take an instance of the former with B replaced by $\bot$. Then using weakening, $A \vee \neg A \vee \neg\neg A$ is derivable, so also $\neg(A \wedge \neg A)$ is derivable. Note, however, that adding (RCAN) to HMTL axiomatizes a different logic to P, known as Product Monoidal t-norm Logic PMTL. Finally, note that two other interesting logics make use of cancellation. An axiomatization HA for Abelian Logic consists of HMALL$^-$ plus (CAN), (e), and (f). For Cross Ratio Logic, the Hilbert system HCRL consists of HMALL plus (RCAN), (e), and (f).

3.4 A Local Deduction Theorem

For axiomatizations of Classical Logic, there is an elegant and useful connection between the left and right hand sides of the derivability turnstile, namely:

$$T \cup \{A\} \vdash B \quad \text{iff} \quad T \vdash A \to B.$$

For substructural logics, this relationship – the so-called *deduction theorem* – usually fails. In particular:

- in weakening-free logics, $\{B\} \vdash A \to A$ but $\nvdash B \to (A \to A)$.
- in contraction-free logics, $\{A\} \vdash A \odot A$ but $\nvdash A \to (A \odot A)$.

These failures hold the key to the problem, and also to its solution. We have to allow the formula A in $T \cup \{A\} \vdash B$ to be used either more than once or not at all in the implication on the right. We do this by combining A into so-called "confusions" (conjunctions and fusions) of A. Since the appropriate confusion depends on the derivation, what we obtain is often called a "local" deduction theorem.

Definition 3.41. A *confusion* of a theory T is defined inductively by:

(1) e, $\top$ (if in the language), and any element of T are confusions of T.
(2) If C_1 and C_2 are confusions of T, then so are $C_1 \odot C_2$ and $C_1 \wedge C_2$.

Lemma 3.42. $T \vdash_{\mathrm{HL}} A$ *for any* HMAILL^-*-extension* HL *and confusion* A *of* T.

Proof. We proceed by induction on $\mathrm{cp}(A)$. For the base case, if $A \in T$, A is e, or A is $\top$, then easily $T \vdash_{\mathrm{HL}} A$. For the inductive step, A is $C_1 \odot C_2$ or $C_1 \wedge C_2$ where both C_1 and C_2 are confusions of T. By the induction hypothesis twice, $T \vdash_{\mathrm{HL}} C_1$ and $T \vdash_{\mathrm{HL}} C_2$. Since $T \vdash_{\mathrm{HL}} C_1 \to (C_2 \to (C_1 \odot C_2))$ by $(\odot 1)$ and Lemma 3.23 (i) twice, $T \vdash_{\mathrm{HL}} C_1 \odot C_2$. Also by Lemma 3.23 (ii), directly $T \vdash_{\mathrm{HL}} C_1 \wedge C_2$. □

Theorem 3.43. *Let* HL *be any* HMAILL^-*-extension:*

(a) $T \cup \{A\} \vdash_{\mathrm{HL}} B$ *iff* $T \vdash_{\mathrm{HL}} C \to B$ *for some confusion* C *of* $\{A\}$.
(b) $T \vdash_{\mathrm{HL}} B$ *iff* $\vdash_{\mathrm{HL}} C \to B$ *for some confusion* C *of* T.

Proof. We show, simultaneously establishing both (a) and (b), that:

$$T_1 \cup T_2 \vdash_{\mathrm{HL}} B \quad \text{iff} \quad T_1 \vdash_{\mathrm{HL}} C \to B \quad \text{for some confusion } C \text{ of } T_2$$

For the right-to-left direction, suppose that $T_1 \vdash_{\mathrm{HL}} C \to B$ for some confusion C of T_2. By Lemma 3.42, $T_2 \vdash_{\mathrm{HL}} C$. So by Proposition 3.7 (a) twice, $T_1 \cup T_2 \vdash_{\mathrm{HL}} C$ and $T_1 \cup T_2 \vdash_{\mathrm{HL}} C \to B$. Hence by Lemma 3.23 (i), $T_1 \cup T_2 \vdash_{\mathrm{HL}} B$. For the opposite direction, we assume that $d; T_1 \cup T_2 \vdash_{\mathrm{HL}} B$ and show that $T_1 \vdash_{\mathrm{HL}} C \to B$ for some confusion C of T_2, proceeding by induction on $\mathrm{ht}(d)$. For the base case, if $B \in T_1$ or B is an axiom, then easily $T_1 \vdash_{\mathrm{HL}} \mathrm{e} \to B$, and if $B \in T_2$, then $T_1 \vdash_{\mathrm{HL}} B \to B$. For the inductive step, there are two possibilities:

- Suppose that B follows by (MP) from $T_1 \cup T_2 \vdash_{\mathrm{HL}} D \to B$ and $T_1 \cup T_2 \vdash_{\mathrm{HL}} D$. Then by the induction hypothesis twice, there exist confusions C_1 and C_2 of T_2 such that $T_1 \vdash_{\mathrm{HL}} C_1 \to (D \to B)$ and $T_1 \vdash_{\mathrm{HL}} C_2 \to D$. Using Lemma 3.23 (vi) and (iii), $T_1 \vdash_{\mathrm{HL}} C_2 \to (C_1 \to B)$. Hence by $(\odot 2)$ and Lemma 3.23 (i), $T_1 \vdash_{\mathrm{HL}} (C_2 \odot C_1) \to B$. So $C_2 \odot C_1$ is the required confusion of T_2.

- Suppose that B is $B_1 \wedge B_2$ and follows by (ADJ) from $T_1 \cup T_2 \vdash_{\mathrm{HL}} B_1$ and $T_1 \cup T_2 \vdash_{\mathrm{HL}} B_2$. Then by the induction hypothesis twice, there exist confusions C_1 and C_2 of T_2 such that $T_1 \vdash_{\mathrm{HL}} C_1 \to B_1$ and $T_1 \vdash_{\mathrm{HL}} C_2 \to B_2$. So by Lemma 3.23 (ii), $T_1 \vdash_{\mathrm{HL}} (C_1 \to B_1) \wedge (C_2 \to B_2)$. But also $T_1 \vdash_{\mathrm{HL}} ((C_1 \to B_1) \wedge (C_2 \to B_2)) \to ((C_1 \wedge C_2) \to (B_1 \wedge B_2))$. Hence by Lemma 3.23 (i), $T_1 \vdash_{\mathrm{HL}} (C_1 \wedge C_2) \to (B_1 \wedge B_2)$. So $C_1 \wedge C_2$ is the required confusion of T_2. □

There may be other local forms of the deduction theorem that hold. E.g. for any HMAILL^--extension HL:

$$T \cup \{A\} \vdash_{\mathrm{HL}} B \quad \text{iff} \quad T \vdash_{\mathrm{HL}} (A \wedge \mathrm{e})^n \to B \text{ for some } n \in \mathbb{N}$$

In the case of logics with weakening, this simplifies to:

$$T \cup \{A\} \vdash_{\mathrm{HL}} B \quad \text{iff} \quad T \vdash_{\mathrm{HL}} A^n \to B \text{ for some } n \in \mathbb{N}$$

We can also characterize substructural logics having the deduction theorem proper.

Proposition 3.44. *The following are equivalent for any* HMAILL^-*-extension* HL:

(a) $T \cup \{A\} \vdash_{\mathrm{HL}} B$ *iff* $T \vdash_{\mathrm{HL}} A \to B$.
(b) $\vdash_{\mathrm{HL}} A \to (A \odot A)$ *and* $\vdash_{\mathrm{HL}} A \to \mathrm{e}$.

Proof. If (a) (i.e. the deduction theorem) holds for HL, then since $\{A\} \vdash_{\mathrm{HL}} A \odot A$ and $\{A\} \vdash_{\mathrm{HL}} \mathrm{e}$, we get $\vdash_{\mathrm{HL}} A \to (A \odot A)$ and $\vdash_{\mathrm{HL}} A \to \mathrm{e}$ as required. Now suppose that (b) holds. Easily if $T \vdash_{\mathrm{HL}} A \to B$, then $T \cup \{A\} \vdash_{\mathrm{HL}} B$. For the other direction suppose that $T \cup \{A\} \vdash_{\mathrm{HL}} B$. By Theorem 3.43, $T \vdash_{\mathrm{HL}} C \to B$ for some confusion C of $\{A\}$, so by Lemma 3.23 (iii), it is sufficient to show:

Claim. $\vdash_{\mathrm{HL}} A \to C$ for every confusion C of $\{A\}$.

Proof of claim. We proceed by induction on $\mathrm{cp}(C)$. If C is A or $\top$, then $\vdash_{\mathrm{HL}} A \to A$ and $\vdash_{\mathrm{HL}} A \to \top$ so we are done. If C is e, then $\vdash_{\mathrm{HL}} A \to \mathrm{e}$ by assumption. For the induction step, C is $C_1 \odot C_2$ or $C_1 \wedge C_2$. Then by the induction hypothesis twice, $\vdash_{\mathrm{HL}} A \to C_1$ and $\vdash_{\mathrm{HL}} A \to C_2$. Hence by Lemma 3.23 (iv), $\vdash_{\mathrm{HL}} A \to (C_1 \wedge C_2)$. Also, $\vdash_{\mathrm{HL}} C_1 \to (C_2 \to (C_1 \odot C_2))$ is an instance of ($\odot$1). So using Lemma 3.23 (iii), $\vdash_{\mathrm{HL}} A \to (C_2 \to (C_1 \odot C_2))$. But then using Lemma 3.23 (vi) and (iii), $\vdash_{\mathrm{HL}} A \to (A \to (C_1 \odot C_2))$, and using ($\odot$2) and Lemma 3.23 (iii), $\vdash_{\mathrm{HL}} (A \odot A) \to (C_1 \odot C_2)$. Hence since $A \to (A \odot A)$, by Lemma 3.23 (iii), $\vdash_{\mathrm{HL}} A \to (C_1 \odot C_2)$. □

This means in particular that any HUL^--extension having the deduction theorem is an extension of some version of Gödel Logic G (possibly with the extra constant f and without $\bot$). Indeed, since any proper HG-extension gives a finite-valued logic, G may be said to be the only fuzzy logic having the deduction theorem proper.

Table 3.1 Common axioms for substructural logics

Label	Axiom
(INV)	$\neg\neg A \to A$
(e)	$\mathrm{f} \to \mathrm{e}$
(f)	$\mathrm{e} \to \mathrm{f}$
(PRL)	$(A \to B) \vee (B \to A)$
(DIS)	$(A \wedge (B \vee C)) \to ((A \wedge B) \vee (A \wedge C))$
(DIV)	$(A \wedge B) \to (A \odot (A \to B))$
(W)	$(A \to \mathrm{e}) \wedge (\mathrm{f} \to A)$
(C_2)	$A \to (A \odot A)$
(M)	$(A \odot A) \to A$
(ID)	$A \leftrightarrow (A \odot A)$
(C_n)	$A^{n-1} \to A^n$
(NC)	$\neg(A \wedge \neg A)$
(EM)	$A \vee \neg A$
(CAN)	$(A \to (A \odot B)) \to B$
(RCAN)	$(A \to \bot) \vee (A \to \top) \vee ((A \to (A \odot B)) \to B)$

Table 3.2 Common substructural and fuzzy logics

Label	Name	Axiomatization
HMALL	Multiplicative Additive Linear Logic	HMAILL + (INV)
HML	Monoidal Logic	HMAILL + (W)
HAMALL	Affine MALL	HML + (INV)
HIL	Intuitionistic Logic	HML + (C_2)
HCL	Classical Logic	HAMALL + (C_2)
HUL	Uninorm Logic	HMAILL + (PRL) + (DIS)
HIUL	Involutive UL	HUL + (INV)
HMTL	Monoidal t-norm Logic	HML + (W)
HIMTL	Involutive MTL	HMTL + (INV)
HG	Gödel Logic	HMTL + (C_2)
HUML	Uninorm Mingle Logic	HUL + (C_2) + (M)
HRM	Relevance Mingle Logic	HIUL + (C_2) + (M)
HIUML	Involutive UML	HRM + (f)
$\mathrm{HMTL_n}$	N-Contractive MTL	HMTL + (C_n)
$\mathrm{HIMTL_n}$	N-Contractive IMTL	HIMTL + (C_n)
HBL	Basic Logic	HMTL + (DIV)
HŁ	Łukasiewicz Logic	HBL + (INV)
HSMTL	Strict MTL	HMTL + (NC)
HSBL	Strict BL	HBL + (NC)
HP	Product Logic	HBL + (RCAN)
HCHL	Cancellative Hoop Logic	HBL^- + (e) + (f) + (CAN)
HPMTL	Product MTL	HMTL + (RCAN)
HA	Abelian Logic	HIUL^- + (e) + (f) + (CAN)
HCRL	Cross Ratio Logic	HIUL + (e) + (f) + (RCAN)

Table 3.3 Classes of L-algebras

Label	Name	Class of algebras
GEN(L)	L-algebras	See Definition 3.45
LIN(L)	L-chains	Linearly ordered L-algebras
DEN(L)	Dense L-chains	Dense linearly ordered L-algebras
STAN(L)	Standard L-algebras	L-algebras with universe $[0,1]$ and the usual ordering

3.5 Soundness and Completeness

The reader will have noticed that there is not really much difference between HMAILL$^-$-extensions and varieties of (bounded) pointed commutative residuated lattices. We simply "translate" axioms into equations and vice versa. Indeed this translation is well behaved enough in general to support a comprehensive theory of "algebraizable logics" (see historical remarks). Here, however, the focus will be on establishing just the elements of this correspondence that we need.

The axiom schema and axiomatizations introduced above are collected together for easy reference in Tables 3.1 and 3.2. Let us assume for now that we are dealing with an HMAILL$^-$-extension HL based on formulas of the language $\mathcal{L}$ (either $\mathcal{L}_F$ or $\mathcal{L}_B$). It is easy to check that the axioms of HMAILL$^-$ (HMAILL) are valid in any pcrl (bpcrl). Hence we can define a class of algebras for HL as follows:

Definition 3.45. An L-*algebra* is a pcrl (bpcrl) in which the axioms of HL are valid.

Example 3.46. Recall that HUL is HMAILL plus (DIS) and (PRL). Then an UL-algebra $\mathbf{A}$ is any bpcrl where for all $\mathbf{A}$-valuations v and formulas A, B, and C:

$$\mathrm{e} \le v((A \to B) \vee (B \to A)) \quad \text{and} \quad \mathrm{e} \le v((A \wedge (B \vee C)) \to ((A \wedge B) \vee (A \wedge C)))$$

That is, UL-algebras are bpcrls satisfying prelinearity and distributivity.

As the previous example illustrates, each algebraic condition listed in Table 2.1 corresponds to an axiom schema in Table 3.1. Integral bpcrls are ML-algebras, involutive bpcrls are MALL-algebras, and so on.

In Table 3.3 we pick out certain classes of these L-algebras as being of particular interest. Standard L-algebras are just the algebras $\mathbf{A}(\odot, \mathrm{f})$ of the previous chapter, where the monoid operation $\odot$ is a residuated uninorm and $\mathrm{f} \in [0,1]$. Our aim is to show that certain choices of HL are "complete" with respect to standard L-algebras. This is often called the "standard completeness problem" for HL.

Our starting point is the following (unsurprising) result:

Theorem 3.47. *If* $T \vdash_{\mathrm{HL}} A$*, then* $T \models_{\mathsf{GEN(L)}} A$.

Proof. Suppose that $d; T \vdash_{\mathrm{HL}} A$. We prove that $T \models_{\mathsf{GEN(L)}} A$ by induction on $\mathrm{ht}(d)$. The base case clearly holds if $A \in T$. Also, if A is an axiom of HL, then A is valid in all L-algebras by definition. For the induction step, let v be an $\mathbf{A}$-model of T for an L-algebra **A**. If $T \vdash_{\mathrm{HL}} A$ follows from $T \vdash_{\mathrm{HL}} B \to A$ and $T \vdash_{\mathrm{HL}} B$ by (MP), then by the induction hypothesis twice, $\mathrm{e} \leq v(B \to A)$ and $\mathrm{e} \leq v(B)$. From the former, we get $\mathrm{e} \leq v(B) \to v(A)$, and by residuation, $v(B) \leq v(A)$. So using the latter, $\mathrm{e} \leq v(A)$. If $T \vdash_{\mathrm{HL}} A$ follows by (ADJ) from $T \vdash_{\mathrm{HL}} A_1$ and $T \vdash_{\mathrm{HL}} A_2$ where $A = A_1 \wedge A_2$, then by the induction hypothesis twice, $\mathrm{e} \leq v(A_1)$ and $\mathrm{e} \leq v(A_2)$. Hence also $\mathrm{e} \leq v(A_1) \wedge v(A_2) = v(A_1 \wedge A_2)$ as required. □

The other direction requires a little more care. For each theory T, we define a special L-algebra $\mathsf{LIND}_T^{\mathrm{L}}$ – the so-called *Lindenbaum algebra* for T – such that a formula A is $\mathsf{LIND}_T^{\mathrm{L}}$-valid iff $T \vdash_{\mathrm{HL}} A$. The basic intuition here is to treat formulas themselves as elements of the algebra. However, this does not quite work. Instead we use sets of formulas that are "provably equivalent with respect to T". Let us write:

$$A \sim B \quad \text{iff} \quad T \vdash_{\mathrm{HL}} A \leftrightarrow B$$

Then $\sim$ is an equivalence relation on the set of formulas; i.e.

(i) $A \sim A$.
(ii) If $A \sim B$, then $B \sim A$.
(iii) If $A \sim B$ and $B \sim C$, then $A \sim C$.

Moreover, $\sim$ is a congruence. For each n-ary connective $\star \in \mathcal{L}$, if $A_i \sim B_i$ for $i = 1 \ldots n$, then, using the equivalences established in Lemma 3.24:

$$\star(A_1, \ldots, A_n) \sim \star(B_1, \ldots, B_n)$$

For example, suppose that $A_1 \sim B_1$ and $A_2 \sim B_2$; i.e. $T \vdash_{\mathrm{HL}} A_1 \leftrightarrow B_1$ and $T \vdash_{\mathrm{HL}} A_2 \leftrightarrow B_2$. To show that $A_1 \wedge A_2 \sim B_1 \wedge B_2$, we need a derivation for $T \vdash_{\mathrm{HL}} (A_1 \wedge A_2) \leftrightarrow (B_1 \wedge B_2)$. By ($\wedge 1$), $T \vdash_{\mathrm{HL}} (A_1 \wedge A_2) \to A_1$ and since $T \vdash_{\mathrm{HL}} A_1 \to B_1$, by Lemma 3.23 (iii), $T \vdash_{\mathrm{HL}} (A_1 \wedge A_2) \to B_1$. Similarly, using ($\wedge 2$), $T \vdash_{\mathrm{HL}} (A_1 \wedge A_2) \to B_2$, so by Lemma 3.23 (iv), $T \vdash_{\mathrm{HL}} (A_1 \wedge A_2) \to (B_1 \wedge B_2)$. The other direction follows symmetrically. We leave the checking of other connectives for the reader's amusement.

These observations allow us to define an algebra of equivalence classes of formulas as follows.

Definition 3.48. For an $\mathrm{Fm}_{\mathcal{L}}$-theory T, let:

$$[A]_T^{\mathrm{L}} =_{\mathrm{def}} \{B \in \mathrm{Fm}_{\mathcal{L}} : T \vdash_{\mathrm{HL}} A \leftrightarrow B\} \quad \text{and} \quad L_T^{\mathrm{L}} =_{\mathrm{def}} \{[A]_T^{\mathrm{L}} : A \in \mathrm{Fm}_{\mathcal{L}}\}$$

Then the *Lindenbaum algebra* of T for HL is defined as:

$$\mathsf{LIND}_T^{\mathrm{L}} =_{\mathrm{def}} \langle L_T^{\mathrm{L}}, \{\star_T^{\mathrm{L}} : \star \in \mathcal{L}\}\rangle \quad \text{where } \star_T^{\mathrm{L}}([A_1]_T^{\mathrm{L}}, \ldots, [A_n]_T^{\mathrm{L}}) =_{\mathrm{def}} [\star(A_1, \ldots, A_n)]_T^{\mathrm{L}}$$

Lemma 3.49. $\mathsf{LIND}_T^{\mathrm{L}}$ *is an* L*-algebra for any* $\mathrm{Fm}_{\mathcal{L}}$*-theory* T.

Proof. The lattice and monoid properties of $\mathsf{LIND}_T^{\mathrm{L}}$ follow easily from the theorems of HMAILL^- collected in Lemma 3.24. For residuation observe that:

$$[A]^{\mathrm{L}}_{T} \leq [B]^{\mathrm{L}}_{T} \quad \text{iff} \quad T \vdash_{\mathrm{HL}} A \leftrightarrow (A \wedge B) \quad \text{iff} \quad T \vdash_{\mathrm{HL}} A \to B$$

Hence $[A]^{\mathrm{L}}_{T} \leq [B]^{\mathrm{L}}_{T} \to [C]^{\mathrm{L}}_{T} = [B \to C]^{\mathrm{L}}_{T}$ iff $T \vdash_{\mathrm{HL}} A \to (B \to C)$ iff $T \vdash_{\mathrm{HL}} (A \odot B) \to C$ iff $[A \odot B]^{\mathrm{L}}_{T} = [A]^{\mathrm{L}}_{T} \odot [B]^{\mathrm{L}}_{T} \leq [C]^{\mathrm{L}}_{T}$. Finally, for each axiom A of HL, since $T \vdash_{\mathrm{HL}} \mathrm{e} \to A$, it follows that $[\mathrm{e}]^{\mathrm{L}}_{T} \leq [A]^{\mathrm{L}}_{T}$. So A is $\mathsf{LIND}^{\mathrm{L}}_{T}$-valid, and hence $\mathsf{LIND}^{\mathrm{L}}_{T}$ is an HL-algebra. □

To show that $\mathsf{LIND}^{\mathrm{L}}_{T}$-validity corresponds to HL-derivability from T, we make use of a specially defined valuation for this algebra that maps each formula to its corresponding equivalence class.

Lemma 3.50. *For any* $\mathrm{Fm}_{\mathcal{L}}$*-theory* T *and* $A \in \mathrm{Fm}_{\mathcal{L}}$*:*

$$T \vdash_{\mathrm{HL}} A \quad \textit{iff} \quad [\mathrm{e}]^{\mathrm{L}}_{T} \leq v^{\mathrm{L}}_{T}(A)$$

where v^{L}_{T} *is the* $\mathsf{LIND}^{\mathrm{L}}_{T}$*-valuation defined by* $v^{\mathrm{L}}_{T}(p) = [p]^{\mathrm{L}}_{T}$ *for each variable* p.

Proof. We prove that $v^{\mathrm{L}}_{T}(B) = [B]^{\mathrm{L}}_{T}$ for all formulas B by induction on $\mathrm{cp}(B)$. The case where B is a variable follows by definition. For the other cases, just note that for any n-ary connective $\star \in \mathcal{L}$ (using the induction hypothesis for the second line):

$$\begin{aligned} v^{\mathrm{L}}_{T}(\star(C_1,\ldots,C_n)) &= \star(v^{\mathrm{L}}_{T}(C_1),\ldots,v^{\mathrm{L}}_{T}(C_n)) \\ &= \star([C_1]^{\mathrm{L}}_{T},\ldots,[C_n]^{\mathrm{L}}_{T}) \\ &= [\star(C_1,\ldots,C_n)]^{\mathrm{L}}_{T} \end{aligned}$$

The result then follows since $[\mathrm{e}]^{\mathrm{L}}_{T} \leq [A]^{\mathrm{L}}_{T}$ iff $T \vdash_{\mathrm{HL}} \mathrm{e} \to A$ iff $T \vdash_{\mathrm{HL}} A$. □

Suppose then that $T \nvdash_{\mathrm{HL}} A$. By the previous lemma, v^{L}_{T} is a $\mathsf{LIND}^{\mathrm{L}}_{T}$-model of T where $[\mathrm{e}]^{\mathrm{L}}_{T} \not\leq v^{\mathrm{L}}_{T}(A)$. So $T \not\models_{\mathsf{GEN(L)}} A$. Hence, putting this together with Theorem 3.47:

Theorem 3.51. $T \vdash_{\mathrm{HL}} A$ *iff* $T \models_{\mathsf{GEN(L)}} A$.

This completeness result holds for any HMAILL^{-}-extension HL. However, if HL includes (DIS) and (PRL) (i.e. is an HUL^{-}-extension), then we can go one step further and obtain completeness with respect to L-*chains*. Suppose as before that $T \nvdash_{\mathrm{HL}} A$. Then there is a $\mathsf{LIND}^{\mathrm{L}}_{T}$-model of T that is not a $\mathsf{LIND}^{\mathrm{L}}_{T}$-model of $\{A\}$. The idea now is that while $\mathsf{LIND}^{\mathrm{L}}_{T}$ might not itself be linearly ordered, we can nevertheless find $\hat{T} \supseteq T$ such that $\mathsf{LIND}^{\mathrm{L}}_{\hat{T}}$ is an L-chain and still $\hat{T} \nvdash_{\mathrm{HL}} A$.

We begin with a useful property of theories.

Definition 3.52. An $\mathrm{Fm}_{\mathcal{L}}$-theory T is L-*linear* if for each pair $A, B \in \mathrm{Fm}_{\mathcal{L}}$:

$$\text{either} \quad T \vdash_{\mathrm{HL}} A \to B \quad \text{or} \quad T \vdash_{\mathrm{HL}} B \to A$$

Notice immediately that T is L-linear iff for all $A, B \in \mathrm{Fm}_{\mathcal{L}}$, either $[A]^{\mathrm{L}}_{T} \leq [B]^{\mathrm{L}}_{T}$ or $[B]^{\mathrm{L}}_{T} \leq [A]^{\mathrm{L}}_{T}$, i.e. iff $\mathsf{LIND}^{\mathrm{L}}_{T}$ is an L-chain.

To extend theories to linear theories, we also make use of a key property of Hilbert systems with distributivity.

Definition 3.53. HL has the *proof-by-cases property* if:

$$\text{whenever } T \cup \{A\} \vdash_{\mathrm{HL}} C \text{ and } T \cup \{B\} \vdash_{\mathrm{HL}} C, \text{ then } T \cup \{A \vee B\} \vdash_{\mathrm{HL}} C.$$

Lemma 3.54. *Any* HMAILL$^-$*-extension plus* (DIS) *has the proof-by-cases property.*

Proof. Let HL be an HMAILL$^-$-extension plus (DIS). Suppose that $T \cup \{A\} \vdash_{\mathrm{HL}} C$ and $T \cup \{B\} \vdash_{\mathrm{HL}} C$. Then $T \vdash_{\mathrm{HL}} A' \to C$ and $T \vdash_{\mathrm{HL}} B' \to C$ for some confusion A' of $\{A\}$ and B' of $\{B\}$. So by Lemma 3.23 (ii), $T \vdash_{\mathrm{HL}} (A' \to C) \wedge (B' \to C)$. But then using (DIS) and Lemma 3.23 (iii), also $T \vdash_{\mathrm{HL}} (A' \vee B') \to C$. Hence it is enough to prove the following:

Claim. If A' is a confusion of $\{A\}$ and B' is a confusion of $\{B\}$, then $\vdash_{\mathrm{HL}} E \to (A' \vee B')$ for some confusion E of $\{A \vee B\}$.

Just notice that in this case, by Lemma 3.42, $T \cup \{A \vee B\} \vdash_{\mathrm{HL}} E$. But then using Lemma 3.23 (iii), $T \cup \{A \vee B\} \vdash_{\mathrm{HL}} C$ as required.

Proof of Claim. By induction on $\mathrm{cp}(A') + \mathrm{cp}(B')$. If A' and B' are just A and B, then the claim is immediate. Also if A' or B' is e or $\top$, then we can just let $E = \mathrm{e}$. For the inductive step, suppose without loss of generality that $A' = A'_1 \wedge A'_2$ or $A' = A'_1 \odot A'_2$ where A'_1 and A'_2 are confusions of $\{A\}$. Then by the induction hypothesis twice:

$$\vdash_{\mathrm{HL}} E_1 \to (A'_1 \vee B') \quad \text{and} \quad \vdash_{\mathrm{HL}} E_2 \to (A'_2 \vee B')$$

for some confusions E_1 and E_2 of $\{A \vee B\}$. But also by (DIS) and Lemma 3.30 (b):

$$\vdash_{\mathrm{HL}} ((A'_1 \vee B') \wedge (A'_2 \vee B')) \to ((A'_1 \wedge A'_2) \vee B').$$
$$\vdash_{\mathrm{HL}} (((A'_1 \vee B') \wedge \mathrm{e}) \odot ((A'_2 \vee B') \wedge \mathrm{e})) \to ((A'_1 \odot A'_2) \vee B').$$

Hence easily, using Lemma 3.23 (various parts):

$$\vdash_{\mathrm{HL}} (E_1 \wedge E_2) \to ((A'_1 \wedge A'_2) \vee B').$$
$$\vdash_{\mathrm{HL}} ((E_1 \wedge \mathrm{e}) \odot (E_2 \wedge \mathrm{e})) \to ((A'_1 \odot A'_2) \vee B').$$

But $E_1 \wedge E_2$ and $(E_1 \wedge \mathrm{e}) \odot (E_2 \wedge \mathrm{e})$ are confusions of $\{A \vee B\}$, so we are done. □

We now show that for any HUL$^-$-extension, theories can be extended to linear theories while still preserving the non-derivability of some formula C. The idea is that for each pair of formulas A and B, we can always choose one of $A \to B$ and $B \to A$ to add without making C derivable.

Lemma 3.55. *Let* HL *be an* HUL$^-$*-extension. If* $T \nvdash_{\mathrm{HL}} C$*, then* $\hat{T} \nvdash_{\mathrm{HL}} C$ *for some* L*-linear theory* $\hat{T} \supseteq T$.

Proof. Suppose that $T \nvdash_{\mathrm{HL}} C$ for some HUL$^-$-extension HL. We enumerate all pairs of formulas $\langle A_n, B_n \rangle$ for $n \in \mathbb{N}$ and define a sequence $(T_n)_{n \in \mathbb{N}}$ of theories as follows:

$$T_0 = T \quad \text{and} \quad T_{n+1} = \begin{cases} T_n \cup \{A_n \to B_n\} & \text{if } T_n \cup \{A_n \to B_n\} \nvdash_{\mathrm{HL}} C \\ T_n \cup \{B_n \to A_n\} & \text{otherwise} \end{cases}$$

We prove that $T_n \nvdash_{\mathrm{HL}} C$ for all $n \in \mathbb{N}$ by induction on n. The base case is immediate. For the inductive step, assume that $T_n \nvdash_{\mathrm{HL}} C$. It is sufficient to show that:

$$T_n \cup \{A_n \to B_n\} \nvdash_{\mathrm{HL}} C \qquad \text{or} \qquad T_n \cup \{B_n \to A_n\} \nvdash_{\mathrm{HL}} C$$

Suppose otherwise. Then by the proof-by-cases property for HL (Lemma 3.54):

$$T_n \cup \{(A_n \to B_n) \vee (B_n \to A_n)\} \vdash_{\mathrm{HL}} C$$

But $\vdash_{\mathrm{HL}} (A_n \to B_n) \vee (B_n \to A_n)$. So $T_n \vdash_{\mathrm{HL}} C$, a contradiction.

Finally, define $\hat{T} = \bigcup_{n \in \mathbb{N}} T_n$, an L-linear theory by construction. Also $\hat{T} \nvdash_{\mathrm{HL}} C$ since otherwise, $T_k \vdash_{\mathrm{HL}} C$ for some $k \in \mathbb{N}$, a contradiction. □

Suppose now that $T \nvdash_{\mathrm{HL}} A$. Then by the previous lemma, $\hat{T} \nvdash_{\mathrm{HL}} A$ for some L-linear theory $\hat{T}$. So using Lemma 3.50, we have the following result.

Theorem 3.56. *Let* HL *be an* HUL^-*-extension. Then* $T \vdash_{\mathrm{HL}} A$ *iff* $T \models_{\mathsf{LIN(L)}} A$.

This theorem is of course only one step towards showing that an HUL^--extension is "fuzzy" in the sense of being complete with respect to standard algebras.

3.6 The Density Rule

We take another step towards fuzziness in this section. We establish completeness of certain Hilbert systems with respect to algebras that are not only linearly ordered but also dense. To do this, we make use of a further rule.

Definition 3.57 (Density Rule).

$$\frac{(A \to p) \vee (p \to B) \vee C}{(A \to B) \vee C} \ (\textsc{density})$$

where p does not occur in A, B, or C

Adding (DENSITY) to a Hilbert system is fine for proving theorems. However, for derivations from a theory T, we have to be a bit careful. We have to ensure that the condition on the new variable p applies also to T. Let us assume as before that HL is an HMAILL^--extension for $\mathrm{Fm}_{\mathcal{L}}(\mathsf{X})$, and that theories and formulas, unless otherwise stated, are also based on $\mathrm{Fm}_{\mathcal{L}}(\mathsf{X})$.

Definition 3.58. HL^{D} is HL + (DENSITY) where HL^{D}-derivations from a finite theory T are restricted so that the new variable p does not occur in T. As before, $T \vdash_{\mathrm{HL}^{\mathrm{D}}} A$ iff there is an HL^{D}-derivation of A from some finite subset of T.

Example 3.59. Observe that extending the axiomatization HCL of Classical Logic with the density rule gives triviality:

$\vdash_{\mathsf{HCL}^{\mathsf{D}}} p \vee (p \to \bot)$	the law of excluded middle
$\vdash_{\mathsf{HCL}^{\mathsf{D}}} (\top \to p) \vee (p \to \bot) \vee \bot$	using $\vdash_{\mathsf{HCL}} p \leftrightarrow ((\top \to p) \vee \bot)$
$\vdash_{\mathsf{HCL}^{\mathsf{D}}} (\top \to \bot) \vee \bot$	by (DENSITY)
$\vdash_{\mathsf{HCL}^{\mathsf{D}}} \bot$	since $\vdash_{\mathsf{HCL}} \bot \leftrightarrow ((\top \to \bot) \vee \bot)$

Several key properties are preserved by the extension with (DENSITY).

Proposition 3.60.
(a) *If $T \vdash_{\mathrm{HL}} A$, then $T \vdash_{\mathrm{HL}^{\mathrm{D}}} A$.*
(b) *If $T \vdash_{\mathrm{HL}^{\mathrm{D}}} A$ and $T^+ \supseteq T$, then $T^+ \vdash_{\mathrm{HL}^{\mathrm{D}}} A$.*
(c) *If $T \vdash_{\mathrm{HL}^{\mathrm{D}}} A$, then $T^F \vdash_{\mathrm{HL}^{\mathrm{D}}} A$ for some finite subset T^F of T.*
(d) *If A is a confusion of T, then $T \vdash_{\mathrm{HL}^{\mathrm{D}}} A$.*

To see that (DENSITY) is sound for dense HL-chains, suppose that for such an algebra $\mathbf{A}$, there is an $\mathbf{A}$-model v of a finite theory T such that $\mathrm{e} \not\leq v((A \to B) \vee C)$. Since $\mathbf{A}$ is a chain, it follows that:

$$v(B) < v(A) \quad \text{and} \quad v(C) < \mathrm{e}$$

But now since $\mathbf{A}$ is dense, there exists an element $x \in L_{\mathbf{A}}$ such that:

$$v(B) < x < v(A)$$

But suppose that p does not occur in T, A, B, or C. Then we can extend the $\mathbf{A}$-model v of T with $v(p) = x$. So $v((A \to p) \vee (p \to B) \vee C) < \mathrm{e}$ as required.

Soundness for HL^{D} with respect to dense L-chains follows: if $T \vdash_{\mathrm{HL}^{\mathrm{D}}} A$, then $T \models_{\mathsf{DEN}(\mathrm{L})} A$. We just proceed as in the proof of Theorem 3.47 by induction on the height of a derivation for $T \vdash_{\mathrm{HL}^{\mathrm{D}}} A$, the only new case being the soundness of the density rule.

Moreover, HUL^--extensions with density still admit the local deduction theorem and the proof-by-cases property.

Theorem 3.61. *Let* HL *be an* HUL^-*-extension:*
(a) *$T \cup \{A\} \vdash_{\mathrm{HL}^{\mathrm{D}}} B$ iff $T \vdash_{\mathrm{HL}^{\mathrm{D}}} C \to B$ for some confusion C of $\{A\}$.*
(b) *$T \vdash_{\mathrm{HL}^{\mathrm{D}}} A$ iff $\vdash_{\mathrm{HL}} C \to A$ for some confusion C of T.*
(c) HL^{D} *has the proof-by-cases property.*

Proof. Note first that (c) follows from (a) exactly as in the proof of Lemma 3.54. For (a) and (b), we proceed as in Theorem 3.43 by showing that $T_1 \cup T_2 \vdash_{\mathrm{HL}^{\mathrm{D}}} B$ iff $T_1 \vdash_{\mathrm{HL}^{\mathrm{D}}} C \to B$ for some confusion C of T_2. Let us concentrate on the trickier left-to-right direction. We can assume by compactness that T_1 and T_2 are both finite. We then prove the claim by induction on the height of an HL^{D}-derivation of B from $T_1 \cup T_2$, the only new case to consider being (DENSITY). Suppose that B is $(B_1 \to B_2) \vee B_3$ and:

$$T_1 \cup T_2 \vdash_{\mathrm{HL}^{\mathrm{D}}} (B_1 \to p) \vee (p \to B_2) \vee B_3$$

where p does not occur in T_1, T_2, B_1, B_2, or B_3. Then by the induction hypothesis, there exists a confusion C of T_2 such that:

$$T_1 \vdash_{\mathrm{HL^D}} C \to ((B_1 \to p) \vee (p \to B_2) \vee B_3)$$

So using the fact that $\vdash_{\mathsf{HUL}} (A \to (B \vee C)) \to ((A \to B) \vee (A \to C))$:

$$T_1 \vdash_{\mathrm{HL^D}} ((C \odot B_1) \to p) \vee (p \to (C \to B_2)) \vee (C \to B_3)$$

Hence by (DENSITY):

$$T_1 \vdash_{\mathrm{HL^D}} ((C \odot B_1) \to (C \to B_2)) \vee (C \to B_3)$$

It then follows easily using HMAILL^--derivabilities that:

$$T_1 \vdash_{\mathrm{HL^D}} ((C \wedge \mathrm{e}) \odot (C \wedge \mathrm{e})) \to ((B_1 \to B_2) \vee B_3)$$

and since C is a confusion of T_2, so also is $(C \wedge \mathrm{e}) \odot (C \wedge \mathrm{e})$. □

We are now ready to give a completeness proof with respect to dense chains. Just as we can extend a theory so that the corresponding Lindenbaum algebra is a chain, so we can extend a theory using (DENSITY) so that this algebra is also dense.

Definition 3.62. An $\mathrm{Fm}_{\mathcal{L}}$-theory T is L-*dense* if for each pair $A, B \in \mathrm{Fm}_{\mathcal{L}}$:

$$T \nvdash_{\mathrm{HL}} A \to B \quad \text{implies} \quad T \nvdash_{\mathrm{HL}} A \to C \text{ and } T \nvdash_{\mathrm{HL}} C \to B \text{ for some } C \in \mathrm{Fm}_{\mathcal{L}}.$$

Suppose now that T is L-linear. It follows that T is L-dense iff whenever $[A]^{\mathrm{L}}_T < [B]^{\mathrm{L}}_T$, then $[A]^{\mathrm{L}}_T < [C]^{\mathrm{L}}_T$ and $[C]^{\mathrm{L}}_T < [B]^{\mathrm{L}}_T$ for some formula C, i.e. iff $\mathsf{LIND}^{\mathrm{L}}_T$ is dense.

Lemma 3.63. *Let* HL *be an* HJL^-*-extension for* $\mathrm{Fm}_{\mathcal{L}}(\mathsf{X})$ *and* $T \cup \{C\} \subseteq \mathrm{Fm}_{\mathcal{L}}(\mathsf{X})$. *If* $T \nvdash_{\mathrm{HL^D}} C$, *then* $\hat{T} \nvdash_{\mathrm{HL^D}} C$ *for some countable* L*-dense* L*-linear* $\mathrm{Fm}_{\mathcal{L}}(\mathsf{X} \cup \mathsf{Y})$*-theory* $\hat{T} \supseteq T$.

Proof. Let HL be an HUL^--extension for $\mathrm{Fm}_{\mathcal{L}}(\mathsf{X})$ and $T \cup \{C\} \subseteq \mathrm{Fm}_{\mathcal{L}}(\mathsf{X})$. Suppose that $T \nvdash_{\mathrm{HL^D}} C$. Now consider HL for $\mathrm{Fm}_{\mathcal{L}}(\mathsf{X} \cup \mathsf{Y})$ where Y is a countably infinite set of variables disjoint from X, noting that in this extended system we still have $T \nvdash_{\mathrm{HL^D}} C$. We enumerate all pairs of formulas in $\mathrm{Fm}_{\mathcal{L}}(\mathsf{X} \cup \mathsf{Y})$ as $\langle A_n, B_n \rangle$ for $n \in \mathbb{N}$. We then define a sequence $(T_n)_{n \in \mathbb{N}}$ of $\mathrm{Fm}_{\mathcal{L}}(\mathsf{X} \cup \mathsf{Y})$-theories and a sequence $(C_n)_{n \in \mathbb{N}}$ of members of $\mathrm{Fm}_{\mathcal{L}}(\mathsf{X} \cup \mathsf{Y})$, starting with $T_0 = T$ and $C_0 = C$.

Suppose that at step n we have defined T_n and C_n such that:

(i) For each $i < n$, either $A_i \to B_i \in T_n$ or $B_i \to A_i \in T_n$.
(ii) For each $i < n$, there is a variable q_i (depending on i but not on n) such that:

1. If $A_i \to B_i \notin T_n$, then $(A_i \to q_i) \vee (q_i \to B_i) \vdash_{\mathrm{HL^D}} C_n$.
2. If $B_i \to A_i \notin T_n$, then $(B_i \to q_i) \vee (q_i \to A_i) \vdash_{\mathrm{HL^D}} C_n$.

(iii) $T_n \nvdash_{\mathrm{HL^D}} C_n$.
(iv) $\vdash_{\mathrm{HL^D}} C_i \to C_n$ for all $i < n$.

Note that (i)...(iv) are satisfied for $n = 0$. We now define T_{n+1} and C_{n+1} such that (i) ... (iv) are satisfied when n is replaced by $n+1$. First, if $T_n \cup \{A_n \to B_n, B_n \to A_n\} \nvdash_{\mathrm{HL^D}} C_n$, then define:

$$T_{n+1} = T_n \cup \{A_n \to B_n, B_n \to A_n\} \text{ and } C_{n+1} = C_n$$

and note that (i) ... (iv) hold for $n+1$.

Suppose that the previous case does not apply. Let q_n be a variable not in $T_n \cup \{A_n, B_n\} \cup \{C_i : i \leq n\}$. We claim that (at least) one of the following conditions holds:

(a) $T_n \cup \{A_n \to B_n\} \nvdash_{\mathrm{HL^D}} C_n \vee (B_n \to q_n) \vee (q_n \to A_n)$.

(b) $T_n \cup \{B_n \to A_n\} \nvdash_{\mathrm{HL^D}} C_n \vee (A_n \to q_n) \vee (q_n \to B_n)$.

Suppose that (a) does not hold. Then by the density rule:

$$T_n \cup \{A_n \to B_n\} \vdash_{\mathrm{HL^D}} C_n \vee (B_n \to A_n)$$

But since $T_n \cup \{A_n \to B_n, B_n \to A_n\} \vdash_{\mathrm{HL^D}} C_n$ and $T_n \cup \{A_n \to B_n, C_n\} \vdash_{\mathrm{HL^D}} C_n$, by the proof-by-cases property, $T_n \cup \{A_n \to B_n, C_n \vee (B_n \to A_n)\} \vdash_{\mathrm{HL^D}} C_n$. So:

$$T_n \cup \{A_n \to B_n\} \vdash_{\mathrm{HL^D}} C_n$$

Similarly, if (b) does not hold, then:

$$T_n \cup \{B_n \to A_n\} \vdash_{\mathrm{HL^D}} C_n$$

Hence by the proof-by-cases property, if neither (a) or (b) holds, then:

$$T_n \cup \{(A_n \to B_n) \vee (B_n \to A_n)\} \vdash_{\mathrm{HL^D}} C_n$$

But $\vdash_{\mathrm{HL^D}} (A_n \to B_n) \vee (B_n \to A_n)$ so $T_n \vdash_{\mathrm{HL^D}} C_n$, a contradiction. Hence:

If (a) holds, let $T_{n+1} = T_n \cup \{A_n \to B_n\}$ and $C_{n+1} = C_n \vee (B_n \to q_n) \vee (q_n \to A_n)$.

If (b) holds, let $T_{n+1} = T_n \cup \{B_n \to A_n\}$ and $C_{n+1} = C_n \vee (A_n \to q_n) \vee (q_n \to B_n)$.

In both cases conditions (i) ... (iv) are preserved.

Now let $\hat{T} = \bigcup_{n \in \mathbb{N}} T_n$. For each $n \in \mathbb{N}$, $\hat{T} \nvdash_{\mathrm{HL^D}} C_n$. Otherwise by compactness there exists k such that $T_k \vdash_{\mathrm{HL^D}} C_n$. Without loss of generality we can assume that $k \geq n$. But since $\vdash_{\mathrm{HL^D}} C_n \to C_k$, we get $T_k \vdash_{\mathrm{HL^D}} C_k$ which contradicts property (iii). Finally, observe that $\hat{T}$ is L-linear and L-dense by construction. □

It follows then as before that if $T \nvdash_{\mathrm{HL^D}} A$, then $\hat{T} \nvdash_{\mathrm{HL^D}} A$ for some L-linear L-dense theory $\hat{T}$. Hence $\hat{T} \nvdash_{\mathrm{HL}} A$ and $\mathsf{LIND}^{\mathrm{L}}_{\hat{T}}$ is a dense chain. Moreover, by Lemma 3.50, $v(A) < \mathrm{e}$ for some $\mathsf{LIND}^{\mathrm{L}}_{\hat{T}}$-model v of T.

Theorem 3.64. *For any* HUL^-*-extension* HL*:* $T \vdash_{\mathrm{HL^D}} A$ *iff* $T \models_{\mathsf{DEN(L)}} A$.

Let us take stock. What we have just shown is that *any* HUL^--extension with (DENSITY) is complete for dense L-chains. Now in many cases, we can take one step further and obtain completeness with respect to standard L-algebras. To do this, we make use of the Dedekind-MacNeille completion defined in the previous chapter. This tells us that for certain classes of pcrls and bpcrls, each countable dense chain from the class can be embedded into a standard algebra of the same class.

Theorem 3.65. *For any extension* HL *of* HUL^- *with axiom schema taken from the set* $\{(\bot),(\top),(\mathrm{e}),(\mathrm{f}),(\mathrm{INV}),(\mathrm{W}),(\mathrm{M}),(\mathrm{EM}),(\mathrm{NC})\} \cup \{(\mathrm{C_n}) : n \geq 3\}$*:*

$$T \vdash_{\mathrm{HL^D}} A \quad \textit{iff} \quad T \models_{\mathsf{STAN(L)}} A$$

Proof. The left-to-right direction follows as above by induction on the height of a derivation for $T \vdash_{\mathrm{HL^D}} A$. For the other direction, suppose that $T \not\vdash_{\mathrm{HL^D}} A$. By Lemma 3.63, $v(A) < \mathrm{e}$ for some countable dense L-chain $\mathbf{A}$ and $\mathbf{A}$-model v of T. Since $\mathbf{A}$ is a countable dense L-chain, we can without loss of generality assume that $L_{\mathbf{A}} = [0,1] \cap \mathbb{Q}$ with the usual ordering of the rationals. Now consider $\mathsf{DM}(\mathbf{A})$, the Dedekind-MacNeille completion of $\mathbf{A}$, where $\mathsf{DM}(L_{\mathbf{A}})$ is the real unit interval $[0,1]$ with the usual ordering of the reals. Moreover, by Theorem 2.58, the characteristic properties of L-algebras corresponding to the axiom schema taken from the set $\{(\bot),(\top),(\mathrm{INV}),(\mathrm{e}),(\mathrm{f}),(\mathrm{W}),(\mathrm{M}),(\mathrm{C_n}),(\mathrm{EM}),(\mathrm{NC})\}$ are all preserved. So $\mathsf{DM}(\mathbf{A})$ is a standard L-algebra. Finally, we can use the embedding Φ of $\mathbf{A}$ into $\mathsf{DM}(\mathbf{A})$ of Theorem 2.58. We define $w(p) = \Phi(v(p))$, where w is a $\mathsf{DM}(\mathbf{A})$-model of T such that $w(A) < \mathrm{e}_{\mathsf{DM}}$. So $T \not\models_{\mathsf{STAN(L)}} A$ as required. □

In particular, extending the Hilbert systems for UL, IUL, MTL, IMTL, UML, IUML, G, $\mathsf{MTL_n}$ ($n \geq 2$), $\mathsf{IMTL_n}$ ($n \geq 3$), and SMTL, with density ensures standard completeness. Note however that since the divisibility condition is not preserved by the Dedekind-MacNeille completion, corresponding systems for logics such as BL are not covered by this method.

These results are nice. However, really we would like to show that the original density-free axiomatizations are standard complete. In later chapters we will give a uniform method that achieves this by eliminating a version of the density rule from corresponding Gentzen systems. In some cases, however, it is possible to show semantically that (DENSITY) is admissible for the axiomatization. Intuitively, what we need to show is that given two elements $a < b$ in a chain with no elements between, we can always add a third element c such that $a < c < b$, while preserving the required properties of the algebra. Let us consider the case of MTL as an example.

Lemma 3.66. (DENSITY) *is* HMTL-*admissible.*

Proof. We proceed by contraposition. Suppose that $\not\vdash_{\mathrm{HL}} (A \to B) \vee C$. Then by Theorem 3.56, $v((A \to B) \vee C) < \mathrm{e}$ for some MTL-chain $\mathbf{A}$ and $\mathbf{A}$-valuation v. It follows easily that:

$$a = v(A) > b = v(B) \qquad \text{and} \qquad \mathrm{e} > v(C)$$

If there exists c in $L_{\mathbf{A}}$ such that $a > c > b$, then we can extend v with $v(p) = c$ and $v((A \to p) \vee (p \to B) \vee C) < \mathrm{e}$ as required. If not, then it is enough to extend $\mathbf{A}$ with a new element between a and b such that $\mathbf{A}$ embeds into the new MTL-chain $\hat{\mathbf{A}}$ since then $v((A \to B) \vee C) < \mathrm{e}$ also in $\hat{\mathbf{A}}$. Let:

$$L_{\hat{\mathbf{A}}} =_{\text{def}} L_{\mathbf{A}} \cup \{c\}$$

and extend the linear order $\leq$ of $L_{\mathbf{A}}$ to $L_{\hat{\mathbf{A}}}$ so that $a > c > b$. We extend $\odot$ to $\hat{\odot}$ first with $c\hat{\odot}c = \min(a \odot a, c)$ and then for all $x \in L_{\mathbf{A}}$ with:

$$c\hat{\odot}x = x\hat{\odot}c = \min(a \odot x, c)$$

We claim that $\langle L_{\hat{\mathbf{A}}}, \hat{\odot}, \mathrm{e}\rangle$ is a commutative monoid. Easily $x\hat{\odot}y = y\hat{\odot}x$ for all $x, y \in \hat{L}_{\mathbf{A}}$ and $c\hat{\odot}\mathrm{e} = \mathrm{e}\hat{\odot}c = \min(a \odot \mathrm{e}, c) = \min(a, c) = c$. The only tricky thing to check here is associativity. Take $x, y \in L_{\mathbf{A}}$ and consider the two expressions:

$$(x\hat{\odot}c)\hat{\odot}y = \min(a \odot x, c)\hat{\odot}y \quad \text{and} \quad x\hat{\odot}(c\hat{\odot}y) = x\hat{\odot}\min(a \odot y, c)$$

Observe that $a \odot x \leq a$ and $a \odot y \leq a$, and, since c is the greatest element below a, either $a \odot x = a$ or $a \odot x \leq c$, and either $a \odot y = a$ or $a \odot y \leq c$. If $a \odot x \leq c$ and $a \odot y \leq c$, then the two expressions are equal using the associativity of $\odot$. If $a \odot x = a$ and $a \odot y = a$, then both expressions reduce to c. If $a \odot x = a$ and $a \odot y \leq c$, then the first expression becomes $c\hat{\odot}y = a \odot y$ and the second becomes $x \odot (a \odot y) = (x \odot a) \odot y = a \odot y$. The situation where $a \odot y = y$ and $a \odot x \leq c$ is symmetrical. Now consider:

$$(x\hat{\odot}y)\hat{\odot}c = \min(a \odot (x \odot y), c) \quad \text{and} \quad x\hat{\odot}(y\hat{\odot}c) = x\hat{\odot}\min(a \odot y, c)$$

If $a \odot y \leq c$, then also $a \odot (x \odot y) \leq c$ and the two expressions are clearly equal. Hence suppose that $a \odot y = a$. Then the first expression becomes $\min(a \odot x, c)$ and the second becomes $x\hat{\odot}c = \min(a \odot x, c)$.

Clearly $\hat{\odot}$ has a residual $\hat{\rightarrow}$ (since we have just added one element), so letting $x\hat{\wedge}y = \min(x, y)$ and $x\hat{\vee}y = \max(x, y)$, $\hat{\mathbf{A}} = \langle L_{\hat{\mathbf{A}}}, \hat{\wedge}, \hat{\vee}, \hat{\odot}, \hat{\rightarrow}, \mathrm{e}, \mathrm{f}, \bot, \top\rangle$ is an MTL-algebra. Moreover, $x\hat{\rightarrow}y = x \rightarrow y$ for all $x, y \in L_{\mathbf{A}}$. This could only fail if $x\hat{\odot}c \leq y$ and $x\hat{\odot}a > y$. But this implies that $x\hat{\odot}c = \min(x \odot a, c) = c$ and $x \odot a = a$. Hence $a = x \odot a > y \geq x\hat{\odot}c = c$ and $y = c$, a contradiction. So the identity function $\Phi(x) = x$ is the required embedding of $\mathbf{A}$ into $\hat{\mathbf{A}}$. □

Theorem 3.67. *$T \vdash_{\mathsf{HMTL}} A$ iff $T \models_{\mathsf{STAN(MTL)}} A$.*

Proof. We have the following chain of reasoning:

$T \vdash_{\mathsf{HMTL}} A$	iff	$\vdash_{\mathsf{HMTL}} C \rightarrow A$ for some confusion C of T	Theorem 3.43
	iff	$\vdash_{\mathsf{HMTL^D}} C \rightarrow A$ for some confusion C of T	Lemma 3.66
	iff	$T \vdash_{\mathsf{HMTL^D}} A$	Theorem 3.61
	iff	$T \models_{\mathsf{STAN(MTL)}} A$	Theorem 3.65 □

With some adaptations, this method can be used to establish standard completeness for HIMTL, HSMTL, and several other fuzzy logics with weakening. However, in cases without weakening such as HUL, it is not at all easy to see how the relevant algebras could be appropriately extended.

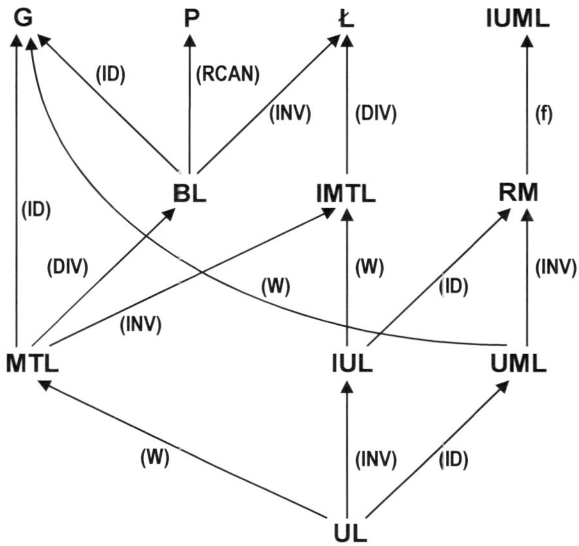

Fig. 3.2 Relationships between fuzzy logics

3.7 Historical Remarks

Hilbert systems take their name from the famous mathematician David Hilbert who popularized the axiomatic approach to Logic in the early part of the twentieth century in works such as [118]. Such systems are also found (with very different notation) in the earlier work of Frege [84] and are hence often called Frege systems.

Hilbert systems are a common feature of the several diverse strands in the literature on substructural logics, including the relevance logics developed by Anderson and Belnap and co-workers in the 1960s and 1970s [6, 7], the contraction-free logics studied by Ono and Komori in their 1985 paper [177], and Linear Logic, introduced by Girard in 1987 [10, 97]. A useful "toolkit" for building axiomatizations for a wide range of substructural logics, including MAILL, MALL, ML, AMALL, and RM, can be found in the 1999 book of Restall [186], as can the notion of a confusion used in this chapter. For fuzzy logics, the main source for the modern axiomatic approach is Hájek's 1998 monograph [105]. This includes Hilbert systems and (standard) completeness proofs for the three fundamental logics Ł, G, and P.[2] Also provided is an axiomatization for Basic Logic BL, proved complete with respect to algebras based on continuous t-norms (using an ordinal sum characterization of BL-chains) a year later by Cignoli et al. [59]. Related work of Esteva, Godo, and co-workers includes extensions of continuous t-norm based logics with

[2] We defer historical remarks for these logics to Chapter 6.

an involutive negation [79], motivating the introduction of the "strict negation" basic logic SBL, and logics combining Łukasiewicz and Product Logics such as ŁP [80],

Monoidal t-norm Logic MTL and extensions such as SMTL and IMTL were introduced via Hilbert systems – essentially by dropping the divisibility axioms from Hájek's axiomatization of BL – by Esteva and Godo in 2001 [77]. However, a fragment of MTL, known as C, was already introduced in Urquhart's 1986 handbook chapter on many-valued logics [212] in the context of defining Kripke semantics for Łukasiewicz Logic. Standard completeness was established for Godo and Esteva's axiomatization by Jenei and Montagna in 2002 [125], the crucial step being an embedding of countable MTL-chains into countable dense MTL-chains. This method was subsequently extended to SMTL and IMTL in [76] and n-contractive logics $\mathsf{MTL_n}$ and $\mathsf{IMTL_n}$ (for $n \geq 3$) in [49]. Other extensions of MTL to have been investigated include logics obeying certain restricted cancellation properties. The logic PMTL, axiomatized by adding the axioms of P to MTL, was proved distinct from P by Hájek in [107], and standard complete by Horčík in [121]. Related "weakly cancellative" logics extending MTL have been considered by Montagna, Noguera, and Horčík in [154]. Finally, an investigation of "hoop" logics – fuzzy logics without negation in the language – by Esteva et al. [78] led to an axiomatization and standard completeness proof for Cancellative Hoop Logic CHL.

Weakening-free Hilbert systems for Uninorm Logic UL and its extensions IUL, UML, and IUML, were introduced by Metcalfe and Montagna in 2007 [144], and for Cross Ratio Logic CRL and other "$[0,1)$-continuous" uninorm logics by Metcalfe and Gabbay the same year [87]. Related work includes the weakly implicative fuzzy logics of Cintula [60]: a broad approach that treats implication as the fundamental connective and characterizes fuzziness as completeness with respect to chains. The systematic approach presented in this chapter using the density rule to axiomatize fuzzy logics, first appeared in Metcalfe and Montagna's [144]. However, the density rule itself was introduced by Takeuti and Titani already in 1984 to axiomatize "Intuitionistic Fuzzy Logic", better known as first-order Gödel Logic [205]. The semantic elimination of this rule given for HMTL is a simplification of the embedding step used in the standard completeness proof of Jenei and Montagna [125].

Finally, we remark that the general approach described here relating Hilbert systems to classes of residuated lattices falls easily into the framework of algebraizable logics developed in particular by Blok and Pigozzi in their 1989 work [36]. In this powerful and very general approach, a Hilbert system is guaranteed a corresponding class of algebras and vice versa by the existence of certain formulas and equations that serve to translate between the two presentations. A detailed exposition of this phenomenon in the context of residuated lattices may be found in [90].

Chapter 4
Gentzen Systems

Hilbert systems are good for presenting a wide range of logics and connecting them with classes of algebras, not so good when it comes to proving theorems in those logics: even simple cases like $A \to A$ might be tricky. The problem is that at each step in a derivation we should guess which axioms or instances of rules like modus ponens to use next. A better option would be proof systems with more restrictions on how to proceed, ideally, systems where derivations are *analytic*: built from the raw material (subformulas) of the formula to be proved.

Several proof frameworks are well-suited to this task: Tableaux, Resolution, and Display Logic, to name just a few, each with their own distinct advantages and disadvantages. In this book, however, we choose to proceed with *Gentzen systems*, a popular framework that offers the twin virtues of simplicity – our systems will be reasonably natural and easy to understand – and flexibility – most of the logics that we are interested in can be covered.

4.1 Sequents and Hypersequents

While Hilbert systems treat formulas directly, Gentzen systems gain flexibility by treating structured collections of formulas: typically sequences, sets, or multisets. Just which structures are appropriate depends on the properties of the logics under consideration. With sets, order and multiplicity are lost: A,A,B is the same as A,B is the same as B,A. Even with sequences, where both order and multiplicity do matter, associativity is implicit (for non-associative logics, *trees* of formulas are needed). In this book, we mostly assume commutativity, so the order of formulas is irrelevant. On the other hand, the number of occurrences of a formula does matter: we will want A,A to be different from A.

These considerations lead us to the following well-known definitions:

Definition 4.1. A *multiset over* α is an ordered pair $\langle \alpha, f \rangle$ where:

1. α is a set.

2. f is a function $f : \alpha \to \mathbb{N}$.

We say that $\langle \alpha, f \rangle$ is *finite* if $\mathrm{Set}(\langle \alpha, f \rangle) =_{\mathrm{def}} \{x \in \alpha : f(x) > 0\}$ is finite.

Definition 4.2. Let $\langle \alpha, f_1 \rangle$ and $\langle \alpha, f_2 \rangle$ be multisets over α:

- $\langle \alpha, f_1 \rangle \uplus \langle \alpha, f_2 \rangle =_{\mathrm{def}} \langle \alpha, f \rangle$ where $f(x) = f_1(x) + f_2(x)$.
- $\langle \alpha, f_1 \rangle \ominus \langle \alpha, f_2 \rangle =_{\mathrm{def}} \langle \alpha, f \rangle$ where $f(x) = \max(f_1(x) - f_2(x), 0)$.
- $\langle \alpha, f_1 \rangle \subseteq \langle \alpha, f_2 \rangle$ iff $f_1(x) \leq f_2(x)$ for all $x \in \alpha$.
- $x \in \langle \alpha, f \rangle$ iff $x \in \alpha$ and $f(x) > 0$.

Multisets are also written using set-like notation with brackets [and] (rather than { and }) where elements can be repeated. E.g. $[a,a,b,b,b]$ stands for a multiset $\langle \alpha, f \rangle$ where $\{a,b\} \subseteq \alpha$; $f(a) = 2$, $f(b) = 3$, and f is 0 elsewhere.

Let us assume again in this chapter that a set of formulas $\mathrm{Fm}_{\mathcal{L}}$ is given by a language $\mathcal{L}$ and a fixed countably infinite set of variables X. Finite multisets of formulas, $\langle \mathrm{Fm}_{\mathcal{L}}, f \rangle$, will be denoted by upper case Greek letters $\Gamma, \Delta, \Pi, \Sigma$. We often write – particularly in Gentzen systems – Γ, Π and Γ, A to denote the multiset sums $\Gamma \uplus \Pi$ and $\Gamma \uplus [A]$, respectively. Sometimes we write A for the multiset $[A]$ and a blank space for the empty multiset $[]$. Also, for any multiset Γ, we let:

$$\Gamma^0 = [] \quad \text{and} \quad \Gamma^{n+1} = \Gamma^n \uplus \Gamma \ \text{ for all } n \in \mathbb{N}$$

For $\star \in \{\wedge, \vee, \odot, \oplus\}$, we implicitly assume an ordering on formulas, and let:

$$\star[A_1, \ldots, A_n] = A_1 \star \ldots \star A_n$$

where $\wedge[] = \top$, $\vee[] = \bot$, $\odot[] = \mathrm{e}$, and $\oplus[] = \mathrm{f}$.

We take the complexity of a finite multiset of formulas to be the multiset of the complexities of its elements, i.e. a finite multiset of natural numbers:

Definition 4.3. For a finite multiset of formulas $\Gamma = \langle \mathrm{Fm}_{\mathcal{L}}, f \rangle$:

$$\mathrm{cp}(\Gamma) =_{\mathrm{def}} \langle \mathbb{N}^+, g \rangle \quad \text{where} \quad g(n) = \sum \{f(A) : A \in \mathrm{Fm}_{\mathcal{L}} \text{ and } \mathrm{cp}(A) = n\}$$

We will also make use of a well-known ordering on sets of finite multisets.

Definition 4.4. For a well-ordered poset $\langle \alpha, \leq \rangle$:

- $\mathsf{m}(\alpha)$ is the set of all finite multisets over α.
- $\langle \alpha, f \rangle \leq_{\mathsf{m}} \langle \alpha, g \rangle$ if $f(x) > g(x)$ implies $y > x$ and $g(y) > f(y)$ for some $y \in \alpha$.

Intuitively, $\Gamma \leq_{\mathsf{m}} \Delta$ holds if Γ can be obtained from Δ by replacing elements with finitely many (possibly zero) smaller elements. For example, for the set of all finite multisets over $\mathbb{N}^+$ and the usual ordering of natural numbers:

$$[1,1,2,2,2] \leq_{\mathsf{m}} [1,1,1,3]$$

Just notice that we can replace 3 on the right hand side with 2,2,2 and 1 with nothing, to get the left hand side.

We will make quite frequent use of the following result:

Theorem 4.5. *If* $\langle \alpha, \leq \rangle$ *is well-ordered, then* $\langle \mathsf{m}(\alpha), \leq_{\mathsf{m}} \rangle$ *is well-ordered.*

Proof. We define the well-ordered part W of $\mathsf{m}(\alpha)$ with respect to $\leq_{\mathsf{m}}$ inductively by $\Gamma \in W$ if $\Delta \in W$ for all $\Delta <_{\mathsf{m}} \Gamma$. Clearly, it is enough to prove that $W = \mathsf{m}(\alpha)$. Let us define $K(\Gamma)$ to mean:

For all $a \in \alpha$, if $\Delta \uplus [b] \in W$ for all $\Delta \in W$ and $b < a$, then $\Gamma \uplus [a] \in W$.

We show that $K(\Gamma)$ holds for all $\Gamma \in W$. By the definition of W, it is sufficient to establish the following:

Claim 1. For $\Gamma \in W$, $K(\Gamma')$ for all $\Gamma' <_m \Gamma$, implies $K(\Gamma)$.

Proof of Claim 1. Let $a \in \alpha$ and suppose that $\Delta \uplus [b] \in W$ for all $\Delta \in W$ and $b < a$. We will show that $\Pi \in W$ for all $\Pi <_{\mathsf{m}} \Gamma \uplus [a]$ and hence that $\Gamma \uplus [a] \in W$ as required. There are two cases:

- $\Pi = \Gamma' \uplus [a]$ for some $\Gamma' <_{\mathsf{m}} \Gamma$. Since $K(\Gamma')$ holds, $\Gamma' \uplus [a] \in W$ as required.
- $\Pi = \Gamma \uplus [b_1, \ldots, b_m]$ where $b_i < a$ for $i = 1 \ldots m$. The result follows by induction on m using the fact that $\Delta \uplus [b] \in W$ for all $\Delta \in W$ and $b < a$.

Claim 2. Let $L(a)$ mean: $\Gamma \uplus [a] \in W$ for all $\Gamma \in W$. Then $L(a)$ holds for all $a \in \alpha$.

Proof of Claim 2. We proceed inductively using the fact that $\langle \alpha, \leq \rangle$ is well-ordered. Suppose that $L(b)$ holds for all $b < a$; i.e. $\Gamma \uplus [b] \in W$ for all $b < a$ and $\Gamma \in W$. But $K(\Gamma)$ holds for all $\Gamma \in W$. So $\Gamma \uplus [a] \in W$ for all $\Gamma \in W$ as required.

Finally, to prove the theorem, we show that each finite multiset $\langle \alpha, f \rangle \in m(\alpha)$ is in W, proceeding by induction on $\sum\{f(x) : x \in \alpha\}$. The empty multiset $[]$ is in W by definition, and the inductive step follows immediately from Claim 2. □

4.1.1 Sequents

In Logic it is often useful to think of one collection of formulas "following from" or being "a consequence of" another collection of formulas. In Gentzen systems, this situation is represented using *sequents*: multisets (or sets, sequences, ...) of formulas separated by the "entails" symbol $\Rightarrow$. Formally:

Definition 4.6. An $\mathcal{L}$*-sequent* for a language $\mathcal{L}$ is an ordered pair of finite multisets of $\mathcal{L}$-formulas, written:

$$A_1, \ldots, A_n \Rightarrow B_1, \ldots, B_m$$

An $\mathcal{L}$-sequent $\Gamma \Rightarrow \Delta$ is:

- *single-conclusion* if Δ contains at most one formula.

- *multiple-conclusion* if it is not single-conclusion.
- *atomic* if Γ and Δ contain only atoms.
- *strictly atomic* if Γ and Δ contain only variables.

We define $\mathrm{cp}(\Gamma \Rightarrow \Delta) =_{\mathrm{def}} \mathrm{cp}(\Gamma) \uplus \mathrm{cp}(\Delta)$.

Intuitively, we might read $(A_1, \ldots, A_n \Rightarrow B_1, \ldots, B_m)$ as:

$$\text{“}A_1 \text{ and } \ldots \text{ and } A_n \text{ entails } B_1 \text{ or } \ldots \text{ or } B_m\text{”}$$

Or algebraically (to look ahead a little), sequents can be read as inequations: the sequent arrow $\Rightarrow$ will correspond to $\leq$, and the comma will correspond to $\odot$ on the left and $\oplus$ on the right.

We refer to $\mathcal{L}$-sequents using the letter S with various subscripts and superscripts, omitting the prefix $\mathcal{L}$ when the language is clear from the context. The notions of sequent rule, proof system, and derivation follow from the general definitions of the previous chapter, but let us recap. A *sequent rule* for a language $\mathcal{L}$ is a set of sequent rule instances: ordered pairs consisting of an $\mathcal{L}$-sequent S called the *conclusion* and a finite set of $\mathcal{L}$-sequents $S_1, \ldots, S_n$ called the *premises*, written $S_1, \ldots, S_n \;/\; S$ or:

$$\frac{S_1 \quad \ldots \quad S_n}{S}$$

Sequent rules where $n = 0$ are often called *initial sequents*. As for Hilbert system rules we usually write sequent rules using schemas involving meta-variables $A, B, C, \ldots$ standing for formulas and here also meta-variables $\Gamma, \Delta, \Pi, \Sigma, \ldots$ standing for finite multisets of formulas. Moreover, we make the following distinction:

Definition 4.7. A sequent rule is *single-conclusion* if all of its instances consist of single-conclusion sequents, *multiple-conclusion* otherwise. The *single-conclusion version* of a sequent rule consists of all of its single-conclusion instances.

A (single-conclusion) $\mathcal{L}$*-sequent calculus* GL for a language $\mathcal{L}$ then consists of all (single-conclusion) sequents for $\mathcal{L}$ and a set of (single-conclusion) sequent rules.[1]

Whereas Hilbert systems typically have many axiom schema with just one or two further rules, sequent calculi usually have just a few initial sequent schema and many other rules. Common initial sequent schema are:

$$\frac{}{A \Rightarrow A}\;(\textsc{id}) \qquad \frac{}{\Gamma, \bot \Rightarrow \Delta}\;(\bot\Rightarrow) \qquad \frac{}{\Gamma \Rightarrow \top, \Delta}\;(\Rightarrow\top)$$

The identity schema (ID) tells us that every formula “follows from itself”. $(\bot\Rightarrow)$ and $(\Rightarrow\top)$ tell us that a sequent always holds if $\bot$ is on the left or $\top$ is on the right.

Other rules can be grouped together into different classes:

[1] We use “G” for Gentzen here to distinguish our calculi from others in the literature characterizing the same logics but with different languages and rules.

- *Logical rules* deal with connectives. Usually, there are one or two rules, labelled $(\star\Rightarrow)$ or $(\Rightarrow\star)$ (with subscripts), for each appearance of the connective $\star$ on the left or right of a sequent. E.g. typical sequent rules for $\to$, $\wedge$, and $\vee$ are:

$$\frac{\Gamma_1 \Rightarrow A, \Delta_1 \quad \Gamma_2, B \Rightarrow \Delta_2}{\Gamma_1, \Gamma_2, A \to B \Rightarrow \Delta_1, \Delta_2}\ (\to\Rightarrow) \qquad \frac{\Gamma, A \Rightarrow B, \Delta}{\Gamma \Rightarrow A \to B, \Delta}\ (\Rightarrow\to)$$

$$\frac{\Gamma, A \Rightarrow \Delta}{\Gamma, A \wedge B \Rightarrow \Delta}\ (\wedge\Rightarrow)_1 \qquad \frac{\Gamma, B \Rightarrow \Delta}{\Gamma, A \wedge B \Rightarrow \Delta}\ (\wedge\Rightarrow)_2$$

$$\frac{\Gamma, A \Rightarrow \Delta \quad \Gamma, B \Rightarrow \Delta}{\Gamma, A \vee B \Rightarrow \Delta}\ (\vee\Rightarrow) \qquad \frac{\Gamma \Rightarrow A, \Delta \quad \Gamma \Rightarrow B, \Delta}{\Gamma \Rightarrow A \wedge B, \Delta}\ (\Rightarrow\wedge)$$

$$\frac{\Gamma \Rightarrow A, \Delta}{\Gamma \Rightarrow A \vee B, \Delta}\ (\Rightarrow\vee)_1 \qquad \frac{\Gamma \Rightarrow B, \Delta}{\Gamma \Rightarrow A \vee B, \Delta}\ (\Rightarrow\vee)_2$$

The formula $\star(A_1, \ldots, A_n)$ in the conclusion of an instance of a logical rule for $\star$ is called the *principal formula* of the instance, the subformulas $A_1, \ldots, A_n$ occurring in the premises are called *active formulas*, and other formulas are called *context formulas*.

- *Structural rules* allow modifications of sequents without reference to the composition of individual formulas. For example, these "weakening" and "contraction" rules allow formulas in a sequent to be added and removed respectively.

$$\frac{\Gamma \Rightarrow \Delta}{\Gamma, A \Rightarrow \Delta}\ (\text{WL}) \qquad \frac{\Gamma \Rightarrow \Delta}{\Gamma \Rightarrow A, \Delta}\ (\text{WR})$$

$$\frac{\Gamma, A, A \Rightarrow \Delta}{\Gamma, A \Rightarrow \Delta}\ (\text{CL}) \qquad \frac{\Gamma \Rightarrow A, A, \Delta}{\Gamma \Rightarrow A, \Delta}\ (\text{CR})$$

Structural rules are often built into the definition of a sequent. In particular, we have defined sequents here as consisting of multisets. To treat sequences of formulas we would also need the "exchange" rules:

$$\frac{\Gamma, B, A, \Pi \Rightarrow \Delta}{\Gamma, A, B, \Pi \Rightarrow \Delta}\ (\text{EL}) \qquad \frac{\Gamma \Rightarrow \Delta, B, A, \Sigma}{\Gamma \Rightarrow \Delta, A, B, \Sigma}\ (\text{ER})$$

The absence or presence of different structural rules in Gentzen systems produces a wide spectrum of "substructural logics". For example, relevant logics omit weakening rules, while Linear Logic does without both weakening and contraction rules.

- *The cut rule* expresses the transitivity of deduction. If B proves A and A proves C, then B proves C:

$$\frac{\Gamma_1, A \Rightarrow \Delta_1 \quad \Gamma_2 \Rightarrow A, \Delta_2}{\Gamma_1, \Gamma_2 \Rightarrow \Delta_1, \Delta_2}\ (\text{CUT})$$

The *cut-formula* A occurring in the premises of an instance of this rule may not occur in the conclusion, making proof search difficult. Crucially, however, it is often possible to do without this rule in a calculus, or even (the topic of Chapter 5) "eliminate" its instances from derivations.

Our choice of examples of initial sequents and rules here is not accidental. Taken together they form a sequent calculus GK for Classical Logic in the language $\mathcal{L}_G = \{\wedge, \vee, \rightarrow, \bot, \top\}$. Note, however, that this is not "the" sequent calculus for Classical Logic, just one among many options. For example, it is possible to define a version with no structural rules in which weakening and contraction rules are "absorbed" into the logical rules and initial sequents.

Using the general definitions in Chapter 3, a GL-*derivation* d for a sequent calculus GL of a sequent S from a finite set of sequents $\mathcal{U}$ is a labelled finite tree (usually written with the root at the bottom) such that:

- S labels the root of the tree and is called the *end-sequent*.
- For each node x labelled S_0, either $S_0 \in \mathcal{U}$ or the child nodes of x are labelled $S_1, \ldots, S_n$ and $S_1, \ldots, S_n \;/\; S_0$ is an instance of a rule of GL.

S is GL-*derivable from* $\mathcal{U}$ if there is a GL-derivation d of S from a finite set $\mathcal{U}^F \subseteq \mathcal{U}$, written $d;\mathcal{U} \vdash_{GL} S$ or just $\mathcal{U} \vdash_{GL} S$.

Example 4.8. Peirce's axioms are derivable in GK as follows:

$$\dfrac{\dfrac{\dfrac{}{A \Rightarrow A}\,(\text{ID}) \qquad \dfrac{\dfrac{\dfrac{}{A \Rightarrow A}\,(\text{ID})}{A \Rightarrow B, A}\,(\text{WR})}{\Rightarrow A \rightarrow B, A}\,(\Rightarrow\rightarrow)}{\dfrac{(A \rightarrow B) \rightarrow A \Rightarrow A, A}{(A \rightarrow B) \rightarrow A \Rightarrow A}\,(\text{CR})}\,(\rightarrow\Rightarrow)}{\Rightarrow ((A \rightarrow B) \rightarrow A) \rightarrow A}\,(\Rightarrow\rightarrow)$$

The sequent framework is remarkably flexible. One of its most interesting features is that by restricting GK (or a calculus like it) to single-conclusion rules, we obtain a calculus for Intuitionistic Logic. For example, the rules $(\Rightarrow\wedge)$ and $(\rightarrow\Rightarrow)$ become:

$$\dfrac{\Gamma \Rightarrow A \quad \Gamma \Rightarrow B}{\Gamma \Rightarrow A \wedge B}\,(\Rightarrow\wedge) \qquad \dfrac{\Gamma_1 \Rightarrow A \quad \Gamma_2, B \Rightarrow \Delta}{\Gamma_1, \Gamma_2, A \rightarrow B \Rightarrow \Delta}\,(\rightarrow\Rightarrow)$$

The rule (CR) is lost altogether since the premise is multiple-conclusion, blocking the derivation in Example 4.8. More generally, the step from single to multiple conclusion permits a derivation of the involution axioms $\neg\neg A \rightarrow A$, and corresponds to the addition of this axiom schema to Hilbert systems.

The flexibility of sequent calculi is illustrated further by removing structural rules. For example, without weakening rules it is no longer possible to prove the classically valid $A \rightarrow (B \rightarrow A)$. Also, alternative rules for logical connectives become available. Compare for example, the rules for $\wedge$ of GK with:

$$\frac{\Gamma, A, B \Rightarrow \Delta}{\Gamma, A \odot B \Rightarrow \Delta}\ (\odot\Rightarrow) \qquad \frac{\Gamma_1 \Rightarrow A, \Delta_1 \quad \Gamma_2 \Rightarrow B, \Delta_2}{\Gamma_1, \Gamma_2 \Rightarrow A \odot B, \Delta_1, \Delta_2}\ (\Rightarrow\odot)$$

With both weakening and contraction rules it is easy to derive $(A \wedge B \Rightarrow A \odot B)$ and $(A \odot B \Rightarrow A \wedge B)$. Removing these rules, however, the derivations of equivalence break down and the connectives "split". Of course this is no surprise to readers familiar with algebraic structures like residuated lattices. In that framework, the split is into lattice (additive) and group (multiplicative) operations.

Removing structural rules and adding new connectives in this way gives characterizations of different logics. In particular, removing the weakening and contraction rules from GK and adding both the rules for $\odot$ above and some more for e and f (which split from $\top$ and $\bot$), we get sequent calculi for Multiplicative Additive Linear Logic MALL, and, restricting to single-conclusion rules, its "intuitionistic" version MAILL. Removing just contraction rules gives calculi for Affine Multiplicative Additive Linear Logic AMALL and Monoidal Logic ML, respectively, while removing just weakening rules gives calculi for distribution-less relevant logics.

4.1.2 Hypersequents

We can do quite a lot with sequents: treat substructural logics, modal logics, and many other logical families. However, for fuzzy logics, sequents are not enough. To see why, consider a potential cut-free derivation of one of the prelinearity axioms. Such a derivation would have as an end-sequent:

$$\Rightarrow (A \to B) \vee (B \to A)$$

The derivation of this in GK uses both contraction and weakening rules:

$$\cfrac{\cfrac{\cfrac{\cfrac{\cfrac{\cfrac{\cfrac{\cfrac{}{A \Rightarrow A}\ (\text{ID})}{A, B \Rightarrow A}\ (\text{WL})}{A, B \Rightarrow A, B}\ (\text{WR})}{A \Rightarrow B, B \to A}\ (\Rightarrow\to)}{\Rightarrow A \to B, B \to A}\ (\Rightarrow\to)}{\Rightarrow A \to B, (A \to B) \vee (B \to A)}\ (\Rightarrow\vee)_2}{\Rightarrow (A \to B) \vee (B \to A), (A \to B) \vee (B \to A)}\ (\Rightarrow\vee)_1}{\Rightarrow (A \to B) \vee (B \to A)}\ (\text{CR})$$

In fact, a "weaker form" of weakening is enough here: the so-called "mix rule". Just replace the top three lines of the above derivation with:

$$\frac{\cfrac{}{A \Rightarrow A}\ (\text{ID}) \qquad \cfrac{}{B \Rightarrow B}\ (\text{ID})}{A, B \Rightarrow A, B}\ (\text{MIX})$$

However, if contraction or weakening is unavailable, we have just two options. We can tinker with different sequent calculi, e.g. by taking non-standard rules or interpretations. This approach is taken for some logics in Chapter 6. Or we can extend the sequent framework itself.

Take another look at $\Rightarrow (A \to B) \vee (B \to A)$. To work further on the constituents of this sequent, it would be helpful to represent in some way:

$$\Rightarrow A \to B \quad \text{“or”} \quad \Rightarrow B \to A$$

Then we could operate on $A \to B$ and $B \to A$ in parallel. Let us take the symbol $|$ to stand for this meta-level “or”.

Definition 4.9. An $\mathcal{L}$-*hypersequent* is a non-empty finite multiset of the form:

$$\Gamma_1 \Rightarrow \Delta_1 \mid \ldots \mid \Gamma_n \Rightarrow \Delta_n$$

where each $(\Gamma_i \Rightarrow \Delta_i)$ is an $\mathcal{L}$-sequent for $i = 1 \ldots n$.

An $\mathcal{L}$-hypersequent $S_1 \mid \ldots \mid S_n$ is:

- *single-conclusion* if $S_1, \ldots, S_n$ are all single-conclusion.
- *multiple-conclusion* if it is not single-conclusion.
- *(strictly) atomic* if $S_1, \ldots, S_n$ are all (strictly) atomic.

We define the *complexity* of $S_1 \mid \ldots \mid S_n$ as $\text{cp}(S_1 \mid \ldots \mid S_n) =_{\text{def}} [\text{cp}(S_1), \ldots, \text{cp}(S_n)]$.

We will drop the $\mathcal{L}$ as before when the language is clear from the context, and use $\mathcal{G}$ and $\mathcal{H}$ to refer to arbitrary hypersequents.

Example 4.10. To get some idea of how the complexity of a hypersequent is calculated, consider:

$$\mathcal{G} = (p, q, (p \vee q) \to r \Rightarrow r \mid q \Rightarrow p \to r \mid q \wedge p, r, \bot \Rightarrow p \to q \mid \Rightarrow)$$

For the sequents, we have:

$$\text{cp}(p, q, (p \vee q) \to r \Rightarrow r) = [1,1,1,5] \qquad \text{cp}(q \Rightarrow p \to r) = [1,3]$$
$$\text{cp}(q \wedge p, r, \bot \Rightarrow p \to q) = [1,1,3,3] \qquad \text{cp}(\Rightarrow) = []$$

So putting this together, we get:

$$\text{cp}(\mathcal{G}) = [[1,1,1,5],[1,3],[1,1,3,3],[]]$$

Recall that the usual well-ordering $\leq$ over $\mathbb{N}^+$ gives rise (via Definition 4.4 and Theorem 4.4) to a well-ordering $\leq_{\mathsf{m}}$ of finite multisets of positive natural numbers. But then this well-ordering also gives rises to a new well-ordering $\leq_{\mathsf{mm}}$ of finite multisets of finite multisets of positive natural numbers. Hence for example:

$$[[1,1,1,1,2,2,2],[1,1,3,3],[2,2,3]] \leq_{\mathsf{mm}} [[1,2,3,3],[1,1,1,2]] \leq_{\mathsf{mm}} [[4]]$$

Just as for sequents where each side is a multiset, here the use of multisets of sequents rather than sets or sequences means that the multiplicity but not the order of sequents is important. Sometimes we find it convenient to write a hypersequent $S_1 \mid \ldots \mid S_n$ directly as a multiset $[S_1, \ldots, S_n]$, possibly using brackets (and) to distinguish the sequents. Also we will frequently use $\mathcal{G}$, $\mathcal{H}$, etc. in rule schema to denote (perhaps empty) hypersequents, called *side-hypersequents*.

Definitions for hypersequent rules, derivations, and calculi follow from the general definitions, and are in any case, very similar to those for sequent calculi. Moreover, there is a simple method for stepping from sequent rules to hypersequent rules:

Definition 4.11. The *hypersequent version* of a (single-conclusion) sequent rule (r) is the set of inferences:

$$\frac{\mathcal{G} \mid S_1 \ \ldots \ \mathcal{G} \mid S_n}{\mathcal{G} \mid S}$$

where $S_1, \ldots, S_n \ / \ S$ is an instance of (r) and $\mathcal{G}$ a (single-conclusion) hypersequent.

For example, the single-conclusion hypersequent versions of $(\rightarrow\Rightarrow)$ and $(\Rightarrow\rightarrow)$ are:

$$\frac{\mathcal{G} \mid \Gamma_1 \Rightarrow A \quad \mathcal{G} \mid \Gamma_2, B \Rightarrow \Delta}{\mathcal{G} \mid \Gamma_1, \Gamma_2, A \rightarrow B \Rightarrow \Delta} \ (\rightarrow\Rightarrow) \qquad \frac{\mathcal{G} \mid \Gamma, A \Rightarrow B}{\mathcal{G} \mid \Gamma \Rightarrow A \rightarrow B} \ (\Rightarrow\rightarrow)$$

There is also a natural way of stepping in the opposite direction:

Definition 4.12. The *sequent version* of a rule consists of all its instances which have sequents as both premises and conclusion.

Except when confusion might occur, we will make use of the same label for a hypersequent rule and also its various single-conclusion and sequent versions.

The fact that $\mid$ is supposed to represent disjunction between sequents is reflected in a couple of key "external" structural rules: external weakening and external contraction.

$$\frac{\mathcal{G}}{\mathcal{G} \mid \mathcal{H}} \ (\text{EW}) \qquad \frac{\mathcal{G} \mid \mathcal{H} \mid \mathcal{H}}{\mathcal{G} \mid \mathcal{H}} \ (\text{EC})$$

These rules add and contract (or, reading upwards, remove and multiply) sequents. However, they do not really add much to the expressivity of sequent calculi. This is achieved only when we add rules that allow interaction between sequents. For example, the following single-conclusion rule "splits" a sequent in two:

$$\frac{\mathcal{G} \mid \Gamma_1, \Gamma_2 \Rightarrow \Delta}{\mathcal{G} \mid \Gamma_1 \Rightarrow \Delta \mid \Gamma_2 \Rightarrow} \ (\text{SPLIT})$$

We can use (SPLIT) to obtain a single-conclusion contraction-free calculus for Classical Logic. Just take single-conclusion hypersequent versions of the initial sequents, logical and weakening rules of GK together with (EW), (EC), and (SPLIT). To see how this works, compare the following derivation of Peirce's axioms in this calculus with the GK-derivation of Example 4.8:

$$
\dfrac{\dfrac{\dfrac{\dfrac{\dfrac{\overline{A \Rightarrow A}\ (\text{ID})}{A \Rightarrow A \mid \Rightarrow A}\ (\text{EW}) \qquad \dfrac{\dfrac{\dfrac{\dfrac{}{A \Rightarrow A}\ (\text{ID})}{A \Rightarrow \mid \Rightarrow A}\ (\text{SPLIT})}{A \Rightarrow B \mid \Rightarrow A}\ (\text{WR})}{\Rightarrow A \to B \mid \Rightarrow A}\ (\Rightarrow\to)}{(A \to B) \to A \Rightarrow A \mid \Rightarrow A}\ (\to\Rightarrow)}{(A \to B) \to A \Rightarrow A \mid (A \to B) \to A \Rightarrow A}\ (\text{WL})}{(A \to B) \to A \Rightarrow A}\ (\text{EC})}{\Rightarrow ((A \to B) \to A) \to A}\ (\Rightarrow\to)
$$

Of course, contraction does occur in this calculus, but at the level of whole sequents rather than formulas.

Finally, we mention a property for hypersequent calculi with schematic rules that will prove very useful in the next chapter. Let us write $\mathcal{G}[p/A]$ for the result of substituting all occurrences of a variable p in a hypersequent $\mathcal{G}$ with a formula A. Then the following is established by an easy induction on the height of a derivation:

Lemma 4.13. *Let* GL *be a hypersequent calculus with schematic rules. If* $\vdash_{\mathrm{GL}} \mathcal{G}$, *then* $\vdash_{\mathrm{GL}} \mathcal{G}[p/A]$.

4.2 Core Systems

Initial sequents or hypersequents – rules with no premises – provide the basic building blocks for derivations in Gentzen systems. The systems presented in this book all contain a stock of such rules that express just "*A* follows from *A*":

Definition 4.14 (Identity Hypersequents).

$$
\dfrac{}{\mathcal{G} \mid A \Rightarrow A}\ (\text{ID})
$$

The core of a Gentzen system, uniform across almost all the systems in this chapter, consists of (ID) plus a fixed set of logical rules characterizing the behaviour of connectives on the left and right of sequents. Let us start with implication.

Definition 4.15 (Rules for Implication).

$$
\dfrac{\mathcal{G} \mid \Gamma_1 \Rightarrow A, \Delta_1 \quad \mathcal{G} \mid \Gamma_2, B \Rightarrow \Delta_2}{\mathcal{G} \mid \Gamma_1, \Gamma_2, A \to B \Rightarrow \Delta_1, \Delta_2}\ (\to\Rightarrow) \qquad \dfrac{\mathcal{G} \mid \Gamma, A \Rightarrow B, \Delta}{\mathcal{G} \mid \Gamma \Rightarrow A \to B, \Delta}\ (\Rightarrow\to)
$$

Note that we can take both single-conclusion and sequent versions of these rules. That is, each rule counts four times (although we use the same label in each case). E.g. for $(\Rightarrow\to)$, we have additionally:

$$
\dfrac{\mathcal{G} \mid \Gamma, A \Rightarrow B}{\mathcal{G} \mid \Gamma \Rightarrow A \to B}\ (\Rightarrow\to) \qquad \dfrac{\Gamma, A \Rightarrow B, \Delta}{\Gamma \Rightarrow A \to B, \Delta}\ (\Rightarrow\to) \qquad \dfrac{\Gamma, A \Rightarrow B}{\Gamma \Rightarrow A \to B}\ (\Rightarrow\to)
$$

Example 4.16. The transitivity (B), permutability (C), and reflexivity (I) axioms are all derivable using the implication rules and (ID). E.g. for (B):

$$\dfrac{\dfrac{\dfrac{\dfrac{\dfrac{}{C \Rightarrow C}\,(\text{ID}) \quad \dfrac{\dfrac{}{B \Rightarrow B}\,(\text{ID}) \quad \dfrac{}{A \Rightarrow A}\,(\text{ID})}{A \to B, A \Rightarrow B}\,(\to\Rightarrow)}{A \to B, B \to C, A \Rightarrow C}\,(\to\Rightarrow)}{A \to B, B \to C \Rightarrow A \to C}\,(\Rightarrow\to)}{A \to B \Rightarrow (B \to C) \to (A \to C)}\,(\Rightarrow\to)}{\Rightarrow (A \to B) \to ((B \to C) \to (A \to C))}\,(\Rightarrow\to)$$

The rule $(\Rightarrow\to)$ also gives us a good idea of how to interpret the "," on the left of sequents. Since $A \to (B \to C)$ is equivalent to $(A \odot B) \to C$ in our logics, reading "$\Rightarrow$" as a meta-level $\to$, it makes sense to read "," as a meta-level $\odot$. Bearing this in mind, the rules for $\odot$ are:

Definition 4.17 (Rules for Multiplicative Conjunction).

$$\dfrac{\mathcal{G} \mid \Gamma, A, B \Rightarrow \Delta}{\mathcal{G} \mid \Gamma, A \odot B \Rightarrow \Delta}\,(\odot\Rightarrow) \qquad \dfrac{\mathcal{G} \mid \Gamma_1 \Rightarrow A, \Delta_1 \quad \mathcal{G} \mid \Gamma_2 \Rightarrow B, \Delta_2}{\mathcal{G} \mid \Gamma_1, \Gamma_2 \Rightarrow A \odot B, \Delta_1, \Delta_2}\,(\Rightarrow\odot)$$

Rules for the multiplicative unit e are based on the idea that e should be both the unit element for $\odot$, and the meaning of the empty multiset on the left. Similarly, f takes the role of the unit for $\oplus$, and the meaning of the empty multiset on the right:

Definition 4.18 (Rules for Multiplicative Constants).

$$\dfrac{\mathcal{G} \mid \Gamma \Rightarrow \Delta}{\mathcal{G} \mid \Gamma, \mathrm{e} \Rightarrow \Delta}\,(\mathrm{e}\Rightarrow) \qquad \dfrac{}{\mathcal{G} \mid\, \Rightarrow \mathrm{e}}\,(\Rightarrow\mathrm{e}) \qquad \dfrac{}{\mathcal{G} \mid \mathrm{f} \Rightarrow}\,(\mathrm{f}\Rightarrow) \qquad \dfrac{\mathcal{G} \mid \Gamma \Rightarrow \Delta}{\mathcal{G} \mid \Gamma \Rightarrow \mathrm{f}, \Delta}\,(\Rightarrow\mathrm{f})$$

The comma on the right of sequents (for multiple-conclusion calculi) is interpreted by the "multiplicative disjunction" connective $\oplus$ defined by $A \oplus B =_{\mathrm{def}} \neg A \to B$ where $\neg A =_{\mathrm{def}} A \to \mathrm{f}$. Derived rules for $\oplus$ and $\neg$ are then:

$$\dfrac{\mathcal{G} \mid \Gamma_1, A \Rightarrow \Delta_1 \quad \mathcal{G} \mid \Gamma_2, B \Rightarrow \Delta_2}{\mathcal{G} \mid \Gamma_1, \Gamma_2, A \oplus B \Rightarrow \Delta_1, \Delta_2}\,(\oplus\Rightarrow) \qquad \dfrac{\mathcal{G} \mid \Gamma \Rightarrow A, B, \Delta}{\mathcal{G} \mid \Gamma \Rightarrow A \oplus B, \Delta}\,(\Rightarrow\oplus)$$

$$\dfrac{\mathcal{G} \mid \Gamma \Rightarrow A, \Delta}{\mathcal{G} \mid \Gamma, \neg A \Rightarrow \Delta}\,(\neg\Rightarrow) \qquad \dfrac{\mathcal{G} \mid \Gamma, A \Rightarrow \Delta}{\mathcal{G} \mid \Gamma \Rightarrow \neg A, \Delta}\,(\Rightarrow\neg)$$

For example, in the case of $(\oplus\Rightarrow)$, we have:

$$\dfrac{\dfrac{\dfrac{\mathcal{G} \mid \Gamma_1, A \Rightarrow \Delta_1}{\mathcal{G} \mid \Gamma_1, A \Rightarrow \mathrm{f}, \Delta_1}\,(\Rightarrow\mathrm{f})}{\mathcal{G} \mid \Gamma_1 \Rightarrow A \to \mathrm{f}, \Delta_1}\,(\Rightarrow\to) \quad \mathcal{G} \mid \Gamma_2, B \Rightarrow \Delta_2}{\mathcal{G} \mid \Gamma_1, \Gamma_2, (A \to \mathrm{f}) \to B \Rightarrow \Delta_1, \Delta_2}\,(\to\Rightarrow)$$

and for $(\Rightarrow\oplus)$:

$$\dfrac{\dfrac{\dfrac{}{\mathcal{G} \mid \mathrm{f} \Rightarrow}\,(\mathrm{f}\Rightarrow) \qquad \mathcal{G} \mid \Gamma \Rightarrow A, B, \Delta}{\mathcal{G} \mid \Gamma, A \to \mathrm{f} \Rightarrow B, \Delta}\,(\to\Rightarrow)}{\mathcal{G} \mid \Gamma \Rightarrow (A \to \mathrm{f}) \to B, \Delta}\,(\Rightarrow\to)$$

It is easy to see how the single-conclusion / multiple-conclusion distinction affects derivations in Gentzen systems. In the multiple-conclusion case for example, we are able to prove the involution axioms (INV):

$$\dfrac{\dfrac{\dfrac{\dfrac{}{A \Rightarrow A}\,(\mathrm{ID})}{\Rightarrow A, \neg A}\,(\Rightarrow\neg)}{\neg\neg A \Rightarrow A}\,(\neg\Rightarrow)}{\Rightarrow \neg\neg A \to A}\,(\Rightarrow\to)$$

But in single-conclusion calculi the crucial application of $(\Rightarrow\neg)$ is blocked.

Taking multiple or single-conclusion sequent versions of the rules introduced so far gives "cut-free" multiplicative fragments of Linear Logic. To introduce the additive connectives $\wedge$ and $\vee$ for these (and other) logics, we have:

Definition 4.19 (Rules for Additive Conjunction).

$$\dfrac{\mathcal{G} \mid \Gamma, A \Rightarrow \Delta}{\mathcal{G} \mid \Gamma, A \wedge B \Rightarrow \Delta}\,(\wedge\Rightarrow)_1 \qquad \dfrac{\mathcal{G} \mid \Gamma, B \Rightarrow \Delta}{\mathcal{G} \mid \Gamma, A \wedge B \Rightarrow \Delta}\,(\wedge\Rightarrow)_2$$

$$\dfrac{\mathcal{G} \mid \Gamma \Rightarrow A, \Delta \quad \mathcal{G} \mid \Gamma \Rightarrow B, \Delta}{\mathcal{G} \mid \Gamma \Rightarrow A \wedge B, \Delta}\,(\Rightarrow\wedge)$$

Definition 4.20 (Rules for Additive Disjunction).

$$\dfrac{\mathcal{G} \mid \Gamma, A \Rightarrow \Delta \quad \mathcal{G} \mid \Gamma, B \Rightarrow \Delta}{\mathcal{G} \mid \Gamma, A \vee B \Rightarrow \Delta}\,(\vee\Rightarrow)$$

$$\dfrac{\mathcal{G} \mid \Gamma \Rightarrow A, \Delta}{\mathcal{G} \mid \Gamma \Rightarrow A \vee B, \Delta}\,(\Rightarrow\vee)_1 \qquad \dfrac{\mathcal{G} \mid \Gamma \Rightarrow B, \Delta}{\mathcal{G} \mid \Gamma \Rightarrow A \vee B, \Delta}\,(\Rightarrow\vee)_2$$

Example 4.21. The axioms $(\wedge 1)$–$(\wedge 3)$ and $(\vee 1)$–$(\vee 3)$ are easily derived using the above rules together with those for implication. For example, the axioms of $(\wedge 3)$ are derived as follows:

$$\dfrac{\dfrac{\dfrac{\dfrac{\dfrac{\dfrac{}{B \Rightarrow B}\,(\mathrm{ID}) \quad \dfrac{}{A \Rightarrow A}\,(\mathrm{ID})}{A \to B, A \Rightarrow B}\,(\to\Rightarrow)}{(A \to B) \wedge (A \to C), A \Rightarrow B}\,(\wedge\Rightarrow)_1 \quad \dfrac{\dfrac{\dfrac{}{C \Rightarrow C}\,(\mathrm{ID}) \quad \dfrac{}{A \Rightarrow A}\,(\mathrm{ID})}{A \to C, A \Rightarrow C}\,(\to\Rightarrow)}{(A \to B) \wedge (A \to C), A \Rightarrow C}\,(\wedge\Rightarrow)_2}{(A \to B) \wedge (A \to C), A \Rightarrow B \wedge C}\,(\Rightarrow\wedge)}{(A \to B) \wedge (A \to C) \Rightarrow A \to (B \wedge C)}\,(\Rightarrow\to)}{\Rightarrow ((A \to B) \wedge (A \to C)) \to (A \to (B \wedge C))}\,(\Rightarrow\to)$$

However, the distributivity axioms (DIS) cannot be proved using these rules.

The top truth $\top$ and bottom falsity $\bot$ are characterized, if required, by the following rules (or, initial hypersequents):

Definition 4.22 (Rules for Additive Constants).

$$\frac{}{\mathcal{G} \mid \Gamma, \bot \Rightarrow \Delta}\ (\bot\Rightarrow) \qquad\qquad \frac{}{\mathcal{G} \mid \Gamma \Rightarrow \top, \Delta}\ (\Rightarrow\top)$$

The rules introduced so far in this section have an important feature. The only formulas occurring in the premises are subformulas of formulas occurring in the conclusion. This is called the "subformula property". It ensures that each derivation using only these rules can be viewed (reading upwards) as a decomposition of formulas in which no new material is added. In fact we can decompose to such an extent that (ID) can be restricted to strictly atomic instances (i.e. containing only variables).

Proposition 4.23. *For* $\mathcal{L} \subseteq \mathcal{L}_3$*, let* GL *consist of the appropriate logical rules for* $\mathcal{L}$ *and the strictly atomic instances of* (ID). *Then* $\vdash_{\mathrm{GL}} A \Rightarrow A$ *for all* $A \in \mathrm{Fm}_{\mathcal{L}}$.

Proof. We proceed by induction on $\mathrm{cp}(A)$. If A is a variable p, then $(A \Rightarrow A)$ is a strictly atomic instance of (ID). If A is a constant, then the cases of $\bot$ and $\top$ follow immediately using $(\bot\Rightarrow)$ and $(\Rightarrow\top)$, respectively, and if A is f or e, then we have the derivations:

$$\dfrac{\dfrac{}{\Rightarrow \mathrm{e}}\ (\Rightarrow\mathrm{e})}{\mathrm{e} \Rightarrow \mathrm{e}}\ (\mathrm{e}\Rightarrow) \qquad\qquad \dfrac{\dfrac{}{\mathrm{f} \Rightarrow}\ (\mathrm{f}\Rightarrow)}{\mathrm{f} \Rightarrow \mathrm{f}}\ (\Rightarrow\mathrm{f})$$

Now suppose that A is $B \star C$ for $\star \in \{\to, \odot, \wedge, \vee\}$. Then by the induction hypothesis twice $\vdash_{\mathrm{GL}} B \Rightarrow B$ and $\vdash_{\mathrm{GL}} C \Rightarrow C$, and we can construct derivations using the left and right rules for $\star$. E.g. for $\to$ and $\vee$, we have:

$$\dfrac{\dfrac{B \Rightarrow B \quad C \Rightarrow C}{B \to C, B \Rightarrow C}\ (\to\Rightarrow)}{B \to C \Rightarrow B \to C}\ (\Rightarrow\to) \qquad \dfrac{\dfrac{B \Rightarrow B}{B \Rightarrow B \vee C}\ (\Rightarrow\vee)_1 \quad \dfrac{C \Rightarrow C}{C \Rightarrow B \vee C}\ (\Rightarrow\vee)_2}{B \vee C \Rightarrow B \vee C}\ (\vee\Rightarrow)$$

□

Our core set of rules is completed by a rule that fails the subformula property, however: the so-called "cut rule", corresponding to the transitivity of deduction or the introduction of lemmas into derivations.

Definition 4.24 (Cut Rule).

$$\frac{\mathcal{G} \mid \Gamma_1, A \Rightarrow \Delta_1 \quad \mathcal{G} \mid \Gamma_2 \Rightarrow A, \Delta_2}{\mathcal{G} \mid \Gamma_1, \Gamma_2 \Rightarrow \Delta_1, \Delta_2}\ (\mathrm{CUT})$$

Example 4.25. The cut rule is closely related to the modus ponens rule used in Hilbert systems. Indeed we can simulate (MP) in systems with (CUT) as follows:

$$\dfrac{\Rightarrow A \qquad \dfrac{\Rightarrow A \to B \qquad \dfrac{\dfrac{}{A \Rightarrow A}\ (\mathrm{ID}) \quad \dfrac{}{B \Rightarrow B}\ (\mathrm{ID})}{A, A \to B \Rightarrow B}\ (\to\Rightarrow)}{A \Rightarrow B}\ (\mathrm{CUT})}{\Rightarrow B}\ (\mathrm{CUT})$$

As we have hinted, (CUT) is not needed for most of the calculi that we consider (at least, the good ones). Everything that we can prove with cuts, we can prove without cuts. However, there are good reasons to have (CUT) as a primitive rule of Gentzen systems. First, it is easy to prove soundness and completeness for such systems. Second, we can investigate the impact of introducing or eliminating cuts from derivations, one of the major themes of Proof Theory.

We now have everything we need to identify some core systems. GMALL, a calculus for Multiplicative Additive Linear Logic, consists of the sequent versions of the initial hypersequents, logical rules, and cut rule presented above (also collected in Fig. 4.1). GMAILL, for the intuitionistic partner of the logic, is obtained as the single-conclusion version of GMALL. As for Hilbert systems, we also introduce notation to denote counterparts of systems such as these without $\bot$ and $\top$.

Definition 4.26. A GL-*extension* for an $\mathcal{L}_F$ or $\mathcal{L}_B$ hypersequent calculus GL consists of GL extended with $\mathcal{L}_B$ hypersequent rule schema. Also, for any GMAILL-extension GL, the Gentzen system GL$^-$ is GL with $(\bot\Rightarrow)$ and $(\Rightarrow\top)$ removed.

Before moving on to study particular extensions of GMAILL$^-$, let us investigate some useful invertibility properties of the logical rules that hold for any such extension GL. Consider an instance of the rule $(\odot\Rightarrow)$. If the conclusion $(\mathcal{G} \mid \Gamma, A \odot B \Rightarrow \Delta)$ is GL-derivable, then the premise $(\mathcal{G} \mid \Gamma, A, B \Rightarrow \Delta)$ is GL-derivable. We just make use of the derivation:

$$\dfrac{\dfrac{\dfrac{}{\mathcal{G} \mid A \Rightarrow A}\,(\text{ID}) \quad \dfrac{}{\mathcal{G} \mid B \Rightarrow B}\,(\text{ID})}{\mathcal{G} \mid A, B \Rightarrow A \odot B}\,(\Rightarrow\odot) \quad \mathcal{G} \mid \Gamma, A \odot B \Rightarrow \Delta}{\mathcal{G} \mid \Gamma, A, B \Rightarrow \Delta}\,(\text{CUT})$$

That is, $(\odot\Rightarrow)$ is GL-invertible. Moreover, the same property can be shown to hold for many other logical rules (typically either for the left rules or right rules for each connective, but not both), left as simple exercises for the reader.

Proposition 4.27. *The logical rules* $(\odot\Rightarrow)$, $(\Rightarrow\to)$, $(\Rightarrow\mathrm{f})$, $(\mathrm{e}\Rightarrow)$, $(\vee\Rightarrow)$, *and* $(\Rightarrow\wedge)$ *are* GL*-invertible for any* GMAILL$^-$*-extension* GL. *Also, if* GL *extends* GMALL$^-$, *then the derived rule* $(\Rightarrow\oplus)$ *is* GL*-invertible.*

4.3 Adding Structural Rules

Structural rules manipulate sequents and formulas with no regard to their internal composition. If we fix the logical rules of our systems, then it is these manipulations that give each calculus its distinctive properties. Here we consider important examples of two kinds of structural rules: *internal rules* that manipulate formulas within individual sequents, and *external rules* that manipulate whole sequents. We also introduce along the way many different sequent and hypersequent calculi, collecting these definitions together for the reader's convenience in Table 4.1.

Initial Sequents

$$\frac{}{\mathcal{G} \mid A \Rightarrow A} \; (\text{ID})$$

Structural Rules

$$\frac{\mathcal{G}}{\mathcal{G} \mid \mathcal{H}} \; (\text{EW}) \qquad \frac{\mathcal{G} \mid \mathcal{H} \mid \mathcal{H}}{\mathcal{G} \mid \mathcal{H}} \; (\text{EC})$$

Logical Rules

$$\frac{}{\mathcal{G} \mid \Gamma, \bot \Rightarrow \Delta} \; (\bot\Rightarrow) \qquad \frac{}{\mathcal{G} \mid \Gamma \Rightarrow \top, \Delta} \; (\Rightarrow\top)$$

$$\frac{\mathcal{G} \mid \Gamma \Rightarrow \Delta}{\mathcal{G} \mid \Gamma, \mathrm{e} \Rightarrow \Delta} \; (\mathrm{e}\Rightarrow) \qquad \frac{}{\mathcal{G} \mid \Rightarrow \mathrm{e}} \; (\Rightarrow\mathrm{e})$$

$$\frac{}{\mathcal{G} \mid \mathrm{f} \Rightarrow} \; (\mathrm{f}\Rightarrow) \qquad \frac{\mathcal{G} \mid \Gamma \Rightarrow \Delta}{\mathcal{G} \mid \Gamma \Rightarrow \mathrm{f}, \Delta} \; (\Rightarrow\mathrm{f})$$

$$\frac{\mathcal{G} \mid \Gamma_1 \Rightarrow A, \Delta_1 \quad \mathcal{G} \mid \Gamma_2, B \Rightarrow \Delta_2}{\mathcal{G} \mid \Gamma_1, \Gamma_2, A \to B \Rightarrow \Delta_1, \Delta_2} \; (\to\Rightarrow) \qquad \frac{\mathcal{G} \mid \Gamma, A \Rightarrow B, \Delta}{\mathcal{G} \mid \Gamma \Rightarrow A \to B, \Delta} \; (\Rightarrow\to)$$

$$\frac{\mathcal{G} \mid \Gamma, A, B \Rightarrow \Delta}{\mathcal{G} \mid \Gamma, A \odot B \Rightarrow \Delta} \; (\odot\Rightarrow) \qquad \frac{\mathcal{G} \mid \Gamma_1 \Rightarrow A, \Delta_1 \quad \mathcal{G} \mid \Gamma_2 \Rightarrow B, \Delta_2}{\mathcal{G} \mid \Gamma_1, \Gamma_2 \Rightarrow A \odot B, \Delta_1, \Delta_2} \; (\Rightarrow\odot)$$

$$\frac{\mathcal{G} \mid \Gamma, A \Rightarrow \Delta}{\mathcal{G} \mid \Gamma, A \wedge B \Rightarrow \Delta} \; (\wedge\Rightarrow)_1 \qquad \frac{\mathcal{G} \mid \Gamma, B \Rightarrow \Delta}{\mathcal{G} \mid \Gamma, A \wedge B \Rightarrow \Delta} \; (\wedge\Rightarrow)_2$$

$$\frac{\mathcal{G} \mid \Gamma, A \Rightarrow \Delta \quad \mathcal{G} \mid \Gamma, B \Rightarrow \Delta}{\mathcal{G} \mid \Gamma, A \vee B \Rightarrow \Delta} \; (\vee\Rightarrow) \qquad \frac{\mathcal{G} \mid \Gamma \Rightarrow A, \Delta \quad \mathcal{G} \mid \Gamma \Rightarrow B, \Delta}{\mathcal{G} \mid \Gamma \Rightarrow A \wedge B, \Delta} \; (\Rightarrow\wedge)$$

$$\frac{\mathcal{G} \mid \Gamma \Rightarrow A, \Delta}{\mathcal{G} \mid \Gamma \Rightarrow A \vee B, \Delta} \; (\Rightarrow\vee)_1 \qquad \frac{\mathcal{G} \mid \Gamma \Rightarrow B, \Delta}{\mathcal{G} \mid \Gamma \Rightarrow A \vee B, \Delta} \; (\Rightarrow\vee)_2$$

Cut Rule

$$\frac{\mathcal{G} \mid \Gamma_1, A \Rightarrow \Delta_1 \quad \mathcal{G} \mid \Gamma_2 \Rightarrow A, \Delta_2}{\mathcal{G} \mid \Gamma_1, \Gamma_2 \Rightarrow \Delta_1, \Delta_2} \; (\text{CUT})$$

Fig. 4.1 The standard rule set

4.3.1 External Weakening and External Contraction

External weakening and external contraction, (EW) and (EC), are core external structural rules that appear in all our hypersequent calculi. Essentially they characterize "|" as an additive disjunction: they add and remove sequents.

Definition 4.28 (External Weakening and External Contraction Rules).

$$\frac{\mathcal{G}}{\mathcal{G} \mid \mathcal{H}} \; (\text{EW}) \qquad \frac{\mathcal{G} \mid \mathcal{H} \mid \mathcal{H}}{\mathcal{G} \mid \mathcal{H}} \; (\text{EC})$$

These rules offer greater flexibility for recording choices in a derivation. In particular, the rules for $\vee$ on the right can be combined, as can those for $\wedge$ on the left. Let us see how.

Definition 4.29 (Hypersequent Rules for $\wedge$ and $\vee$).

$$\frac{\mathcal{G} \mid \Gamma, A \Rightarrow \Delta \mid \Gamma, B \Rightarrow \Delta}{\mathcal{G} \mid \Gamma, A \wedge B \Rightarrow \Delta}\ (\wedge\Rightarrow) \qquad \frac{\mathcal{G} \mid \Gamma \Rightarrow A, \Delta \mid \Gamma \Rightarrow B, \Delta}{\mathcal{G} \mid \Gamma \Rightarrow A \vee B, \Delta}\ (\Rightarrow\vee)$$

Lemma 4.30. *Let* GL *be any calculus with* (EW) *and* (EC)*. Then* $(\wedge\Rightarrow)$ *and* $(\Rightarrow\vee)$ *are derivable in* GL *extended with* $(\wedge\Rightarrow)_1$, $(\wedge\Rightarrow)_2$, $(\vee\Rightarrow)_1$, *and* $(\vee\Rightarrow)_2$, *and vice versa: these rules are derivable in* GL *extended with* $(\wedge\Rightarrow)$ *and* $(\Rightarrow\vee)$.

Proof. We just consider the rules for $\vee$. To show that $(\Rightarrow\vee)$ is derivable using $(\Rightarrow\vee)_1$ and $(\Rightarrow\vee)_2$, we have:

$$\cfrac{\cfrac{\cfrac{\mathcal{G} \mid \Gamma \Rightarrow A, \Delta \mid \Gamma \Rightarrow B, \Delta}{\mathcal{G} \mid \Gamma \Rightarrow A, \Delta \mid \Gamma \Rightarrow A \vee B, \Delta}\ (\Rightarrow\vee)_2}{\mathcal{G} \mid \Gamma \Rightarrow A \vee B, \Delta \mid \Gamma \Rightarrow A \vee B, \Delta}\ (\Rightarrow\vee)_1}{\mathcal{G} \mid \Gamma \Rightarrow A \vee B, \Delta}\ (\text{EC})$$

and for the other direction:

$$\cfrac{\cfrac{\mathcal{G} \mid \Gamma \Rightarrow A, \Delta}{\mathcal{G} \mid \Gamma \Rightarrow A, \Delta \mid \Gamma \Rightarrow B, \Delta}\ (\text{EW})}{\mathcal{G} \mid \Gamma \Rightarrow A \vee B, \Delta}\ (\Rightarrow\vee) \qquad \cfrac{\cfrac{\mathcal{G} \mid \Gamma \Rightarrow B, \Delta}{\mathcal{G} \mid \Gamma \Rightarrow A, \Delta \mid \Gamma \Rightarrow B, \Delta}\ (\text{EW})}{\mathcal{G} \mid \Gamma \Rightarrow A \vee B, \Delta}\ (\Rightarrow\vee)$$

□

However, adding (EW) and (EC) to the hypersequent version of a sequent calculus has no effect on which sequents are derivable. To get more, we will need to add rules that allow sequents to interact.

Proposition 4.31. *Let* GL^{H} *consist of the hypersequent versions of the rules of a sequent calculus* GL *plus* (EW) *and* (EC)*. Then* $\vdash_{\mathsf{GL}^{\mathsf{H}}} \mathcal{G}$ *iff* $\vdash_{\mathrm{GL}} S$ *for some* $S \in \mathcal{G}$.

Proof. Clearly if $\vdash_{\mathrm{GL}} S$ for some $S \in \mathcal{G}$, then using (EW), $\vdash_{\mathsf{GL}^{\mathsf{H}}} \mathcal{G}$. For the other direction, we prove that if $d \vdash_{\mathsf{GL}^{\mathsf{H}}} \mathcal{G}$, then $\vdash_{\mathrm{GL}} S$ for some $S \in \mathcal{G}$, proceeding by induction on $\mathrm{ht}(d)$. For the base case, some $S \in \mathcal{G}$ is an initial sequent of GL. For the inductive step, consider first $\mathcal{G} = \mathcal{H}_1 \mid \mathcal{H}_2$. If the last step in d is an application of (EW) to $\mathcal{H}_1$, then by the induction hypothesis, $\vdash_{\mathrm{GL}} S$ for some $S \in \mathcal{H}_1$ as required. Similarly, for (EC), we can apply the induction hypothesis to $\mathcal{H}_1 \mid \mathcal{H}_2 \mid \mathcal{H}_2$, to get $\vdash_{\mathrm{GL}} S$ for some $S \in \mathcal{H}_1 \uplus \mathcal{H}_2$. Suppose finally that d ends with:

$$\frac{\mathcal{H} \mid S_1 \ \ldots\ \mathcal{H} \mid S_n}{\mathcal{H} \mid S}$$

where $S_1, \ldots, S_n \,/\, S$ is an instance of a rule of GL. By the induction hypothesis n times, either $\vdash_{\mathrm{GL}} S'$ for some $S' \in \mathcal{H}$ or $\vdash_{\mathrm{GL}} S_i$ for $i = 1 \ldots n$. The first case is immediate and for the second we just apply the sequent rule to get $\vdash_{\mathrm{GL}} S$. □

Example 4.32. Even when not strictly necessary, hypersequents can still be useful for recording options in a derivation, e.g.

$$\dfrac{\dfrac{\dfrac{\dfrac{}{A \Rightarrow B \quad B \Rightarrow B \mid A \Rightarrow C \mid B \Rightarrow C}\ (\text{ID})}{A \wedge B \Rightarrow B \mid A \Rightarrow C \mid B \Rightarrow C}\ (\wedge\Rightarrow)}{A \wedge B \Rightarrow B \mid A \wedge B \Rightarrow C}\ (\wedge\Rightarrow)}{A \wedge B \Rightarrow B \vee C}\ (\Rightarrow\vee)$$

Here, we do not have to guess which conjunct or disjunct to work on first. We just decompose them all and see which one is most useful; no backtracking is required.

It is often convenient (in particular, for fitting derivations onto the page) to use rules combined with applications of (EW) and (EC). We will denote such combinations with a $*$. For example, we can make use of a version of (CUT) where the context side-hypersequents are added rather than merged:

$$\dfrac{\mathcal{G}_1 \mid \Gamma, A \Rightarrow \Delta \quad \mathcal{G}_2 \mid \Pi \Rightarrow A, \Sigma}{\mathcal{G}_1 \mid \mathcal{G}_2 \mid \Gamma, \Pi \Rightarrow \Sigma, \Delta}\ (\text{CUT})^*$$

4.3.2 *Communication and Split*

For Hilbert systems, it is the prelinearity and distributivity axioms that are key for characterizing linearity. For Gentzen systems, it is the communication rule, so-called because formulas – pieces of information – are "communicated" between different sequents.

Definition 4.33 (Communication Rule).

$$\dfrac{\mathcal{G} \mid \Gamma_1, \Pi_1 \Rightarrow \Sigma_1, \Delta_1 \quad \mathcal{G} \mid \Gamma_2, \Pi_2 \Rightarrow \Sigma_2, \Delta_2}{\mathcal{G} \mid \Gamma_1, \Gamma_2 \Rightarrow \Delta_1, \Delta_2 \mid \Pi_1, \Pi_2 \Rightarrow \Sigma_1, \Sigma_2}\ (\text{COM})$$

The single-conclusion version of (COM) simplifies a little to:

$$\dfrac{\mathcal{G} \mid \Gamma_1, \Pi_1 \Rightarrow \Delta \quad \mathcal{G} \mid \Gamma_2, \Pi_2 \Rightarrow \Sigma}{\mathcal{G} \mid \Gamma_1, \Gamma_2 \Rightarrow \Delta \mid \Pi_1, \Pi_2 \Rightarrow \Sigma}\ (\text{COM})$$

Example 4.34. The best way to understand communication is by returning to the tricky prelinearity axioms $(A \rightarrow B) \vee (B \rightarrow A)$. The following derivation uses (COM) and (EC) (implicit in the derived rule $(\Rightarrow\vee)$) but no other structural rules:

$$\dfrac{\dfrac{\dfrac{\dfrac{\dfrac{}{A\Rightarrow A}\,(\mathrm{ID}) \qquad \dfrac{}{B\Rightarrow B}\,(\mathrm{ID})}{A\Rightarrow B \mid B\Rightarrow A}\,(\mathrm{COM})}{A\Rightarrow B \mid\, \Rightarrow B\to A}\,(\Rightarrow\to)}{\Rightarrow A\to B \mid\, \Rightarrow B\to A}\,(\Rightarrow\to)}{\Rightarrow (A\to B)\vee(B\to A)}\,(\Rightarrow\vee)$$

Notice that the hypersequent $(A\Rightarrow B \mid B\Rightarrow A)$ two lines down might be read as just a "hypersequent translation" of $(A\to B)\vee(B\to A)$.

Example 4.35. Consider also these derivations of a helpful property of disjunction:

$$\dfrac{\dfrac{}{A\Rightarrow A \mid A\vee B\Rightarrow B}\,(\mathrm{ID}) \qquad \dfrac{\dfrac{\dfrac{}{A\Rightarrow A}\,(\mathrm{ID}) \quad \dfrac{}{B\Rightarrow B}\,(\mathrm{ID})}{B\Rightarrow A \mid A\Rightarrow B}\,(\mathrm{COM}) \qquad \dfrac{}{A\Rightarrow B \mid B\Rightarrow B}\,(\mathrm{ID})}{B\Rightarrow A \mid A\vee B\Rightarrow B}\,(\vee\Rightarrow)}{A\vee B\Rightarrow A \mid A\vee B\Rightarrow B}\,(\vee\Rightarrow)$$

and conjunction:

$$\dfrac{\dfrac{}{A\Rightarrow A \mid B\Rightarrow A\wedge B}\,(\mathrm{ID}) \qquad \dfrac{\dfrac{\dfrac{}{A\Rightarrow A}\,(\mathrm{ID}) \quad \dfrac{}{B\Rightarrow B}\,(\mathrm{ID})}{A\Rightarrow B \mid B\Rightarrow A}\,(\mathrm{COM}) \qquad \dfrac{}{A\Rightarrow B \mid B\Rightarrow B}\,(\mathrm{ID})}{A\Rightarrow B \mid B\Rightarrow A\wedge B}\,(\vee\Rightarrow)}{A\Rightarrow A\wedge B \mid B\Rightarrow A\wedge B}\,(\Rightarrow\wedge)$$

Example 4.36. The communication rule is crucial for deriving the prelinearity axioms. But also the distributivity axioms are derivable in any extension of the core rules with (EC), (EW), and (COM). In fact, it is not easy to see how hypersequent calculi can be defined that derive only one of (PRL) and (DIS). For convenience, let us first consider the following rule, easily derived using the rules for $\wedge$:

$$\dfrac{\mathcal{G} \mid B\Rightarrow C}{\mathcal{G} \mid A\wedge B\Rightarrow A\wedge C}\,(\wedge\mathrm{CAN})$$

Then we obtain the following derivation, where the top hypersequent is derived as in Example 4.35:

$$\dfrac{\dfrac{\dfrac{\dfrac{B\vee C\Rightarrow B \mid B\vee C\Rightarrow C}{B\vee C\Rightarrow B \mid A\wedge(B\vee C)\Rightarrow A\wedge C}\,(\wedge\mathrm{CAN})}{A\wedge(B\vee C)\Rightarrow A\wedge B \mid A\wedge(B\vee C)\Rightarrow A\wedge C}\,(\wedge\mathrm{CAN})}{A\wedge(B\vee C)\Rightarrow (A\wedge B)\vee(A\wedge C)}\,(\Rightarrow\vee)}{\Rightarrow (A\wedge(B\vee C))\to((A\wedge B)\vee(A\wedge C))}\,(\Rightarrow\to)$$

We are ready now to identify Gentzen systems for the most elementary of our fuzzy logics. We let GIUL consist of the initial hypersequents, logical rules, and cut rule of the previous section (i.e. the hypersequent version of GMALL), together with

(EC), (EW), and (COM). GUL is then the single-conclusion version of GIUL (i.e. the hypersequent version of GMAILL).

The step up to hypersequents also gains us a couple of helpful extra invertibilities to add to those in Proposition 4.27. Consider first the defined rule ($\vee\Rightarrow$). In any GUL$^-$-extension, we can derive the premise of an instance of this rule from its conclusion:

$$\dfrac{\dfrac{A\vee B\Rightarrow A\mid A\vee B\Rightarrow B \quad \mathcal{G}\mid\Gamma\Rightarrow A\vee B,\Delta}{\mathcal{G}\mid A\vee B\Rightarrow A\mid\Gamma\Rightarrow B,\Delta}\,(\text{CUT})^* \qquad \mathcal{G}\mid\Gamma\Rightarrow A\vee B,\Delta}{\mathcal{G}\mid\Gamma\Rightarrow A,\Delta\mid\Gamma\Rightarrow B,\Delta}\,(\text{CUT})^*$$

where again the leftmost hypersequent is derived as in Example 4.35. A similar derivation works for the defined rule ($\wedge\Rightarrow$), so we get the following result.

Proposition 4.37. ($\wedge\Rightarrow$) *and* ($\Rightarrow\vee$) *are* GL*-invertible for any* GUL$^-$*-extension* GL.

Finally, a related rule that also allows interaction between sequents is the "split rule" which divides one sequent in the premise into two sequents in the conclusion.

Definition 4.38 (Split Rule).

$$\dfrac{\mathcal{G}\mid\Gamma_1,\Gamma_2\Rightarrow\Delta_1,\Delta_2}{\mathcal{G}\mid\Gamma_1\Rightarrow\Delta_1\mid\Gamma_2\Rightarrow\Delta_2}\,(\text{SPLIT})$$

Example 4.39. Just as (COM) corresponds to (PRL) and (DIS), so (SPLIT) corresponds to the law of excluded middle (EM):

$$\dfrac{\dfrac{\dfrac{\dfrac{}{A\Rightarrow A}\,(\text{ID})}{\Rightarrow A\mid A\Rightarrow}\,(\text{SPLIT})}{\Rightarrow A\mid\Rightarrow\neg A}\,(\Rightarrow\neg)}{\Rightarrow A\vee\neg A}\,(\Rightarrow\vee)$$

Note that there are no sequent instances of (COM) and (SPLIT), or indeed of (EW) and (EC), so sequent versions of these rules are simply empty.

4.3.3 Weakening

Weakening axioms (or theorems) are a common feature of fuzzy logics based on t-norms, including MTL and the fundamental logics Ł, G, and P. In Gentzen systems, weakening rules typically come in pairs, introducing new formulas on the left and right of sequents. E.g.

$$\dfrac{\mathcal{G}\mid\Gamma\Rightarrow\Delta}{\mathcal{G}\mid\Gamma,A\Rightarrow\Delta}\,(\text{WL}) \qquad \dfrac{\mathcal{G}\mid\Gamma\Rightarrow\Delta}{\mathcal{G}\mid\Gamma\Rightarrow A,\Delta}\,(\text{WR})$$

Here, for simplicity, we combine these left and right rules into one:

Definition 4.40 (Weakening Rule).

$$\frac{\mathcal{G} \mid \Gamma \Rightarrow \Delta}{\mathcal{G} \mid \Gamma, \Pi \Rightarrow \Sigma, \Delta}\ (\textsc{w})$$

Example 4.41. Unsurprisingly, the weakening rule allows us to derive weakening axioms:

$$\frac{\dfrac{\dfrac{\dfrac{}{\Rightarrow \mathrm{e}}\ (\textsc{id})}{A \Rightarrow \mathrm{e}}\ (\textsc{w})}{\Rightarrow A \to \mathrm{e}}\ (\Rightarrow\to) \qquad \dfrac{\dfrac{\dfrac{}{\mathrm{f} \Rightarrow}\ (\mathrm{f}\Rightarrow)}{\mathrm{f} \Rightarrow A}\ (\textsc{w})}{\Rightarrow \mathrm{f} \to A}\ (\Rightarrow\to)}{\Rightarrow (A \to \mathrm{e}) \wedge (\mathrm{f} \to A)}\ (\Rightarrow\wedge)$$

In fact, we can use the weakening rule to prove that f and $\bot$ and $\top$ and e collapse: that is, $\mathrm{f} \leftrightarrow \bot$ and $\mathrm{e} \leftrightarrow \top$ are derivable using $(\textsc{w})$ (assuming that $\bot$ and $\top$ are in the language).

Gentzen systems for Affine Multiplicative Additive Linear Logic and its intuitionistic version, Monoidal Logic, are obtained by adding the multiple and single-conclusion versions of $(\textsc{w})$ to GMALL and GMAILL, respectively. The systems GIMTL and GMTL are then obtained by extending the hypersequent versions of these calculi with $(\textsc{com})$. Or put another way: GIMTL and GMTL are GIUL and GUL plus $(\textsc{w})$.

A further (weaker) kind of weakening allows two sequents to be combined or "mixed" into one. There is also an extreme case. The empty sequent is a "nullary mix" where there are no formulas to combine.

Definition 4.42 (Mix and Nullary Mix Rules).

$$\frac{\mathcal{G} \mid \Gamma_1 \Rightarrow \Delta_1 \quad \mathcal{G} \mid \Gamma_2 \Rightarrow \Delta_2}{\mathcal{G} \mid \Gamma_1, \Gamma_2 \Rightarrow \Delta_1, \Delta_2}\ (\textsc{mix}) \qquad\qquad \frac{}{\mathcal{G} \mid \Rightarrow}\ (\textsc{emp})$$

Example 4.43. $(\textsc{mix})$ and $(\textsc{emp})$ provide a link between the multiplicative constants e and f. The former allows us to prove (not unreasonably) that f is less true than e. The latter (more bizarrely) allows us to prove the converse.

$$\frac{\dfrac{\dfrac{\dfrac{}{\Rightarrow}\ (\textsc{emp})}{\mathrm{e} \Rightarrow}\ (\mathrm{e}\Rightarrow)}{\mathrm{e} \Rightarrow \mathrm{f}}\ (\Rightarrow\mathrm{f})}{\Rightarrow \mathrm{e} \to \mathrm{f}}\ (\Rightarrow\to) \qquad\qquad \frac{\dfrac{\dfrac{}{\mathrm{f} \Rightarrow}\ (\mathrm{f}\Rightarrow) \qquad \dfrac{}{\Rightarrow \mathrm{e}}\ (\Rightarrow\mathrm{e})}{\mathrm{f} \Rightarrow \mathrm{e}}\ (\textsc{mix})}{\Rightarrow \mathrm{f} \to \mathrm{e}}\ (\Rightarrow\to)$$

Notice also that $(\textsc{com})$ is derivable using $(\textsc{split})$ and $(\textsc{mix})$ (and so is redundant in Gentzen systems where these are present):

$$\frac{\dfrac{\mathcal{G} \mid \Gamma_1, \Pi_1 \Rightarrow \Sigma_1, \Delta_1 \quad \mathcal{G} \mid \Gamma_2, \Pi_2 \Rightarrow \Sigma_2, \Delta_2}{\mathcal{G} \mid \Gamma_1, \Gamma_2, \Pi_1, \Pi_2 \Rightarrow \Sigma_1, \Sigma_2, \Delta_1, \Delta_2}\ (\textsc{mix})}{\mathcal{G} \mid \Gamma_1, \Gamma_2 \Rightarrow \Delta_1, \Delta_1 \mid \Pi_1, \Pi_2 \Rightarrow \Sigma_1, \Sigma_2}\ (\textsc{split})$$

Conversely, it is easy to see that (SPLIT) is derivable using (COM) and (EMP). Note finally that in the presence of weakening, (EMP) is a route to triviality. For any formula A, the sequent $(\Rightarrow A)$ is derivable from the empty sequent using (W). This fits of course, since the corresponding axioms would allow us to derive $\mathrm{e} \to \mathrm{f}$ and $\mathrm{f} \to \bot$, and hence $\mathrm{e} \to \bot$ and $\bot$.

4.3.4 Contraction

The other (alongside weakening) core structural manipulation for Gentzen systems is contraction: reduction of the number of occurrences of a formula in a sequent. Typically, we encounter rules that contract single formulas on the left and right:

$$\frac{\mathcal{G} \mid \Gamma, A, A \Rightarrow \Delta}{\mathcal{G} \mid \Gamma, A \Rightarrow \Delta}\ (\text{CL}) \qquad \frac{\mathcal{G} \mid \Gamma \Rightarrow A, A, \Delta}{\mathcal{G} \mid \Gamma \Rightarrow A, \Delta}\ (\text{CR})$$

Again however, as for weakening, we will consider here a more general version:

Definition 4.44 (Contraction Rule).

$$\frac{\mathcal{G} \mid \Gamma, \Pi, \Pi \Rightarrow \Sigma, \Sigma, \Delta}{\mathcal{G} \mid \Gamma, \Pi \Rightarrow \Sigma, \Delta}\ (\text{C})$$

Example 4.45. This rule helps us derive its Hilbert counterpart (C_2) as follows:

$$\dfrac{\dfrac{\dfrac{\dfrac{}{A \Rightarrow A}\ (\text{ID}) \quad \dfrac{}{A \Rightarrow A}\ (\text{ID})}{A, A \Rightarrow A \odot A}\ (\Rightarrow \odot)}{A \Rightarrow A \odot A}\ (\text{C})}{\Rightarrow A \to (A \odot A)}\ (\Rightarrow \to)$$

Adding the single-conclusion version of (C) to GMTL gives the hypersequent calculus GG for Gödel Logic. Or, from a different perspective, GG is the hypersequent version of a calculus for Intuitionistic Logic plus (EW), (EC), and (COM). In the multiple-conclusion case, adding (C) to GIMTL or GAMALL leads directly to calculi for Classical Logic.

The situation for calculi without weakening is more complicated. Adding (C) and (MIX) to GMALL gives a calculus for the non-distributive relevant logic $\mathrm{RM}^{\mathrm{ND}}$. The calculus GRM (which does prove distributivity) is defined at the hypersequent level by adding (C) and (MIX) to GIUL. However, as we will see later, it is useful to consider also a calculus GIUML defined as the extension of GRM with (EMP).

Example 4.46. Sometimes structural rules interact in unexpected ways. For example, in the presence of (W) and (SPLIT), we can derive the contraction rule (C):

$$\cfrac{\cfrac{\cfrac{\mathcal{G} \mid \Gamma,\Pi,\Pi \Rightarrow \Sigma,\Sigma,\Delta}{\mathcal{G} \mid \Pi \Rightarrow \Sigma \mid \Gamma,\Pi \Rightarrow \Sigma,\Delta}\ (\text{SPLIT})}{\mathcal{G} \mid \Gamma,\Pi \Rightarrow \Sigma,\Delta \mid \Gamma,\Pi \Rightarrow \Sigma,\Delta}\ (\text{W})}{\mathcal{G} \mid \Gamma,\Pi \Rightarrow \Sigma,\Delta}\ (\text{EC})$$

In fact, GMTL extended with (SPLIT) is a single-conclusion hypersequent calculus for Classical Logic.

Contraction rules are sometimes matched with "anti-contraction" rules that multiply (or selectively weaken) formulas:

$$\frac{\mathcal{G} \mid \Gamma,A \Rightarrow \Delta}{\mathcal{G} \mid \Gamma,A,A \Rightarrow \Delta}\ (\text{ML}) \qquad \frac{\mathcal{G} \mid \Gamma \Rightarrow A,\Delta}{\mathcal{G} \mid \Gamma \Rightarrow A,A,\Delta}\ (\text{MR})$$

However, these rules often cause problems for cut elimination (see Chapter 5), and so will not be considered any further here.

A more complicated rule, which contracts formulas selectively, is "mingle" where (as the name suggests) elements from two sequents are combined into one:

Definition 4.47 (Mingle Rule).

$$\frac{\mathcal{G} \mid \Gamma_1,\Pi \Rightarrow \Sigma,\Delta_1 \quad \mathcal{G} \mid \Gamma_2,\Pi \Rightarrow \Sigma,\Delta_2}{\mathcal{G} \mid \Gamma_1,\Gamma_2,\Pi \Rightarrow \Sigma,\Delta_1,\Delta_2}\ (\text{MINGLE})$$

The single-conclusion version of this rule (which includes (MIX) as a special case) is particularly useful:

$$\frac{\mathcal{G} \mid \Gamma_1,\Pi \Rightarrow \Sigma \quad \mathcal{G} \mid \Gamma_2,\Pi \Rightarrow \Sigma}{\mathcal{G} \mid \Gamma_1,\Gamma_2,\Pi \Rightarrow \Sigma}\ (\text{MINGLE})$$

The calculus GUL extended with single-conclusion (MINGLE) and (C) is called GUML. The multiple-conclusion version provides an alternative calculus for the relevant logic RM.

Example 4.48. Consider the following derivation in a single-conclusion calculus with (MINGLE):

$$\cfrac{\cfrac{\cfrac{\cfrac{\cfrac{\cfrac{\overline{A \Rightarrow A}\ (\text{ID}) \quad \overline{C \Rightarrow C}\ (\text{ID})}{A \rightarrow C, A \Rightarrow C}\ (\rightarrow\Rightarrow)}{(A \rightarrow C) \wedge (B \rightarrow C), A \Rightarrow C}\ (\wedge\Rightarrow)_1 \quad \cfrac{\cfrac{\overline{B \Rightarrow B}\ (\text{ID}) \quad \overline{C \Rightarrow C}\ (\text{ID})}{B \rightarrow C, B \Rightarrow C}\ (\rightarrow\Rightarrow)}{(A \rightarrow C) \wedge (B \rightarrow C), B \Rightarrow C}\ (\wedge\Rightarrow)_2}{(A \rightarrow C) \wedge (B \rightarrow C), A, B \Rightarrow C}\ (\text{MINGLE})}{(A \rightarrow C) \wedge (B \rightarrow C), A \odot B \Rightarrow C}\ (\odot\Rightarrow)}{(A \rightarrow C) \wedge (B \rightarrow C) \Rightarrow (A \odot B) \rightarrow C}\ (\Rightarrow\rightarrow)}{\Rightarrow ((A \rightarrow C) \wedge (B \rightarrow C)) \rightarrow ((A \odot B) \rightarrow C)}\ (\Rightarrow\rightarrow)$$

To obtain a derivation of the same sequent using (C) and (MIX) would require a multiple-conclusion calculus.

The contraction rule (C) can also be generalized to allow contraction from n copies of a formula to $n-1$ copies. The idea is simple, but the rule required to take care of this is a little complicated.

Definition 4.49 (N-Contraction Rule).

$$\dfrac{\mathcal{G} \mid \Gamma, \Pi_1^n \Rightarrow \Sigma_1^n, \Delta \quad \ldots \quad \mathcal{G} \mid \Gamma, \Pi_{n-1}^n \Rightarrow \Sigma_{n-1}^n, \Delta}{\mathcal{G} \mid \Gamma, \Pi_1, \ldots, \Pi_{n-1} \Rightarrow \Sigma_1, \ldots, \Sigma_{n-1}, \Delta}\ (\text{C}_\text{n}) \qquad n = 2, 3, \ldots$$

Adding this rule to GMTL and GIMTL gives calculi GMTL_n and GIMTL_n, respectively, for n-contractive logics.

Example 4.50. The n-contraction rule (C_n) corresponds to the axiom schema (C_n):

$$\dfrac{\dfrac{\dfrac{\dfrac{\dfrac{\dfrac{\dfrac{}{A \Rightarrow A}\,(\text{ID})}{\vdots}\,(\Rightarrow\odot) \quad \cdots \quad \dfrac{\dfrac{}{A \Rightarrow A}\,(\text{ID})}{\vdots}\,(\Rightarrow\odot)}{[A]^n \Rightarrow A^n}\,(\Rightarrow\odot) \quad \cdots \quad \dfrac{\dfrac{\dfrac{}{A \Rightarrow A}\,(\text{ID})}{\vdots}\,(\Rightarrow\odot) \quad \cdots \quad \dfrac{\dfrac{}{A \Rightarrow A}\,(\text{ID})}{\vdots}\,(\Rightarrow\odot)}{[A]^n \Rightarrow A^n}\,(\Rightarrow\odot)}{[A]^{n-1} \Rightarrow A^n}\,(\text{C}_\text{n})}{\vdots}\,(\odot\Rightarrow)}{A^{n-1} \Rightarrow A^n}\,(\odot\Rightarrow)}{\Rightarrow A^{n-1} \to A^n}\,(\Rightarrow\to)$$

Notice that in the case where $n = 2$, (C_n) is just $A \to (A \odot A)$, and (C_n) is (C).

Finally, let us consider one more form of contraction: the contraction of a whole sequent n times.

Definition 4.51 (Global N-Contraction Rules).

$$\dfrac{\mathcal{G} \mid \Gamma^n \Rightarrow \Delta^n}{\mathcal{G} \mid \Gamma \Rightarrow \Delta}\ (\text{SC}_\text{n}) \qquad n = 2, 3, \ldots$$

The single-conclusion versions of these rules are sometimes useful. Notice that in such cases Δ must necessarily be empty, i.e. we have:

$$\dfrac{\mathcal{G} \mid \Gamma^n \Rightarrow}{\mathcal{G} \mid \Gamma \Rightarrow}\ (\text{SC}_\text{n}) \qquad n = 2, 3, \ldots$$

Example 4.52. It is not immediately obvious which axioms correspond to the global n-contraction rules. Here, we just point out that in the case where $n = 2$, we can derive the non-contradiction axioms (NC):

$$
\cfrac{\cfrac{\cfrac{\cfrac{\cfrac{\cfrac{}{A \Rightarrow A}\;(\text{ID})}{A, \neg A \Rightarrow}\;(\neg\Rightarrow)}{A, A \wedge \neg A \Rightarrow}\;(\wedge\Rightarrow)_2}{A \wedge \neg A, A \wedge \neg A \Rightarrow}\;(\wedge\Rightarrow)_1}{A \wedge \neg A \Rightarrow}\;(\text{SC}_2)}{\Rightarrow \neg(A \wedge \neg A)}\;(\Rightarrow\neg)
$$

In particular, adding (SC_2) to GMTL gives a calculus GSMTL for Strict Monoidal t-norm Logic.

4.3.5 Cancellation

Not all properties are as easy as weakening and contraction to characterize via structural rules. In particular, no rule has yet been discovered that corresponds to divisibility (see Chapter 9 for further comments). Even in cases where an obvious rule is available, it may not be so benign. Consider the following rule permitting the "cancellation" of formulas occurring on both sides of a sequent.

Definition 4.53 (Cancellation Rule).

$$
\frac{\mathcal{G} \mid \Gamma, A \Rightarrow A, \Delta}{\mathcal{G} \mid \Gamma \Rightarrow \Delta}\;(\text{CAN})
$$

The cancellation rule is useful for defining calculi for varieties of cancellative residuated lattices. For example, it is not hard to show that GMALL^- plus (MIX) and (CAN) where $(\Rightarrow\rightarrow)$ is single-conclusion is a calculus for the variety of cancellative pcrls. We can derive the cancellation axioms (CAN) in such a calculus as follows:

$$
\cfrac{\cfrac{\cfrac{\cfrac{}{A \Rightarrow A}\;(\text{ID}) \qquad \cfrac{\cfrac{\cfrac{}{A \Rightarrow A}\;(\text{ID}) \quad \cfrac{}{B \Rightarrow B}\;(\text{ID})}{A, B \Rightarrow A, B}\;(\text{MIX})}{A \odot B \Rightarrow A, B}\;(\odot\Rightarrow)}{A \rightarrow (A \odot B), A \Rightarrow A, B}\;(\rightarrow\Rightarrow)}{A \rightarrow (A \odot B) \Rightarrow B}\;(\text{CAN})}{\Rightarrow (A \rightarrow (A \odot B)) \rightarrow B}\;(\Rightarrow\rightarrow)
$$

Like the cut rule, however, (CAN) does not have the subformula property. Any formula A can be cancelled. In fact, in the presence of (MIX), the cut rule is even derivable using (CAN):

$$
\cfrac{\cfrac{\mathcal{G} \mid \Gamma, A \Rightarrow \Delta \quad \mathcal{G} \mid \Pi \Rightarrow A, \Sigma}{\mathcal{G} \mid \Gamma, \Pi, A \Rightarrow A, \Sigma, \Delta}\;(\text{MIX})}{\mathcal{G} \mid \Gamma, \Pi \Rightarrow \Sigma, \Delta}\;(\text{CAN})
$$

We will have more to say on this topic in Chapter 5.

4.4 Non-Standard Logical Rules

One of the nice things about the systems discussed so far is their uniformity: essentially we fix the rules for connectives and tinker with the structural rules. However, sometimes it is useful (essential even) to be more flexible. In Chapter 5 we will see that by making some fairly radical changes to the logical rules, we obtain calculi for the fundamental Łukasiewicz and Product logics. Our first example, however, requires only minor modifications. We alter slightly the rules $(\to\Rightarrow)$ and $(\Rightarrow\odot)$ to obtain a calculus for Abelian Logic A, the logic of lattice-ordered abelian groups:

Definition 4.54 (Abelian Logic Rules).

$$\frac{\mathcal{G} \mid \Gamma, B \Rightarrow A, \Delta}{\mathcal{G} \mid \Gamma, A \to B \Rightarrow \Delta}\ (\to\Rightarrow)_{\mathsf{A}} \qquad \frac{\mathcal{G} \mid \Gamma \Rightarrow A, B, \Delta}{\mathcal{G} \mid \Gamma \Rightarrow A \odot B, \Delta}\ (\Rightarrow\odot)_{\mathsf{A}}$$

The calculus GA is obtained by extending GIUL with (EMP) and (MIX), and replacing $(\to\Rightarrow)$ and $(\Rightarrow\odot)$ with $(\to\Rightarrow)_{\mathsf{A}}$ and $(\Rightarrow\odot)_{\mathsf{A}}$.

Example 4.55. Consider this GA-derivation of the so-called "axioms of relativity":

$$\dfrac{\dfrac{\dfrac{\dfrac{\overline{B \Rightarrow B}\ (\text{ID}) \quad \overline{A \Rightarrow A}\ (\text{ID})}{B, A \Rightarrow B, A}\ (\text{MIX})}{B \Rightarrow A \to B, A}\ (\Rightarrow\to)}{(A \to B) \to B \Rightarrow A}\ (\to\Rightarrow)_{\mathsf{A}}}{\Rightarrow ((A \to B) \to B) \to A}\ (\Rightarrow\to)$$

This derivation fails in all of the other systems that we have seen up to now, and rightly so; the formula $((A \to B) \to B) \to A$ is not even classically valid.

One way of looking at the Abelian Logic rules is as building cancellation properties into the connectives. It is easy to see for instance that both $(\to\Rightarrow)_{\mathsf{A}}$ and $(\Rightarrow\odot)_{\mathsf{A}}$ are derivable using the standard logical rules and (CAN). E.g. for $(\to\Rightarrow)_{\mathsf{A}}$:

$$\dfrac{\dfrac{\overline{\mathcal{G} \mid A \Rightarrow A}\ (\text{ID}) \quad \mathcal{G} \mid \Gamma, B \Rightarrow A, \Delta}{\mathcal{G} \mid \Gamma, A \to B, A \Rightarrow A, \Delta}\ (\to\Rightarrow)}{\mathcal{G} \mid \Gamma, A \to B \Rightarrow \Delta}\ (\text{CAN})$$

Moreover, in the presence of $(\to\Rightarrow)_{\mathsf{A}}$ it is possible to derive (CAN) using (CUT).

$$\dfrac{\dfrac{\mathcal{G} \mid \Gamma, A \Rightarrow A, \Delta}{\mathcal{G} \mid \Gamma, A \to A \Rightarrow \Delta}\ (\to\Rightarrow)_{\mathsf{A}} \quad \dfrac{\overline{\mathcal{G} \mid A \Rightarrow A}\ (\text{ID})}{\mathcal{G} \mid \Rightarrow A \to A}\ (\Rightarrow\to)}{\mathcal{G} \mid \Gamma \Rightarrow \Delta}\ (\text{CUT})$$

To whet the appetite for Chapter 6, let us mention now that the rule $(\rightarrow\Rightarrow)_{\mathsf{A}}$ can also be used to define a calculus for Łukasiewicz Logic. In this case, however, we will need a further non-standard implication rule on the right:

$$\frac{\mathcal{G} \mid \Gamma \Rightarrow \Delta \quad \mathcal{G} \mid \Gamma, A \Rightarrow B, \Delta}{\mathcal{G} \mid \Gamma \Rightarrow A \rightarrow B, \Delta} \ (\Rightarrow\rightarrow)_{Ł}$$

These implication rules, together with the initial sequents (EMP), (ID), and $(\bot\Rightarrow)$ (single-conclusion) and the structural rules (EW), (EC), (SPLIT), and (MIX), provide a calculus for Ł in the (fully expressive) language $\mathcal{L}_{\mathrm{L}} = \{\rightarrow, \bot\}$.

4.5 Density Again

In the previous chapter we introduced a special "density rule" for Hilbert systems with the nice property that adding it to any HUL^--extension HL gives a system HL^{D} complete with respect to dense L-chains. We can do something similar here for Gentzen systems by defining a hypersequent density rule, understandable as a "hypersequent translation" of the Hilbert-style density rule.[2]

Definition 4.56 (Density Rule).

$$\frac{\mathcal{G} \mid \Gamma_1, p \Rightarrow \Delta_1 \mid \Gamma_2 \Rightarrow p, \Delta_2}{\mathcal{G} \mid \Gamma_1, \Gamma_2 \Rightarrow \Delta_1, \Delta_2} \ (\mathrm{DENSITY})$$

where p does not occur in $\mathcal{G}$, Γ_1, Γ_2, Δ_1, or Δ_2

As in the Hilbert system case, we can rework our definition of a derivation to account for systems with (DENSITY). However, since the only derivations we will consider for such systems are from the empty set of hypersequents, the standard definition will do well enough here.

Example 4.57. Adding the density rule to a calculus can have a dramatic effect. Let GL be any multiple-conclusion calculus with (DENSITY), (COM), and the contraction rule (C). Then the empty sequent is derivable in GL:

$$\dfrac{\dfrac{\dfrac{\dfrac{\overline{p \Rightarrow p}\ (\mathrm{ID}) \qquad \overline{p \Rightarrow p}\ (\mathrm{ID})}{p, p \Rightarrow \mid \Rightarrow p, p}\ (\mathrm{COM})}{p, p \Rightarrow \mid \Rightarrow p}\ (\mathrm{C})}{p \Rightarrow \mid \Rightarrow p}\ (\mathrm{C})}{\Rightarrow}\ (\mathrm{DENSITY})$$

---

[2] In the single-conclusion case, this can be seen by interpreting $\mid$ as $\vee$, $\Rightarrow$ as $\rightarrow$, and the comma on the left as $\odot$. For multiple-conclusion calculi, the rule seems a little stronger (it is not) than its Hilbert counterpart since there may be extra formulas in Δ_2 occurring with p on the right.

This shows that adding (DENSITY) to Classical Logic gives inconsistency, since any sequent is derivable from the empty sequent using (W). It also shows that (DENSITY) is not admissible for the calculus GRM (GIUL plus (C) and (MIX)) since (EMP) is derivable in GRM with (DENSITY) but not without.

Like (CUT), the density rule does not have the subformula property, by definition in fact, since p in the premise should always be a new variable. On the other hand, a judicious use of the rule can be beneficial for proof search. It can be used to shorten derivations and reduce the size of hypersequents.

Example 4.58. Suppose that we want to try to prove a sequent of the form $(A \wedge B \Rightarrow C \vee D)$ in the Gentzen system GUL. We might begin our attempted derivation with:

$$\dfrac{\dfrac{\dfrac{A \Rightarrow C \mid B \Rightarrow C \mid A \Rightarrow D \mid B \Rightarrow D}{A \Rightarrow C \mid B \Rightarrow C \mid A \wedge B \Rightarrow D}\,(\wedge\Rightarrow)}{A \wedge B \Rightarrow C \mid A \wedge B \Rightarrow D}\,(\wedge\Rightarrow)}{A \wedge B \Rightarrow C \vee D}\,(\Rightarrow\vee)$$

Notice that each subformula A, B, C, and D occurs twice in the top hypersequent, which could be very costly if these formulas are large. Using (DENSITY), on the other hand, we obtain:

$$\dfrac{\dfrac{\dfrac{p \Rightarrow C \mid p \Rightarrow D \mid A \Rightarrow p \mid B \Rightarrow p}{p \Rightarrow C \mid p \Rightarrow D \mid A \wedge B \Rightarrow p}\,(\wedge\Rightarrow)}{p \Rightarrow C \vee D \mid A \wedge B \Rightarrow p}\,(\Rightarrow\vee)}{A \wedge B \Rightarrow C \vee D}\,(\text{DENSITY})$$

Here we need to apply $(\wedge\Rightarrow)$ only once, and, depending on the complexity of A, B, C, and D, we may have a much smaller top hypersequent.

Moreover, reckless applications of (DENSITY) can always be retracted using communication. That is, the density rule is invertible for any calculus with (COM). Just consider the derivation:

$$\dfrac{\dfrac{}{\mathcal{G} \mid p \Rightarrow p}\,(\text{ID}) \qquad \mathcal{G} \mid \Gamma_1, \Gamma_2 \Rightarrow \Delta_1, \Delta_2}{\mathcal{G} \mid \Gamma_1, p \Rightarrow \Delta_1 \mid \Gamma_2 \Rightarrow p, \Delta_2}\,(\text{COM})$$

4.6 Soundness and Completeness

With rules and systems proliferating, it is high time that we made precise the connections between this and the preceding chapters. What we want to show is that Gentzen systems and Hilbert systems (and hence varieties of pcrls and bpcrls)

Table 4.1 Some sequent and hypersequent calculi

Calculus	Initial sequents and rules	Sequent	Single-conclusion
GMAILL	Standard Rules	yes	yes
GMALL	Standard Rules	yes	no
GML	GMAILL + (W)	yes	yes
GAMALL	GMALL + (W)	yes	no
GIL	GML + (C)	yes	yes
GCL	GAMALL + (C)	yes	no
GUL	Standard Rules + (COM)	no	yes
GIUL	Standard Rules + (COM)	no	no
GMTL	GUL + (W)	no	yes
GIMTL	GIUL + (W)	no	no
GMTL_n	GMTL + (C_n)	no	yes
GIMTL_n	GIMTL + (C_n)	no	no
GSMTL	GMTL + (SC_2)	no	yes
GG	GMTL + (C)	no	yes
GUML	GUL + (C) + (MINGLE)	no	yes
GRM	GIUL + (C) + (MIX)	no	no
GIUML	GRM + (EMP)	no	no

characterize the same logics. Our starting point for this task is a "standard translation" of sequents and hypersequents into formulas, recalling from earlier that $\star[A_1,\ldots,A_n] = (A_1 \star \ldots \star A_n)$ for $\star \in \{\odot, \oplus\}$, where $\odot[] = \mathrm{e}$ and $\oplus[] = \mathrm{f}$.

Definition 4.59 (Standard Interpretation).

$$\mathrm{I}(\Gamma \Rightarrow \Delta) =_{\mathrm{def}} \odot\Gamma \to \oplus\Delta$$
$$\mathrm{I}(S_1 \mid \ldots \mid S_n) =_{\mathrm{def}} \mathrm{I}(S_1) \vee \ldots \vee \mathrm{I}(S_n)$$

Example 4.60. Consider the following hypersequent:

$$\mathcal{G} = (A, A \to C \Rightarrow B \mid \Rightarrow A, B \mid A, B, C \Rightarrow)$$

To find the standard interpretation of $\mathcal{G}$, we first interpret the sequents:

$$\mathrm{I}(A, A \to C \Rightarrow B) = (A \odot (A \to C)) \to B$$
$$\mathrm{I}(\Rightarrow A, B) = \mathrm{e} \to (A \oplus B)$$
$$\mathrm{I}(A, B, C \Rightarrow) = (A \odot B \odot C) \to \mathrm{f}$$

and then take the disjunction, to get:

$$\mathrm{I}(\mathcal{G}) = ((A \odot (A \to C)) \to B) \vee (\mathrm{e} \to (A \oplus B)) \vee ((A \odot B \odot C) \to \mathrm{f})$$

Let us assume for the rest of this section that GL is a GMAILL^--extension. It is easy to see that if a single-conclusion sequent S is GL-derivable, then so is $\mathrm{I}(S)$. We just use $(\Rightarrow\to)$, $(\odot\Rightarrow)$, $(\mathrm{e}\Rightarrow)$, and $(\Rightarrow \mathrm{f})$ to derive the latter from the former. Moreover, since by Proposition 4.27, these rules are GL-invertible, the opposite direction also

Table 4.2 Matching rules and axioms

Structural rule	Matching axioms
$\dfrac{\mathcal{G} \mid \Gamma_1, \Pi_1 \Rightarrow \Sigma_1, \Delta_1 \quad \mathcal{G} \mid \Gamma_2, \Pi_2 \Rightarrow \Sigma_2, \Delta_2}{\mathcal{G} \mid \Gamma_1, \Gamma_2 \Rightarrow \Delta_1, \Delta_2 \mid \Pi_1, \Pi_2 \Rightarrow \Sigma_1, \Sigma_2}$ (COM)	(PRL) $(A \to B) \vee (B \to A)$ (DIS) $(A \wedge (B \vee C)) \to ((A \wedge B) \vee (A \wedge C))$
$\dfrac{\mathcal{G} \mid \Gamma_1, \Gamma_2 \Rightarrow \Delta_1, \Delta_2}{\mathcal{G} \mid \Gamma_1 \Rightarrow \Delta_1 \mid \Gamma_2 \Rightarrow \Delta_2}$ (SPLIT)	(EM) $A \vee \neg A$
$\dfrac{\mathcal{G} \mid \Gamma \Rightarrow \Delta}{\mathcal{G} \mid \Gamma, \Pi \Rightarrow \Sigma, \Delta}$ (W)	(W) $(A \to \mathrm{e}) \wedge (\mathrm{f} \to A)$
$\dfrac{\mathcal{G} \mid \Gamma_1 \Rightarrow \Delta_1 \quad \mathcal{G} \mid \Gamma_2 \Rightarrow \Delta_2}{\mathcal{G} \mid \Gamma_1, \Gamma_2 \Rightarrow \Delta_1, \Delta_2}$ (MIX)	(e) $\mathrm{f} \to \mathrm{e}$
$\dfrac{}{\mathcal{G} \mid \Rightarrow}$ (EMP)	(f) $\mathrm{e} \to \mathrm{f}$
$\dfrac{\mathcal{G} \mid \Gamma, \Pi, \Pi \Rightarrow \Sigma, \Sigma, \Delta}{\mathcal{G} \mid \Gamma, \Pi \Rightarrow \Sigma, \Delta}$ (C)	(C_2) $A \to (A \odot A)$
$\dfrac{\mathcal{G} \mid \Gamma_1, \Pi \Rightarrow \Sigma, \Delta_1 \quad \mathcal{G} \mid \Gamma_2, \Pi \Rightarrow \Sigma, \Delta_2}{\mathcal{G} \mid \Gamma_1, \Gamma_2, \Pi \Rightarrow \Sigma, \Delta_1, \Delta_2}$ (MINGLE)	(M) $(A \odot A) \to A$
$\dfrac{\mathcal{G} \mid \Gamma, \Pi_1^n \Rightarrow \Sigma_1^n, \Delta \quad \ldots \quad \mathcal{G} \mid \Gamma, \Pi_{n-1}^n \Rightarrow \Sigma_{n-1}^n, \Delta}{\mathcal{G} \mid \Gamma, \Pi_1, \ldots, \Pi_{n-1} \Rightarrow \Sigma_1, \ldots, \Sigma_{n-1}, \Delta}$ (C_n)	(C_n) $A^{n-1} \to A^n$
$\dfrac{\mathcal{G} \mid \Gamma, \Gamma \Rightarrow \Delta, \Delta}{\mathcal{G} \mid \Gamma \Rightarrow \Delta}$ (SC_2)	(NC) $\neg(A \wedge \neg A)$

holds. Similarly, if GL is a GMALL$^-$-extension, we can use $(\Rightarrow\oplus)$ to get $\vdash_{\mathrm{GL}} S$ iff $\vdash_{\mathrm{GL}} \mathrm{I}(S)$ for any sequent S. For the case where GL is a GUL$^-$-extension, we can use the invertible rule $(\Rightarrow\vee)$ to extend the correspondence also to hypersequents.

Let us say that a hypersequent $\mathcal{G}$ or a rule (r) is "appropriate for GL" if it is a sequent or sequent rule when GL is not a GUL$^-$-extension, and single-conclusion when GL is not a GMALL$^-$-extension. We have argued that:

Proposition 4.61. $\vdash_{\mathrm{GL}} \mathcal{G}$ *iff* $\vdash_{\mathrm{GL}} \mathrm{I}(\mathcal{G})$ *where* $\mathcal{G}$ *is appropriate for* GL.

Soundness and completeness for a Gentzen system GL with respect to a Hilbert system HL now takes the following form. We want to show that $\vdash_{\text{GL}} \mathcal{G}$ iff $\vdash_{\text{HL}} \text{I}(\mathcal{G})$ for all hypersequents $\mathcal{G}$ appropriate for GL. We will establish this result for a wide range of Gentzen systems, taking as a basis the standard rule set displayed in Fig. 4.1, with popular calculi from the literature listed in Table 4.1. In general, a correspondence with Hilbert systems can be established by "matching" rules with axioms. Intuitively, the rule should derive the axioms (in a basic system) and vice versa.

Definition 4.62. A hypersequent rule (r) and set of axioms $\mathcal{A}$ are L-*matching* if:

(1) For every instance $\mathcal{G}_1, \ldots, \mathcal{G}_n \;/\; \mathcal{G}$ of (r):

$$\vdash_{\text{HL}+\mathcal{A}} C \to \text{I}(\mathcal{G}) \quad \text{for some confusion } C \text{ of } \{\text{I}(\mathcal{G}_1), \ldots, \text{I}(\mathcal{G}_n)\}.$$

(2) $\vdash_{\text{GL}+(\text{r})} \Rightarrow A$ for every axiom A in $\mathcal{A}$.

Example 4.63. Let GL be the standard rule set of Fig. 4.1, and let HL be just HMALL. Then the set of axioms (DIS) and (PRL) and the rule (COM) are L-matching. We have already established (2) earlier in the chapter: the axioms are derivable in GL + (COM) (i.e. GIUL). For (1), we define:

$$A_i = \text{I}(\Gamma_i, \Pi_i \Rightarrow \Delta_i, \Sigma_i) \text{ for } i = 1,2 \quad \text{and} \quad B = \text{I}(\Gamma_1, \Gamma_2 \Rightarrow \Delta_1, \Delta_2 \mid \Pi_1, \Pi_2 \Rightarrow \Sigma_1, \Sigma_2)$$

It is enough to prove that $\vdash_{\text{HL}} (A_1 \wedge A_2) \to B$, since then by Lemma 3.30, $\vdash_{\text{HL}} ((\text{I}(\mathcal{G}) \vee A_1) \wedge (\text{I}(\mathcal{G}) \vee A_2)) \to (\text{I}(\mathcal{G}) \vee B)$ as required. First, we can show, using various parts of Lemma 3.23, that:

$$\vdash_{\text{HL}} (\text{I}(\Pi_1 \Rightarrow \Sigma_1) \to \text{I}(\Gamma_2 \Rightarrow \Delta_2)) \to ((A_1 \wedge A_2) \to B)$$
$$\vdash_{\text{HL}} (\text{I}(\Gamma_2 \Rightarrow \Delta_2) \to \text{I}(\Pi_1 \Rightarrow \Sigma_1)) \to ((A_1 \wedge A_2) \to B)$$

So by Lemma 3.23 (v):

$$\vdash_{\text{HL}} ((\text{I}(\Pi_1 \Rightarrow \Sigma_1) \to \text{I}(\Gamma_2 \Rightarrow \Delta_2)) \vee (\text{I}(\Gamma_2 \Rightarrow \Delta_2) \to \text{I}(\Pi_1 \Rightarrow \Sigma_1))) \to ((A_1 \wedge A_2) \to B)$$

But then $\vdash_{\text{HL}} (A_1 \wedge A_2) \to B$, using (PRL) and (MP).

Other matching relationships between axioms from Chapter 3 and hypersequent rules introduced above are displayed in Table 4.2.

Lemma 4.64. *For each rule* (r) *and set of axioms* $\mathcal{A}$ *matched in Table 4.2:*

(i) (r) *and* $\mathcal{A}$ *are* IUL$^-$*-matching.*
(ii) *the sequent version of* (r) *(if non-empty) and* $\mathcal{A}$ *are* MALL$^-$*-matching.*
(iii) *the single-conclusion version of* (r) *and* $\mathcal{A}$ *are* UL$^-$*-matching.*
(iv) *the single-conclusion sequent version of* (r) *(if non-empty) and* $\mathcal{A}$ *are* MAILL$^-$*-matching.*

Proof. Let (r) be a rule and $\mathcal{A}$ a set of axioms matched in Table 4.2. We have already shown in examples scattered throughout the text that the matching axioms are derivable in the appropriate extended systems. So condition (2) holds for (i)–(iv), noting that the sequent versions of (SPLIT) and (COM) are empty. Establishing condition (1) is more tedious. It involves finding several Hilbert system derivations. However, we can simplify the cases (i) and (iii) slightly for each rule (where n can be zero):

$$\frac{\mathcal{G} \mid \mathcal{H}_1 \quad \dots \quad \mathcal{G} \mid \mathcal{H}_n}{\mathcal{G} \mid \mathcal{H}}$$

Suppose that we can find appropriate derivations of:

$$\text{either} \quad (\mathrm{I}(\mathcal{H}_1) \odot \dots \odot \mathrm{I}(\mathcal{H}_n)) \to \mathrm{I}(\mathcal{H}) \quad \text{or} \quad (\mathrm{I}(\mathcal{H}_1) \wedge \dots \wedge \mathrm{I}(\mathcal{H}_n)) \to \mathrm{I}(\mathcal{H})$$

It then follows, using Lemma 3.30, that we have the required derivations of:

$$\text{either} \quad ((\mathrm{I}(\mathcal{G} \mid \mathcal{H}_1) \wedge \mathrm{e}) \odot \dots \odot (\mathrm{I}(\mathcal{G} \mid \mathcal{H}_n) \wedge \mathrm{e})) \to \mathrm{I}(\mathcal{G} \mid \mathcal{H})$$

$$\text{or} \quad (\mathrm{I}(\mathcal{G} \mid \mathcal{H}_1) \wedge \dots \wedge \mathrm{I}(\mathcal{G} \mid \mathcal{H}_n)) \to \mathrm{I}(\mathcal{G} \mid \mathcal{H})$$

Let us give some examples, leaving other cases for the interested reader:

- (C). Let HL be $\mathsf{HMALL}^- + (\mathrm{C}_2)$. Let $A = \odot(\Pi \uplus [\neg C : C \in \Sigma])$ and $B = \mathrm{I}(\Gamma \Rightarrow \Delta)$. Note that $\vdash_{\mathrm{HL}} (A \to (A \to B)) \to (A \to B)$ (see Example 3.35). But also, using various parts of Lemma 3.23, $\vdash_{\mathrm{HL}} \mathrm{I}(\Gamma, \Pi, \Pi \Rightarrow \Sigma, \Sigma, \Delta) \to (A \to (A \to B))$ and $\vdash_{\mathrm{HL}} (A \to B) \to \mathrm{I}(\Gamma, \Pi \Rightarrow \Sigma, \Delta)$. So, using Lemma 3.23 (iii), $\vdash_{\mathrm{HL}} \mathrm{I}(\Gamma, \Pi, \Pi \Rightarrow \Sigma, \Sigma, \Delta) \to \mathrm{I}(\Gamma, \Pi \Rightarrow \Sigma, \Delta)$ as required. The case where HL is $\mathsf{HMAILL}^- + (\mathrm{C}_2)$ and $\Sigma = []$ is very similar.
- (MIX). Let HL be $\mathsf{HMALL}^- + (\mathrm{f})$. Using Lemma 3.23, we get that $(A \odot (A \to \mathrm{f})) \to \mathrm{f}$, $\mathrm{f} \to \mathrm{e}$, and $\mathrm{e} \to (B \to B)$ are HL-derivable. Hence also, using Lemma 3.23 (iii), $\vdash_{\mathrm{HL}} (A \odot (A \to \mathrm{f})) \to (B \to B)$. It then follows easily that $(A \odot B) \to ((A \to \mathrm{f}) \to B)$ is derivable. So $\vdash_{\mathrm{HL}} (\mathrm{I}(\Gamma_1 \Rightarrow \Delta_1) \odot \mathrm{I}(\Gamma_2 \Rightarrow \Delta_2)) \to \mathrm{I}(\Gamma_1, \Gamma_2 \Rightarrow \Delta_1, \Delta_2)$ is derivable. □

Putting everything together, we arrive at the following characterization.

Definition 4.65. Let $\mathrm{L} \in \{\mathsf{MAILL}, \mathsf{MALL}, \mathsf{UL}, \mathsf{IUL}\}$. If (r_i) and $\mathcal{A}_i$ are L-matching for $i = 1 \dots n$, then $\mathrm{HL} + \mathcal{A}_1 + \dots + \mathcal{A}_n$ and $\mathrm{GL} + (\mathrm{r}_1) + \dots + (\mathrm{r}_n)$ (also $\mathsf{HL}^- + \mathcal{A}_1 + \dots + \mathcal{A}_n$ and $\mathsf{GL}^- + (\mathrm{r}_1) + \dots + (\mathrm{r}_n)$) are called *matching*.

Theorem 4.66. *For any* HMAILL^-*-extension* HL *and matching calculus* GL*:*

$$\vdash_{\mathrm{GL}} \mathcal{G} \ \textit{iff} \ \vdash_{\mathrm{HL}} \mathrm{I}(\mathcal{G}) \quad \textit{and} \quad \vdash_{\mathrm{GL^D}} \mathcal{G} \ \textit{iff} \ \vdash_{\mathrm{HL^D}} \mathrm{I}(\mathcal{G})$$

where $\mathcal{G}$ *is a hypersequent appropriate for* GL.

Proof. We tackle the right-to-left directions first. Suppose that $\vdash_{\mathrm{HL}} \mathrm{I}(\mathcal{G})$. All the axioms of HMAILL^- are GMAILL^--derivable and the involution axioms are GMALL^--derivable. Also the axioms for $\bot$ and $\top$ are GMAILL-derivable. Moreover, since HL and GL are matching, all the extra axioms of HL are GL-derivable. But also the GL-derivable formulas are closed under (ADJ) and (MP), using $(\Rightarrow\wedge)$ and Example 4.25, respectively. So $\vdash_{\text{GL}} \Rightarrow \mathrm{I}(\mathcal{G})$. But then by Proposition 4.61, $\vdash_{\text{GL}} \mathcal{G}$.

If $\vdash_{\mathsf{HL}^{\mathsf{D}}} \mathrm{I}(\mathcal{G})$, then we proceed in the same way. We just need to establish the extra claim that (DENSITY) is GL^{D}-admissible. Suppose that $\vdash_{\mathsf{GL}^{\mathsf{D}}} \Rightarrow (A \to p) \vee (p \to B) \vee C$ where p does not occur in A, B, or C. It follows using Proposition 4.61 that $\vdash_{\mathsf{GL}^{\mathsf{D}}} A \Rightarrow p \mid p \Rightarrow B \mid \Rightarrow C$. Hence by (DENSITY), $\vdash_{\mathsf{GL}^{\mathsf{D}}} A \Rightarrow B \mid \Rightarrow C$. So by (EC) and $(\Rightarrow\to)$, $\vdash_{\mathsf{GL}^{\mathsf{D}}} \Rightarrow (A \to B) \vee C$ as required.

For the other direction, suppose that $d \vdash_{\text{GL}} \mathcal{G}$. We prove that $\vdash_{\text{HL}} \mathrm{I}(\mathcal{G})$, proceeding by induction on $\mathrm{ht}(d)$. Suppose that $\mathcal{G}$ follows by some rule of GL from $\mathcal{G}_1, \ldots, \mathcal{G}_n$ (this includes the case of initial hypersequents when $n = 0$). By the induction hypothesis n times, $\vdash_{\text{HL}} \mathrm{I}(\mathcal{G}_1), \ldots, \vdash_{\text{HL}} \mathrm{I}(\mathcal{G}_n)$. But now $\vdash_{\text{HL}} C \to \mathrm{I}(\mathcal{G})$ for some confusion C of $\{\mathrm{I}(\mathcal{G}_1), \ldots, \mathrm{I}(\mathcal{G}_n)\}$. This is easy (if tedious) to show for any rule of GL$'$ for L$' \in \{\mathsf{MAILL}, \mathsf{MALL}, \mathsf{UL}, \mathsf{IUL}\}$ such that GL extends GL$'$. For the other rules of GL, this follows from the assumption that HL and GL are matching, But using Theorem 3.43 (since C is a confusion of derivable formulas), $\vdash_{\text{HL}} C$. So by (MP), $\vdash_{\text{HL}} \mathrm{I}(\mathcal{G})$ as required.

If $d \vdash_{\mathsf{GL}^{\mathsf{D}}} \mathcal{G}$, then we proceed in the same way. We just need to show that (DENSITY) is HL^{D}-admissible. Suppose that:

$$\vdash_{\mathsf{HL}^{\mathsf{D}}} \mathrm{I}(\mathcal{G}) \vee (\odot(\Gamma_1 \uplus [p]) \to \oplus\Delta_1) \vee (\odot\Gamma_2 \to \oplus(\Delta_2 \uplus [p]))$$

where p does not occur in $\mathcal{G}$, Γ_1, Γ_2, Δ_1, or Δ_2. It follows easily that:

$$\vdash_{\mathsf{HL}^{\mathsf{D}}} \mathrm{I}(\mathcal{G}) \vee (p \to (\odot\Gamma_1 \to \oplus\Delta_1)) \vee (\odot(\Gamma_2 \uplus [\neg A : A \in \Delta_2]) \to p)$$

Hence by the Hilbert system rule (DENSITY):

$$\vdash_{\mathsf{HL}^{\mathsf{D}}} \mathrm{I}(\mathcal{G}) \vee (\odot(\Gamma_2 \uplus [\neg A : A \in \Delta_2]) \to (\odot\Gamma_1 \to \oplus\Delta_1))$$

It follows easily that $\vdash_{\mathsf{HL}^{\mathsf{D}}} \mathrm{I}(\mathcal{G}) \vee (\odot(\Gamma_1 \uplus \Gamma_2) \to \oplus(\Delta_1 \uplus \Delta_2))$ as required. □

4.7 Historical Remarks

Gentzen systems have a long and distinguished history. The sequent calculi LK and LJ for first-order Classical Logic and Intuitionistic Logic were introduced by Gentzen in the 1930s [93] to aid investigations of another proof-theoretic formalism, Natural Deduction. These systems and core topics of the proof theory of Classical Logic and Intuitionistic Logic are treated in detail in e.g. the monographs [163, 204, 209] and the handbook [41].

Gentzen systems for substructural logics have many sources. In Linguistics, a sequent calculus – the so-called "Lambek calculus" – was defined by Lambek in the 1950s to model the assignment of types such as "adjective" or "verb phrase" to strings of words in natural language [130]. Since in this context, order (of words or types) as well as multiplicity is crucial, sequents consist (as for Gentzen) of sequences of formulas, and the calculus lacks not only weakening and contraction but also exchange rules. Gentzen systems for other non-commutative logics have been investigated by many authors, including the full Lambek Calculus FL studied by Ono in [176] and the non-commutative Linear Logic of Ruet and Abrusci [1, 190].

Another important source of substructural logics is Philosophy and the relevance logics developed by Anderson and Belnap and co-workers (in particular Dunn and Meyer) from the 1960s onwards [6, 7]. In these logics, formulas such as $B \to (A \to A)$ are disallowed as theorems because B is not "used" in the proof of $A \to A$ (or B does not "relevantly imply" $A \to A$). This is mirrored proof-theoretically by dropping weakening rules (e.g. from LK or LJ) to prevent the addition of irrelevant formulas to sequents. However, sequent calculi obtained in this way, investigated by Brady in the 1990s [38, 39], do not derive the distributivity axioms assumed by most relevance logics. Gentzen systems for distributive relevance logics were defined independently by Dunn [74] and Minc [151] in the 1970s, making use of sequents with two structural connectives "," and ";" corresponding to $\wedge$ and $\vee$, and $\odot$ and $\oplus$, respectively. This method of replacing logical symbols with structural connectives was developed even further by Belnap in his 1982 paper [34]. The result, Display Logic, is a formalism capable of capturing a wide range of systems (also some hypersequent calculi [216]) but at the cost of introducing a great deal of extra structure.

A further source of substructural logics is Set Theory. Logics (sequent calculi) without contraction were introduced by Grishin in the 1980s [103] with the aim of avoiding paradoxes involving the comprehension principle, and developed further by Ono and Komori in [173, 177]. Finally, in Computer Science, Girard's Linear Logic [97], introduced in 1987, drops weakening and contraction rules from LK but allows these to be recovered for certain formulas using special modal operators ! and ?. One motivation for Linear Logic is to provide a more careful analysis of constructive proofs: formulas can be interpreted as resources that are available to be used once exactly. A good exposition of Linear Logic is given by Troelstra in [208]. More generally, introductions to sequent calculi and other features of substructural logics are provided in the books [90, 180, 186].

The step up from sequents to hypersequents was first taken by Avron [9] in 1987 in order to provide a Gentzen system for the relevance logic RM. However, hypersequents were also used implicitly (and independently) by Pottinger [182] in 1983 to provide a calculus for the modal logic S5. The first hypersequent calculus for a fuzzy logic was defined for Gödel Logic by Avron in 1991 [11] (earlier sequent calculi for the logic are discussed in Chapter 6), and many others have since been developed for this family. In particular, the calculi for MTL and other fuzzy logics with weakening were defined by Baaz, Ciabattoni, and co-authors in [17, 49], and calculi for UL and related logics without weakening, by Metcalfe and Montagna in [144]. Hypersequent calculi were introduced for Łukasiewicz and Product logics, the subject of Chapter 6, by Metcalfe, Olivetti, and Gabbay in [146, 148], and for many other logics, fuzzy and otherwise, in the papers [51, 52, 54, 143]. Finally, in a 2008 paper by Ciabattoni, Galatos, and Terui [53], an algorithm is provided for transforming axiom schema in certain syntactic classes into either sequent or hypersequent (depending on the class) structural rules. As a nice example, a rather complicated structural rule is obtained for the fuzzy logic based on the nilpotent minimum t-norm.

Finally, we remark that the useful multiset ordering of Definition 4.4 was introduced by Dershowitz and Manna in their 1979 paper [70]. The proof that this is a well-ordering is due to Buchholz, written up by Nipkow as [164].

Chapter 5
Syntactic Eliminations

The Gentzen systems defined in Chapter 4 provide a uniform and natural presentation of a wide range of (fuzzy) logics. But are they useful? Proof search in Hilbert systems is rendered tedious and difficult by the need to guess formulas A and $A \to B$ as premises when applying modus ponens. The same situation seems to occur for Gentzen Systems: we have to guess which formula A to use when applying (CUT). Certainly finding derivations would be much simpler if we could do without this rule. Then we could just apply rules where formulas in the premises are subformulas of formulas in the conclusion. Indeed, this "subformula property" is useful for, among other things, establishing decidability and conservative extension results. Here we show that this happy situation occurs for most of the Gentzen systems introduced in the previous chapter. For calculi satisfying certain properties we can algorithmically transform derivations in the calculus with (CUT) into derivations without this rule, that is, "eliminate" (CUT) from derivations.

We will also treat eliminations of other key rules: (CAN), the "cut rule" of Abelian Logic, and (DENSITY). Density elimination in particular has an important application. Since we know from earlier chapters that Gentzen systems extended with (DENSITY) are complete with respect to dense chains, density elimination implies that the same holds for these calculi without (DENSITY). From this we are able to deduce standard completeness results for wide classes of Hilbert systems and Gentzen systems for fuzzy logics.

Since in this chapter we will often be dealing with quite complicated structures, let us recall some notational conveniences:

- $\lambda, \mu, m, n, i, j, k$ denote natural numbers.
- $\Gamma, \Delta, \Pi, \Sigma$ denote multisets of formulas, with $\Gamma^0 = []$ and $\Gamma^{n+1} = \Gamma \uplus \Gamma^n$.
- S denotes a sequent and $\mathcal{G}, \mathcal{H}$ denote hypersequents.
- $[\mathcal{G}_i]_{i=1}^n$ denotes the hypersequent $\mathcal{G}_1 \mid \ldots \mid \mathcal{G}_n$.
- $\{\mathcal{G}_i\}_{i=1}^n$ denotes a set of hypersequents $\{\mathcal{G}_1, \ldots, \mathcal{G}_n\}$ (perhaps the premises of some rule application).

5.1 Cut Elimination

Let us start with an example. Suppose that we want to eliminate an application of (CUT) from a derivation in a single-conclusion sequent calculus without structural rules, e.g. the sequent version of GMAILL. We are confronted by:

$$\dfrac{\dfrac{\vdots}{\Gamma, A \Rightarrow \Delta} \quad \dfrac{\vdots}{\Pi \Rightarrow A}}{\Gamma, \Pi \Rightarrow \Delta}\ (\text{CUT})$$

The cut-formula A occurs on the left in one premise, and on the right in the other. A natural strategy for eliminating this application of (CUT) is to look at the derivations of these premises. If one of the premises is an instance of (ID), then it must be $A \Rightarrow A$ and the other premise must be $\Gamma, \Pi \Rightarrow \Delta$, derived with one fewer applications of (CUT). Otherwise, we have two possibilities. The first is that one of the premises ends with an application of a rule where A is not the principal formula, e.g. letting $\Gamma = \Gamma_1 \uplus \Gamma_2 \uplus [B \to C]$:

$$\dfrac{\dfrac{\dfrac{\vdots}{\Gamma_1, C, A \Rightarrow \Delta} \quad \dfrac{\vdots}{\Gamma_2 \Rightarrow B}}{\Gamma_1, \Gamma_2, B \to C, A \Rightarrow \Delta}\ (\to\Rightarrow) \quad \dfrac{\vdots}{\Pi \Rightarrow A}}{\Gamma_1, \Gamma_2, B \to C, \Pi \Rightarrow \Delta}\ (\text{CUT})$$

In this case, we can "push the cut upwards" in the derivation to get:

$$\dfrac{\dfrac{\dfrac{\vdots}{\Gamma_1, C, A \Rightarrow \Delta} \quad \dfrac{\vdots}{\Pi \Rightarrow A}}{\Gamma_1, C, \Pi \Rightarrow \Delta}\ (\text{CUT}) \quad \dfrac{\vdots}{\Gamma_2 \Rightarrow B}}{\Gamma_1, \Gamma_2, B \to C, \Pi \Rightarrow \Delta}\ (\to\Rightarrow)$$

That is, we have a derivation where the left premise in the new application of (CUT) has a shorter derivation than the application in the original derivation.

The second possibility is that the last application of a rule in both premises involves A as the principal formula, e.g. with $\Gamma = \Gamma_1 \uplus \Gamma_2$ and $A = B \to C$:

$$\dfrac{\dfrac{\dfrac{\vdots}{\Gamma_1 \Rightarrow B} \quad \dfrac{\vdots}{\Gamma_2, C \Rightarrow \Delta}}{\Gamma_1, \Gamma_2, B \to C \Rightarrow \Delta}\ (\to\Rightarrow) \quad \dfrac{\dfrac{\vdots}{\Pi, B \Rightarrow C}}{\Pi \Rightarrow B \to C}\ (\Rightarrow\to)}{\Gamma_1, \Gamma_2, \Pi \Rightarrow \Delta}\ (\text{CUT})$$

Here we rearrange our derivation in a different way: we replace the application of (CUT) with applications of (CUT) with cut-formulas B and C:

$$\dfrac{\dfrac{\dfrac{\vdots}{\Gamma_2, C \Rightarrow \Delta} \quad \dfrac{\vdots}{\Pi, B \Rightarrow C}}{\Gamma_2, \Pi, B \Rightarrow \Delta}\ (\text{CUT}) \quad \dfrac{\vdots}{\Gamma_1 \Rightarrow B}}{\Gamma_1, \Gamma_2, \Pi \Rightarrow \Delta}\ (\text{CUT})$$

We now have two applications of (CUT) but with cut-formulas of a smaller complexity than the original application.

This procedure, formalized using a double induction on cut-formula complexity and the combined height of derivations of the premises, eliminates applications of (CUT) for many sequent calculi. However, it encounters a problem with rules that contract formulas in one or more of the premises. Consider the following situation:

$$\dfrac{\dfrac{\dfrac{\vdots}{\Gamma, A, A \Rightarrow \Delta}}{\Gamma, A \Rightarrow \Delta}\ (\text{C}) \quad \dfrac{\vdots}{\Pi \Rightarrow A}}{\Gamma, \Pi \Rightarrow \Delta}\ (\text{CUT})$$

In this case we need to perform several cuts at once, using a rule something like:

$$\dfrac{\Gamma, [A]^{\lambda} \Rightarrow \Delta \quad \Pi \Rightarrow A}{\Gamma, \Pi^{\lambda} \Rightarrow \Delta}$$

For hypersequent calculi, the situation is further complicated by the fact that whole sequents may be contracted using (EC). This means that a cut-formula occurring in the premises of an application of (CUT) may appear in several sequents in a hypersequent higher up in the derivation, e.g.

$$\dfrac{\dfrac{\dfrac{\vdots}{\Gamma, A \Rightarrow \Delta \mid \Gamma, A \Rightarrow \Delta}}{\Gamma, A \Rightarrow \Delta}\ (\text{EC}) \quad \dfrac{\vdots}{\Pi \Rightarrow A}}{\Gamma, \Pi \Rightarrow \Delta}\ (\text{CUT})$$

To cope with this situation, we use even more general versions of (CUT) that perform multiple cuts in different sequents, e.g.

$$\dfrac{\Gamma_1, [A]^{\lambda_1} \Rightarrow \Delta_1 \mid \ldots \mid \Gamma_n, [A]^{\lambda_n} \Rightarrow \Delta_n \quad \Pi \Rightarrow A}{\Gamma_1, \Pi^{\lambda_1} \Rightarrow \Delta_1 \mid \ldots \Gamma_n, \Pi^{\lambda_n} \Rightarrow \Delta_n}$$

Then, similarly to the sequent calculi case, we can give inductive proofs of the eliminations of such rules.

One of our main goals is to apply this method of cut elimination to a general class of hypersequent calculi satisfying "substitutivity" and "reductivity" conditions. Intuitively, substitutivity ensures that we can move cuts upwards in derivations, and reductivity ensures that when both premises introduce the cut-formula, we can replace the cut with cuts on its subformulas. The general approach avoids repeating work for different systems and allows cut elimination for new systems to be decided in a systematic manner. Moreover, we can think of substitutivity and reductivity as supplying a characterization of calculi that admit cut elimination.

5.1.1 Regular Calculi

In general, a hypersequent calculus is just a set of hypersequent rules. Here we want to be more precise, while still allowing a wide range of logical and structural rules, and covering all (or most) of the examples in the previous chapter.

Recall that a "typical" (to be made more precise below) sequent or hypersequent calculus consists of the axioms (ID), the cut rule (CUT), (internal and external) structural rules, and logical rules. In particular, the latter consist of possibly empty sets of rules for each n-ary connective $\star$ where each instance has a distinguished *principal formula* $\star(A_1,\ldots A_n)$ in the conclusion, and premises containing *active formulas* from $A_1,\ldots,A_n$, all other formulas being *context formulas*. For instances of hypersequent rules, we also speak of the distinguished *active sequents* in the premises and conclusion, and other *context sequents*.

To identify rules suitable for cut elimination, we first need some way of indicating a distinguished formula in hypersequents, either the cut-formula or the principal formula of a logical rule.

Definition 5.1. A *marked hypersequent* is a hypersequent with exactly one occurrence of a formula A distinguished, written $\mathcal{G} \mid \Gamma,\underline{A} \Rightarrow \Delta$ or $\mathcal{G} \mid \Gamma \Rightarrow \underline{A},\Delta$. A *marked rule instance* is a rule instance with the principal formula, if there is one, marked.

We can now define the result of applying (CUT) multiple times, assuming that all the usual notions for hypersequents apply also to marked hypersequents.

Definition 5.2. For a (marked) hypersequent $\mathcal{G}$ and a marked hypersequent $\mathcal{H}$:

(1) $(\mathcal{G} \mid \mathcal{H}') \in \text{CUT}(\mathcal{G},\mathcal{H})$ if $\mathcal{H} = (\mathcal{H}' \mid \Pi \Rightarrow \underline{A},\Sigma)$ or $\mathcal{H} = (\mathcal{H}' \mid \Pi,\underline{A} \Rightarrow \Sigma)$.

(2) $(\mathcal{G}' \mid \Gamma,\Pi \Rightarrow \Sigma,\Delta) \in \text{CUT}(\mathcal{G},\mathcal{H})$ if:

$$\begin{aligned}\text{either } & (\mathcal{G}' \mid \Gamma,A \Rightarrow \Delta) \in \text{CUT}(\mathcal{G},\mathcal{H}) \text{ and } \mathcal{H} = (\mathcal{H}' \mid \Pi \Rightarrow \underline{A},\Sigma)\\ \text{or } & (\mathcal{G}' \mid \Gamma \Rightarrow A,\Delta) \in \text{CUT}(\mathcal{G},\mathcal{H}) \text{ and } \mathcal{H} = (\mathcal{H}' \mid \Pi,\underline{A} \Rightarrow \Sigma)\end{aligned}$$

noting that the occurrence of A in $(\mathcal{G}' \mid \Gamma,A \Rightarrow \Delta)$ or $(\mathcal{G}' \mid \Gamma \Rightarrow A,\Delta)$ is unmarked.

Hypersequents obtained in this way have a particular form. Suppose that A does not occur unmarked in $\biguplus_{i=1}^{n} \Gamma_i$ with:

$$\mathcal{G} = (\Gamma_1,[A]^{\lambda_1} \Rightarrow \Delta_1 \mid \ldots \mid \Gamma_n,[A]^{\lambda_n} \Rightarrow \Delta_n) \quad \text{and} \quad \mathcal{H} = (\mathcal{H}' \mid \Pi \Rightarrow \underline{A},\Sigma)$$

Then $\text{CUT}(\mathcal{G},\mathcal{H})$ contains, for all $0 \leq \mu_i \leq \lambda_i$ for $i = 1\ldots n$:

$$\mathcal{H}' \mid \Gamma_1, \Pi^{\mu_1}, [A]^{\lambda_1 - \mu_1} \Rightarrow \Sigma^{\mu_1}, \Delta_1 \mid \ldots \mid \Gamma_n, \Pi^{\mu_n}, [A]^{\lambda_n - \mu_n} \Rightarrow \Sigma^{\mu_n}, \Delta_n$$

Similarly, suppose that A does not occur unmarked in $\biguplus_{i=1}^{n} \Delta_i$ with:

$$\mathcal{G} = (\Gamma_1 \Rightarrow [A]^{\lambda_1}, \Delta_1 \mid \ldots \mid \Gamma_n \Rightarrow [A]^{\lambda_n}, \Delta_n) \quad \text{and} \quad \mathcal{H} = (\mathcal{H}' \mid \Pi, \underline{A} \Rightarrow \Sigma)$$

Then $\text{CUT}(\mathcal{G}, \mathcal{H})$ contains, for all $0 \leq \mu_i \leq \lambda_i$ for $i = 1 \ldots n$:

$$\mathcal{H}' \mid \Gamma_1, \Pi^{\mu_1} \Rightarrow [A]^{\lambda_1 - \mu_1}, \Sigma^{\mu_1}, \Delta_1 \mid \ldots \mid \Gamma_n, \Pi^{\mu_n} \Rightarrow [A]^{\lambda_n - \mu_n}, \Sigma^{\mu_n}, \Delta_n$$

One of the crucial steps for our method of cut elimination will be shifting applications of (CUT) upwards over applications of other rules. For example, the derivation:

$$\dfrac{\dfrac{\dfrac{\vdots}{\Gamma, A, A \Rightarrow \Delta}}{\Gamma, A \Rightarrow \Delta}\,(\text{C}) \qquad \dfrac{\vdots}{\Pi \Rightarrow A, \Sigma}}{\Gamma, \Pi \Rightarrow \Sigma, \Delta}\,(\text{CUT})$$

can be transformed into:

$$\dfrac{\dfrac{\dfrac{\dfrac{\vdots}{\Gamma, A, A \Rightarrow \Delta} \quad \dfrac{\vdots}{\Pi \Rightarrow A, \Sigma}}{\Gamma, \Pi, A \Rightarrow \Sigma, \Delta}\,(\text{CUT}) \qquad \dfrac{\vdots}{\Pi \Rightarrow A, \Sigma}}{\Gamma, \Pi, \Pi \Rightarrow \Sigma, \Sigma, \Delta}\,(\text{CUT})}{\Gamma, \Pi \Rightarrow \Sigma, \Delta}\,(\text{C})$$

To ensure that this is possible, we require the following condition. Extending the convention of the previous chapter, let us say that a hypersequent $\mathcal{G}$ is "appropriate for a rule (r)" if it is a sequent when (r) is a sequent rule, and single-conclusion when (r) is single-conclusion.

Definition 5.3. A rule (r) is *substitutive* if for any:

1. marked instance $\mathcal{G}_1, \ldots, \mathcal{G}_n \;/\; \mathcal{G}$ of (r);
2. marked hypersequent $\mathcal{H}$ appropriate for (r);
3. $\mathcal{G}' \in \text{CUT}(\mathcal{G}, \mathcal{H})$;

there exist $\mathcal{G}_i' \in \text{CUT}(\mathcal{G}_i, \mathcal{H})$ for $i = 1 \ldots n$ such that:

$$\dfrac{\mathcal{G}_1' \quad \ldots \quad \mathcal{G}_n'}{\mathcal{G}'} \quad \text{is an instance of (r).}$$

The name "substitutive" is apt because the condition implies that substituting occurrences of A with Π on the left and Σ on the right, in both the conclusion of a rule instance and suitably in its premises, gives another instance of the rule.[1]

Example 5.4. Consider a marked instance of the single-conclusion hypersequent rule $(\rightarrow\Rightarrow)$ of the form:

$$\frac{B \vee C \Rightarrow B \quad C \Rightarrow D}{B \vee C, \underline{B \rightarrow C} \Rightarrow D}$$

Suppose that $\mathcal{H}$ is any marked hypersequent of the form:

$$\mathcal{G} \mid \Pi \Rightarrow \underline{B \vee C} \quad \text{or} \quad \mathcal{G} \mid \Pi, \underline{D} \Rightarrow \Sigma$$

Then the following are instances of $(\rightarrow\Rightarrow)$:

$$\frac{\mathcal{G} \mid \Pi \Rightarrow B \quad \mathcal{G} \mid C \Rightarrow D}{\mathcal{G} \mid \Pi, \underline{B \rightarrow C} \Rightarrow D} \quad \text{and} \quad \frac{\mathcal{G} \mid B \vee C \Rightarrow B \quad \mathcal{G} \mid C, \Pi \Rightarrow \Sigma}{\mathcal{G} \mid B \vee C, \underline{B \rightarrow C}, \Pi \Rightarrow \Sigma}$$

However, the following "anti-contraction" rule is not substitutive:

$$\frac{\mathcal{G} \mid \Gamma, \Pi \Rightarrow \Sigma, \Delta}{\mathcal{G} \mid \Gamma, \Pi, \Pi \Rightarrow \Sigma, \Sigma, \Delta}$$

Just consider an instance of the form $A \Rightarrow \ / \ A, A \Rightarrow$, and a sequent $B \Rightarrow \underline{A}$. Neither $A \Rightarrow \ / \ A, B \Rightarrow$ nor $B \Rightarrow \ / \ A, B \Rightarrow$ is an instance of the rule.

It is easy to see that the logical and structural rules introduced in Table 4.2 are substitutive. Apart from the principal formula and active formulas in the logical rules, each sequent rule schema involves just multiset variables $\Gamma, \Pi, \Delta, \ldots$. Call these multiset variables "parameters". All parameters in the multiple-conclusion versions of these rules occur in pairs in the premises and the conclusion: one variable Γ on the left and an accompanying variable Δ on the right. Any parameter occurs just once in the conclusion and any parameter occurring in the premises occurs in the conclusion. Now for an instance of the rule, suppose that we replace a formula A in the conclusion with Π on the left and Σ on the right. But A occurs in the instance of some pair of parameters Γ or Δ. Hence we can also replace the occurrence of A in the instances of Γ and Δ in the premises with Π on the left and Σ on the right, to obtain another instance of the rule. The cases for single-conclusion and hypersequent rules work in a similar way.

Let us now take a closer look at the logical rules. A crucial aspect of our cut elimination proof is the reduction of a cut with a complex formula A to cuts with the subformulas of A. Suppose that the premises of the application of (CUT) end with A introduced on the left in one premise, and on the right in the other. Then we should

[1] A more general approach would be to define substitutivity relative to a calculus, and allow the required derivations to use any rules of the calculus. However, it is always possible to "complete" the rules of such calculi to satisfy the condition given here, essentially by closing under (CUT).

be able to use cuts on the premises of those introductions with subformulas of A to obtain the original conclusion.[2] We formalize this as follows:

Definition 5.5. A set of logical rules for a connective $\star$ is *reductive* if for all instances of left and right rules for $\star$:

$$\frac{\mathcal{G} \mid S_1 \quad \dots \quad \mathcal{G} \mid S_k}{\mathcal{G} \mid \Gamma, \star(A_1,\dots,A_n) \Rightarrow \Delta} \quad \text{and} \quad \frac{\mathcal{G} \mid S'_1 \quad \dots \quad \mathcal{G} \mid S'_m}{\mathcal{G} \mid \Pi \Rightarrow \star(A_1,\dots,A_n), \Sigma}$$

$\mathcal{G} \mid \Gamma, \Pi \Rightarrow \Sigma, \Delta$ is derivable from $(\mathcal{G} \mid S_1), \dots, (\mathcal{G} \mid S_k), (\mathcal{G} \mid S'_1), \dots, (\mathcal{G} \mid S'_m)$ using only (CUT) with cut-formulas from $A_1, \dots, A_n$.

The logical rules for connectives collected in Table 4.2 are all reductive. Consider for example, instances of $(\rightarrow\Rightarrow)$ and $(\Rightarrow\rightarrow)$:

$$\frac{\mathcal{G} \mid \Gamma_1 \Rightarrow B, \Delta_1 \quad \mathcal{G} \mid \Gamma_2, C \Rightarrow \Delta_2}{\mathcal{G} \mid \Gamma_1, \Gamma_2, B \rightarrow C \Rightarrow \Delta_1, \Delta_2} \quad \text{and} \quad \frac{\mathcal{G} \mid \Pi, B \Rightarrow C, \Sigma}{\mathcal{G} \mid \Pi \Rightarrow B \rightarrow C, \Sigma}$$

By applying (CUT) to the premises we obtain the following derivation:

$$\frac{\dfrac{\mathcal{G} \mid \Gamma_1 \Rightarrow B, \Delta_1 \quad \mathcal{G} \mid \Pi, B \Rightarrow C, \Sigma}{\mathcal{G} \mid \Gamma_1, \Pi \Rightarrow C, \Sigma, \Delta_1}\,(\text{CUT}) \quad \mathcal{G} \mid \Gamma_2, C \Rightarrow \Delta_2}{\mathcal{G} \mid \Gamma_1, \Gamma_2, \Pi \Rightarrow \Sigma, \Delta_1, \Delta_2}\,(\text{CUT})$$

For $\wedge$ and $\vee$ there are more rules to consider, e.g.

$$\frac{\mathcal{G} \mid \Gamma, A \Rightarrow \Delta}{\mathcal{G} \mid \Gamma, A \wedge B \Rightarrow \Delta} \qquad \frac{\mathcal{G} \mid \Gamma, B \Rightarrow \Delta}{\mathcal{G} \mid \Gamma, A \wedge B \Rightarrow \Delta} \qquad \frac{\mathcal{G} \mid \Pi \Rightarrow A, \Sigma \quad \mathcal{G} \mid \Pi \Rightarrow B, \Sigma}{\mathcal{G} \mid \Pi \Rightarrow A \wedge B, \Sigma}$$

Hence we need to provide two derivations:

$$\frac{\mathcal{G} \mid \Gamma, A \Rightarrow \Delta \quad \mathcal{G} \mid \Pi \Rightarrow A, \Sigma}{\mathcal{G} \mid \Gamma, \Pi \Rightarrow \Sigma, \Delta}\,(\text{CUT}) \qquad \frac{\mathcal{G} \mid \Gamma, B \Rightarrow \Delta \quad \mathcal{G} \mid \Pi \Rightarrow B, \Sigma}{\mathcal{G} \mid \Gamma, \Pi \Rightarrow \Sigma, \Delta}\,(\text{CUT})$$

Cases for other connectives are very similar and left as exercises.

Example 5.6. Reductivity as defined here is quite sensitive to the form of the rules. Consider, for example, implication rules:

$$\frac{\mathcal{G} \mid \Gamma \Rightarrow B, \Delta \quad \mathcal{G} \mid \Gamma, C \Rightarrow \Delta}{\mathcal{G} \mid \Gamma, B \rightarrow C \Rightarrow \Delta}\,(\rightarrow\Rightarrow)^c \qquad \frac{\mathcal{G} \mid \Pi, B \Rightarrow C, \Sigma}{\mathcal{G} \mid \Pi \Rightarrow B \rightarrow C, \Sigma}\,(\Rightarrow\rightarrow)$$

We cannot in general derive instances of $(\mathcal{G} \mid \Gamma, \Pi \Rightarrow \Sigma, \Delta)$ from $(\mathcal{G} \mid \Gamma \Rightarrow B, \Delta)$, $(\mathcal{G} \mid \Gamma, C \Rightarrow \Delta)$, and $(\mathcal{G} \mid \Pi, B \Rightarrow C, \Sigma)$ using just (CUT) with cut-formulas B and C. So this set of rules for $\rightarrow$ is not reductive. However, note that the derivations go through if we allow extra structural rules such as contraction.

---

[2] As with substitutivity, this condition could be broadened to allow applications of other rules as well as cuts on subformulas.

We are now able to characterize a class of Gentzen systems suitable for our cut elimination method, noting that this includes all the calculi defined in Table 4.1.

Definition 5.7. A *regular hypersequent calculus* consists of

1. the identity axioms (ID) and cut rule (CUT);
2. schematic substitutive reductive logical rules;
3. schematic substitutive structural rules;

restricted possibly to single-conclusion and/or sequent versions.

In particular, any extension of GMAILL, GMALL, GUL, or GIUL (perhaps dropping rules for some connectives) with substitutive structural rules – such as those in Table 4.2 – is regular.

5.1.2 The Main Theorem

Before embarking on our general cut elimination proof, let us review some of the main ideas. First, note that it is enough just to consider an uppermost application of (CUT) in a derivation. If we can remove such an application without introducing any new applications, then we can eliminate applications of (CUT) one by one. Let us define:

Definition 5.8. GL$^\circ$ is the calculus GL without (CUT).

Then to establish cut elimination for GL it is enough to show constructively that if the premises of an instance of (CUT) are GL$^\circ$-derivable, then the conclusion is GL$^\circ$-derivable.

Now recall from our earlier discussion that to make this proof work, we need to consider a generalization of (CUT). In particular, we show that if two hypersequents $\mathcal{G}$ and $\mathcal{H}$ are derivable, the first with a marked occurrence of A, the second with occurrences of A on the opposite side, then repeated applications of (CUT) give derivable hypersequents; i.e. any hypersequent in CUT$(\mathcal{G},\mathcal{H})$ is derivable. The proof of this is by induction. Suppose that the marked cut-formula A occurs on the right in $\mathcal{H}$. We consider a derivation of $\mathcal{G}$, using substitutivity to push applications of (CUT) upwards until the last rule application is a logical rule applied to an occurrence of A on the left. We apply the induction hypothesis to the premises, then turn our attention to $\mathcal{H}$. Again, applications of (CUT) are pushed upwards until the last rule application is a logical rule applied to an occurrence of A, this time on the left. Finally, we use reductivity to replace the application of (CUT) with applications of (CUT) to the subformulas of A occurring in the logical rules.

Theorem 5.9. *Cut elimination holds for any regular hypersequent calculus.*

Proof. Let GL be a regular hypersequent calculus. As observed above, it is sufficient to show that an "uppermost" application of (CUT) in any GL-derivation (i.e. where the premises are GL$^\circ$-derivable) can be eliminated without introducing new applications of (CUT). Hence it is enough to prove the following:

Claim. For any hypersequent $\mathcal{G}$ and hypersequent $\mathcal{H}$ with marked formula A:

$$\text{If } d_{\mathcal{G}} \vdash_{\mathrm{GL}^\circ} \mathcal{G} \text{ and } d_{\mathcal{H}} \vdash_{\mathrm{GL}^\circ} \mathcal{H}, \text{ then } \vdash_{\mathrm{GL}^\circ} \mathrm{CUT}(\mathcal{G},\mathcal{H}).$$

We prove the claim by a triple induction on the lexicographically ordered triple:

$$\langle \mathrm{cp}(A), \mathrm{e}(d_{\mathcal{H}}), \mathrm{ht}(d_{\mathcal{G}}) \rangle$$

$$\text{where} \quad \mathrm{e}(d) = \begin{cases} 0 & \text{if } d \text{ ends with a logical rule applied to a marked formula} \\ 1 & \text{otherwise} \end{cases}$$

We begin by considering the last application of a rule (r) in $d_{\mathcal{G}}$. If (r) is (ID), then $\mathcal{G}$ is of the form $(\mathcal{G}' \mid C \Rightarrow C)$. So every member of $\mathrm{CUT}(\mathcal{G},\mathcal{H})$ is of the form $(\mathcal{H}' \mid C \Rightarrow C)$ or $(\mathcal{H}' \mid \mathcal{H})$, and the claim follows using (EW). Otherwise, there are two cases:

(a) The application of (r) is of the form:

$$\frac{\mathcal{G}_1 \quad \ldots \quad \mathcal{G}_n}{\mathcal{G}}$$

and the principal formula (if there is one) is *not* an occurrence of A on the opposite side to the marked occurrence in $\mathcal{H}$. Pick $\mathcal{G}' \in \mathrm{CUT}(\mathcal{G},\mathcal{H})$. By the substitutivity of (r), there exist $\mathcal{G}_i' \in \mathrm{CUT}(\mathcal{G}_i,\mathcal{H})$ for $i = 1 \ldots n$ such that:

$$\frac{\mathcal{G}_1' \quad \ldots \quad \mathcal{G}_n'}{\mathcal{G}'} \quad \text{is an instance of (r).}$$

But by the induction hypothesis $\vdash_{\mathrm{GL}^\circ} \mathcal{G}_i'$ for $i = 1 \ldots n$, so also $\vdash_{\mathrm{GL}^\circ} \mathcal{G}'$ as required.

(b) The application of (r) is of the form:

$$\frac{\mathcal{G}_1 \quad \ldots \quad \mathcal{G}_n}{\mathcal{G}' \mid \Gamma, [A]^\lambda \Rightarrow \Delta} \quad \text{or} \quad \frac{\mathcal{G}_1 \quad \ldots \quad \mathcal{G}_n}{\mathcal{G}' \mid \Gamma \Rightarrow [A]^\lambda, \Delta}$$

where the principal formula is an occurrence of A on the opposite side to the marked occurrence in $\mathcal{H}$, and $A \notin \Gamma$ or $A \notin \Delta$ as appropriate. Pick $\mathcal{G}^{\mathcal{H}} \in \mathrm{CUT}(\mathcal{G},\mathcal{H})$ where $\mathcal{H}$ is of the form:

$$\mathcal{H}' \mid \Pi \Rightarrow \underline{A}, \Sigma \quad \text{or} \quad \mathcal{H}' \mid \Pi, \underline{A} \Rightarrow \Sigma$$

The only tricky case (others follow as above using substitutivity) is when $\mathcal{G}^{\mathcal{H}}$ is of the form:

$$\mathcal{H}' \mid \mathcal{G}'' \mid \Gamma, \Pi^\lambda \Rightarrow \Sigma^\lambda, \Delta$$

where $\mathcal{G}'' \in \mathrm{CUT}(\mathcal{G}',\mathcal{H})$. Notice that also:

$$\mathcal{H}' \mid \mathcal{G}'' \mid \Gamma, \Pi^{\lambda-1}, A \Rightarrow \Sigma^{\lambda-1}, \Delta \quad \text{or} \quad \mathcal{H}' \mid \mathcal{G}'' \mid \Gamma, \Pi^{\lambda-1} \Rightarrow A, \Sigma^{\lambda-1}, \Delta$$

is a member of $\mathrm{CUT}(\mathcal{G},\mathcal{H})$. So by the substitutivity of (r), there exist $\mathcal{G}_i' \in \mathrm{CUT}(\mathcal{G}_i,\mathcal{H})$ for $i = 1 \ldots n$ such that:

$$\frac{\mathcal{G}'_1 \quad \ldots \quad \mathcal{G}'_n}{\mathcal{H}' \mid \mathcal{G}'' \mid \Gamma, \Pi^{\lambda-1}, A \Rightarrow \Sigma^{\lambda-1}, \Delta} \quad \text{or} \quad \frac{\mathcal{G}'_1 \quad \ldots \quad \mathcal{G}'_n}{\mathcal{H}' \mid \mathcal{G}'' \mid \Gamma, \Pi^{\lambda-1} \Rightarrow A, \Sigma^{\lambda-1}, \Delta}$$

is an instance of (r). Moreover, by the induction hypothesis, $\vdash_{\mathrm{GL}^\circ} \mathcal{G}'_i$ for $i = 1 \ldots n$, so we have a derivation d ending with such a rule application.

Now we consider two subcases:

(1) $\mathrm{e}(d_{\mathcal{H}}) = 1$: i.e. $d_{\mathcal{H}}$ does not end with the application of a logical rule to the marked occurrence of A. Mark the remaining occurrence of A on the left or right as appropriate in d and remove the underlining in $d_{\mathcal{H}}$. So $\mathrm{e}(d) = 0$ and:

$$\langle \mathrm{cp}(A), \mathrm{e}(d), \mathrm{ht}(d_{\mathcal{H}}) \rangle <_d \langle \mathrm{cp}(A), \mathrm{e}(d_{\mathcal{H}}), \mathrm{ht}(d_{\mathcal{G}}) \rangle$$

Hence by the induction hypothesis and a further application of (EC), $\vdash_{\mathrm{GL}^\circ} \mathcal{G}^{\mathcal{H}}$.

(2) $\mathrm{e}(d_{\mathcal{H}}) = 0$: i.e. $d_{\mathcal{H}}$ ends with the application of a logical rule to the marked occurrence of A, and is of the form:

$$\frac{\mathcal{H}_1 \quad \ldots \quad \mathcal{H}_m}{\mathcal{H}' \mid \Pi \Rightarrow A, \Sigma} \quad \text{or} \quad \frac{\mathcal{H}_1 \quad \ldots \quad \mathcal{H}_m}{\mathcal{H}' \mid \Pi, A \Rightarrow \Sigma}$$

By reductivity, $\mathcal{G}^{\mathcal{H}}$ is derivable from $\mathcal{G}'_1, \ldots, \mathcal{G}'_n, \mathcal{H}_1, \ldots, \mathcal{H}_m$ by cuts on subformulas $A_1, \ldots, A_k$ of A. But:

$$\langle \mathrm{cp}(A_i), \mathrm{e}(d_{\mathcal{H}}), \mathrm{ht}(d) \rangle <_d \langle \mathrm{cp}(A), \mathrm{e}(d_{\mathcal{H}}), \mathrm{ht}(d_{\mathcal{G}}) \rangle \quad \text{for } i = 1 \ldots k.$$

So by the induction hypothesis and a further application of (EC), $\vdash_{\mathrm{GL}^\circ} \mathcal{G}^{\mathcal{H}}$. □

5.1.3 Conservative Extensions

One immediate consequence of cut elimination for regular calculi is that cut-free derivations have the *subformula property*. Any formula occurring in such a derivation must occur as a subformula in the end-hypersequent. Or put another way, cut-free derivations do not make use of any "extra materials" to achieve their goal.

The subformula property is key for developing automated reasoning methods based on Gentzen systems. It is also useful in other ways. Notice for example that the empty sequent $\Rightarrow$ (which leads to inconsistency in logics with weakening) can, by the subformula property, only be derivable if it is an initial sequent of the calculus. More generally, a hypersequent $\mathcal{G}$ is derivable iff it is derivable when the logical rules are restricted to those for connectives occurring in $\mathcal{G}$.

Let us make this more precise. A system $\mathbb{C}_1$ for a set of structures $\mathcal{S}_1$ is a *conservative extension* of a system $\mathbb{C}_2$ for $\mathcal{S}_2 \subseteq \mathcal{S}_1$ if for all $X \in \mathcal{S}_2$: $\vdash_{\mathbb{C}_1} X$ iff $\vdash_{\mathbb{C}_2} X$. In this case, $\mathbb{C}_2$ is often called the *$\mathcal{S}_2$-fragment* of $\mathbb{C}_1$. Our first (easy) example will be fragments of regular Gentzen systems based on a reduced language. Let $\mathrm{GL}^{\mathcal{L}}$ be GL where the logical rules are restricted to the connectives in $\mathcal{L}$. Then by cut elimination for regular calculi and the subformula property:

Proposition 5.10. *Let* GL *be any regular extension of* GUL *and* $\mathcal{L}_1 \subseteq \mathcal{L}_2 \subseteq \mathcal{L}_F$. *Then* $\mathsf{GL}^{\mathcal{L}_2}$ *is a conservative extension of* $\mathsf{GL}^{\mathcal{L}_1}$.

However, this is cheating slightly. Although rules for $\vee$ may not be available in the restricted calculus, we still have plenty of rules for the external disjunction "|", namely, (EW), (EC), and (COM). It is therefore more interesting to ask if we can obtain conservative extension results for Hilbert systems, where only formulas are involved in derivations. Or to put this another way, can we find axiomatizations for fragments of Hilbert systems for fuzzy logics? Let us just consider a pertinent example: the implicational fragment of Monoidal t-norm Logic.

The Hilbert system BCK consists of the modus ponens rule (MP) and the axioms:

$$\begin{array}{lll} \text{(B)} & (A \to B) \to ((B \to C) \to (A \to C)) & \text{(transitivity)} \\ \text{(C)} & (A \to (B \to C)) \to (B \to (A \to C)) & \text{(permutation)} \\ \text{(K)} & A \to (B \to A) & \text{(weakening)} \end{array}$$

Then let $\mathsf{HMTL}^{\to}$ be BCK extended with the axiom schema:

$$\text{(IPRL)} \quad ((A \to B) \to C) \to (((B \to A) \to C) \to C)$$

Proposition 5.11. HMTL *is a conservative extension of* $\mathsf{HMTL}^{\to}$.

Proof. Suppose that $\vdash_{\mathsf{HMTL}} A$ for some implicational formula A. Then $\vdash_{\mathsf{GMTL}} \Rightarrow A$ and by Proposition 5.10, $\vdash_{\mathsf{GMTL}^{\to}} \Rightarrow A$. So it remains to show that $\vdash_{\mathsf{GMTL}^{\to}} \Rightarrow A$ implies $\vdash_{\mathsf{HMTL}^{\to}} A$. For this we need to eliminate mention of any connective other than $\to$ from our interpretation of hypersequents. Let us define:

$$\begin{aligned} \mathrm{I_q}(A_1,\ldots,A_n \Rightarrow C) &=_{\mathrm{def}} A_1 \to \ldots \to A_n \to C \\ \mathrm{I_q}(A_1,\ldots,A_n \Rightarrow) &=_{\mathrm{def}} A_1 \to \ldots \to A_n \to q \\ \mathrm{I_q}(S_1 \mid \ldots \mid S_n) &=_{\mathrm{def}} (\mathrm{I_q}(S_1) \to q) \to \ldots \to (\mathrm{I_q}(S_n) \to q) \to q \end{aligned}$$

where $A_1 \to \ldots \to A_n \to C$ is short for $A_1 \to (A_2 \to (\ldots \to (A_n \to C)\ldots)$. It is then enough to prove the following:

Claim. If $\vdash_{\mathsf{GMTL}^{\to}} \mathcal{G}$, then $\vdash_{\mathsf{HMTL}^{\to}} \mathrm{I_q}(\mathcal{G})$ for any q not occurring in $\mathcal{G}$.

Just notice that if $\vdash_{\mathsf{GMTL}^{\to}} \Rightarrow A$, then $\vdash_{\mathsf{HMTL}^{\to}} (A \to q) \to q$ for some q not occurring in A. So by Lemma 3.10, substituting A for q, $\vdash_{\mathsf{HMTL}^{\to}} (A \to A) \to A$. Hence $\vdash_{\mathsf{HMTL}^{\to}} A$.

Proof of claim. We proceed by induction on the height of a cut-free $\mathsf{GMTL}^{\to}$-derivation of $\mathcal{G}$. For the base case, $\mathcal{G}$ is of the form $\mathcal{G}' \mid C \Rightarrow C$ and we note that $\vdash_{\mathsf{HMTL}^{\to}} D \to (((C \to C) \to q) \to q)$ for any formula D as required. The inductive step involves a number of tedious Hilbert derivations. For rules of the form $(\mathcal{G} \mid S_1)\ldots(\mathcal{G} \mid S_n) \;/\; (\mathcal{G} \mid S)$, it is sufficient to show that:

$$\vdash_{\mathsf{HMTL}^{\to}} \mathrm{I_q}(S_1) \to \ldots \to \mathrm{I_q}(S_n) \to \mathrm{I_q}(S)$$

For example, in the case of ($\rightarrow\Rightarrow$), let $\Gamma_1 = [C_1, \ldots, C_k]$ and $D = \mathrm{I_q}(\Gamma_2 \Rightarrow \Delta)$. Then we can show that the following formula is derivable in $\mathsf{HMTL}^{\rightarrow}$:

$$(C_1 \rightarrow \ldots \rightarrow C_k \rightarrow A) \rightarrow (B \rightarrow D) \rightarrow (C_1 \rightarrow \ldots \rightarrow C_k \rightarrow (A \rightarrow B) \rightarrow D)$$

For cases not of this form, we proceed a little differently. E.g. for (EC), suppose that $\vdash_{\mathsf{GMTL}^{\rightarrow}} \mathcal{G} \mid S \mid S$. Then by the induction hypothesis, for suitable $D_1, \ldots, D_n$:

$$\vdash_{\mathsf{HMTL}^{\rightarrow}} (D_1 \rightarrow q) \rightarrow \ldots (D_n \rightarrow q) \rightarrow (\mathrm{I_q}(S) \rightarrow q) \rightarrow (\mathrm{I_q}(S) \rightarrow q) \rightarrow q$$

But then by Lemma 3.10, substituting $(D_1 \rightarrow q) \rightarrow \ldots (D_n \rightarrow q) \rightarrow (\mathrm{I_q}(S) \rightarrow q) \rightarrow q$ for q above, we easily get $\vdash_{\mathsf{HMTL}^{\rightarrow}} (D_1 \rightarrow q) \rightarrow \ldots (D_n \rightarrow q) \rightarrow (\mathrm{I_q}(S) \rightarrow q) \rightarrow q$ as required. □

To obtain axiomatizations for other fragments, we let $\mathsf{HMTL}^{\mathcal{L}}$ be $\mathsf{HMTL}^{\rightarrow}$ extended with the various axioms for $\odot$, $\wedge$, $\vee$, $\bot$, and $\top$ given in Chapter 3, only replacing ($\vee$3) with the schema (acceptable since we have weakening):

$$(\vee 3)' \ (A \rightarrow C) \rightarrow ((B \rightarrow C) \rightarrow ((A \vee B) \rightarrow C))$$

So, courtesy of a lot of tedious Hilbert-style derivations of the interpretations of logical rules (as sketched above), we end up with:

Proposition 5.12. *Let* $\{\rightarrow\} \subseteq \mathcal{L}_1 \subseteq \mathcal{L}_2 \subseteq \{\rightarrow, \odot, \wedge, \vee, \bot, \top\}$. *Then* $\mathsf{HMTL}^{\mathcal{L}_2}$ *is a conservative extension of* $\mathsf{HMTL}^{\mathcal{L}_1}$.

These results can be extended to Hilbert systems for other logics with weakening such as HG and HIMTL. However, in the case of weakening-free logics, we can no longer simulate the connective $\vee$ using $\rightarrow$ or the other multiplicative connectives. An interesting open question is therefore to provide an axiomatization (if one exists) of the implicational fragment of HUL. Could it be that HUL is a conservative extension of BCI? I.e. is the implicational fragment of HUL just the corresponding fragment of Linear Logic?

5.1.4 Decidability

For sequent calculi, cut elimination can be a key tool for establishing the decidability of the *validity problem* for a logic L: the problem of determining whether a formula of the appropriate language belongs to the set of L-valid formulas. As an easy case, consider any regular extension of GMAILL or GMALL with "non-expansive" structural rules, that is, having instances where the (multiset) complexity of each premise is strictly less than the complexity of the conclusion. Cut-free proof search proceeding by applying rules of the cut-free calculus backwards must terminate. So derivability in the calculus is decidable.

This argument fails in the presence of contraction rules. In such cases, premises can be bigger than the conclusion. Sometimes, e.g. for fragments of the relevance

Initial S-Hypersequents

$$\frac{}{\mathcal{G} \mid A \Rightarrow A}\ (\text{ID}) \qquad \frac{}{\mathcal{G} \mid \Rightarrow}\ (\text{EMP})$$

Structural Rules

$$\frac{\mathcal{G}' \mid \Pi_1 \Rightarrow \Sigma_1 \quad \mathcal{G}' \mid \Pi_2 \Rightarrow \Sigma_2}{\mathcal{G} \mid \Gamma_1 \Rightarrow \Delta_1 \mid \Gamma_2 \Rightarrow \Delta_2}\ (\text{COM})_s$$

where $\mathcal{G}' = (\mathcal{G} \mid \Gamma_1 \Rightarrow \Delta_1 \mid \Gamma_2 \Rightarrow \Delta_2)$; $\Gamma_1 \cup \Gamma_2 = \Pi_1 \cup \Pi_2$; and $\Delta_1 \cup \Delta_2 = \Sigma_1 \cup \Sigma_2$.

Logical Rules

$$\frac{\mathcal{G}' \mid \Gamma_1, B \Rightarrow \Delta_1 \quad \mathcal{G}' \mid \Gamma_2 \Rightarrow A, \Delta_2}{\mathcal{G} \mid \Gamma, A \to B \Rightarrow \Delta}\ (\to\Rightarrow)_s \qquad \frac{\mathcal{G} \mid \Gamma, A \Rightarrow B, \Delta}{\mathcal{G} \mid \Gamma \Rightarrow A \to B, \Delta}\ (\Rightarrow\to)$$

where $\Gamma_1 \cup \Gamma_2 \cup \{A \to B\} = \Gamma \cup \{A \to B\}$; $\Delta_1 \cup \Delta_2 = \Delta$; and $\mathcal{G}' = (\mathcal{G} \mid \Gamma, A \to B \Rightarrow \Delta)$.

Fig. 5.1 The hypersequent calculus $\mathsf{GIUML}_s^{\to}$

logic R, this can be dealt with by using restricted rules and some kind of loop-checking mechanism. For hypersequent calculi, however, the approach is further complicated by the presence of the external contraction rule (EC) that can duplicate whole sequents. Nevertheless, with both internal and external contraction rules, as e.g. for GG, GIUML, and GUML, we can again obtain terminating proof search fairly easily. In such cases, we can replace multisets by sets, and have only a finite number of possible structures occurring in a derivation.

Let us define *s-sequents* as expressions $\Gamma \Rightarrow \Delta$ where Γ and Δ are sets rather than hypersequents of formulas and *s-hypersequents* as sets of s-sequents (for familiarity, we retain the same notation). Now consider the set of formulas α occurring in some s-hypersequent. There is only a finite number of s-sequents containing subformulas of the formulas in α, and hence also only a finite number of s-hypersequents built from such s-sequents. E.g. let:

$$\mathcal{G} = (p \to q \Rightarrow q \to p \mid q \to p \Rightarrow q)$$

Then $\alpha = \{p \to q, q \to p, q\}$ and the number of different (multiple-conclusion) sequents with subsets of $\{p \to q, q \to p, p, q\}$ on the left and right is $2^4 \times 2^4 = 256$. So the number of different hypersequents is 2^{256}, a very big number indeed, but still finite.

What this means is that for any s-hypersequent calculus with the subformula property, applying the rules backwards can lead only to a finite number of possible s-hypersequents. So using loop-checking, e.g. checking that the premises of a rule instance have not already occurred lower in the derivation, we get:

Lemma 5.13. *Any s-hypersequent calculus with the subformula property is decidable.*

Hence to show decidability for a logic or proof system, it is sufficient to exhibit a corresponding (sound and complete) s-hypersequent calculus with the subformula property. In general we can do this for any regular hypersequent calculus with contraction rules. The idea is to build contraction into the other rules of the calculus. Consider for concreteness, Fig. 5.1, a cut-free s-hypersequent calculus $\mathsf{GIUML}_s^{\rightarrow}$ for the implicational fragment of IUML. Notice that we have removed (EW) as well as the internal and external contraction rules. Also, the (MIX) of GIUML is subsumed by the revised communication rule $(\text{COM})_s$.

This s-hypersequent calculus has the subformula property, so it is decidable. The question then is whether this calculus really characterizes the implicational fragment of IUML. To answer this affirmatively, we just have to show that $\vdash_{\mathsf{GIUML}} \mathcal{G}$ iff $\vdash_{\mathsf{GIUML}_s^{\rightarrow}} \mathcal{G}_s$ for any implicational hypersequent $\mathcal{G}$, where $\mathcal{G}_s$ is the s-hypersequent version of $\mathcal{G}$ obtained by removing duplications at the sequent level and then the hypersequent level. Both directions can be established by induction on the heights of a derivation. More generally, we can use this method to show that:

Proposition 5.14. *The validity problem for* UML, IUML, *and* G *is decidable.*

5.2 Cancellation Elimination

We now consider a case – the calculus GA for Abelian Logic – where cut elimination is better treated via an elimination of the "cancellation" rule (CAN). To see how the usual cut elimination procedure fails for this calculus, consider:

$$\dfrac{\dfrac{\dfrac{\vdots}{\Gamma, B \Rightarrow A, \Delta}}{\Gamma, A \rightarrow B \Rightarrow \Delta}(\rightarrow\Rightarrow)_{\mathsf{A}} \quad \dfrac{\dfrac{\vdots}{\Pi, A \Rightarrow B, \Sigma}}{\Pi \Rightarrow A \rightarrow B, \Sigma}(\Rightarrow\rightarrow)}{\Gamma, \Pi \Rightarrow \Sigma, \Delta}(\text{CUT})$$

The usual trick, applying (CUT) to the premises of $(\rightarrow\Rightarrow)_{\mathsf{A}}$ and $(\Rightarrow\rightarrow)$ with cut-formula A or B, gives $(\Gamma, \Pi, A \Rightarrow A, \Sigma, \Delta)$ or $(\Gamma, \Pi, B \Rightarrow B, \Sigma, \Delta)$, neither of much use here. Instead, let us rearrange this derivation a little using (CAN):

$$\dfrac{\dfrac{\dfrac{\dfrac{\vdots}{\Gamma, B \Rightarrow A, \Delta}}{\Gamma, A \rightarrow B \Rightarrow \Delta}(\rightarrow\Rightarrow)_{\mathsf{A}} \quad \dfrac{\dfrac{\vdots}{\Pi, A \Rightarrow B, \Sigma}}{\Pi \Rightarrow A \rightarrow B, \Sigma}(\Rightarrow\rightarrow)}{\Gamma, \Pi, A \rightarrow B \Rightarrow A \rightarrow B, \Sigma, \Delta}(\text{MIX})}{\Gamma, \Pi \Rightarrow \Sigma, \Delta}(\text{CAN})$$

Then making some further changes, we apply (CAN) to A and B rather than $A \rightarrow B$:

Initial Sequents

$$\frac{}{\mathcal{G} \mid A \Rightarrow A}\ (\text{ID}) \qquad \frac{}{\mathcal{G} \mid \Rightarrow}\ (\text{EMP})$$

Structural Rules

$$\frac{\mathcal{G}}{\mathcal{G} \mid \mathcal{H}}\ (\text{EW}) \qquad \frac{\mathcal{G} \mid \mathcal{H} \mid \mathcal{H}}{\mathcal{G} \mid \mathcal{H}}\ (\text{EC})$$

$$\frac{\mathcal{G} \mid \Gamma_1 \Rightarrow \Delta_1 \quad \mathcal{G} \mid \Gamma_2 \Rightarrow \Delta_2}{\mathcal{G} \mid \Gamma_1, \Gamma_2 \Rightarrow \Delta_1, \Delta_2}\ (\text{MIX}) \qquad \frac{\mathcal{G} \mid \Gamma_1, \Gamma_2 \Rightarrow \Delta_1, \Delta_2}{\mathcal{G} \mid \Gamma_1 \Rightarrow \Delta_1 \mid \Gamma_2 \Rightarrow \Delta_2}\ (\text{SPLIT})$$

Logical Rules

$$\frac{\mathcal{G} \mid \Gamma, B \Rightarrow A, \Delta}{\mathcal{G} \mid \Gamma, A \to B \Rightarrow \Delta}\ (\to\Rightarrow)_{\mathsf{A}} \qquad \frac{\mathcal{G} \mid \Gamma, A \Rightarrow B, \Delta}{\mathcal{G} \mid \Gamma \Rightarrow A \to B, \Delta}\ (\Rightarrow\to)$$

$$\frac{\mathcal{G} \mid \Gamma, A \Rightarrow \Delta \mid \Gamma, B \Rightarrow \Delta}{\mathcal{G} \mid \Gamma, A \wedge B \Rightarrow \Delta}\ (\wedge\Rightarrow) \qquad \frac{\mathcal{G} \mid \Gamma \Rightarrow A, \Delta \quad \mathcal{G} \mid \Gamma \Rightarrow B, \Delta}{\mathcal{G} \mid \Gamma \Rightarrow A \wedge B, \Delta}\ (\Rightarrow\wedge)$$

Fig. 5.2 The hypersequent calculus GA*

$$\cfrac{\cfrac{\cfrac{\cfrac{\vdots}{\Gamma, B \Rightarrow A, \Delta} \quad \cfrac{\vdots}{\Pi, A \Rightarrow B, \Sigma}}{\Gamma, \Pi, A, B \Rightarrow A, B, \Sigma, \Delta}\ (\text{MIX})}{\Gamma, \Pi, A \Rightarrow A, \Sigma, \Delta}\ (\text{CAN})}{\Gamma, \Pi \Rightarrow \Sigma, \Delta}\ (\text{CAN})$$

The idea of cancellation elimination (as with cuts) is first to reduce all applications of (CAN) to variables, and then to perform "atomic cancellation elimination". Since applications of (CUT) in GA can be replaced by applications of (CAN) (see Section 4.3.5), this also provides a cut elimination procedure for GA.

5.2.1 *The Proof*

For convenience (saving a lot of repetition), we will use a restricted but still fully expressive language $\mathcal{L}_{\mathsf{A}} = \{\to, \wedge\}$ for Abelian Logic, defining:

$$\begin{aligned} \mathrm{e} &=_{\text{def}} q \to q & \mathrm{f} &=_{\text{def}} \mathrm{e} \\ \neg A &=_{\text{def}} A \to \mathrm{f} & A \vee B &=_{\text{def}} \neg(\neg A \wedge \neg B) \\ A \odot B &=_{\text{def}} -A \to B & A \oplus B &=_{\text{def}} A \odot B \end{aligned}$$

A hypersequent calculus GA* for $\mathcal{L}_{\mathsf{A}}$ is given in Fig. 5.2: essentially the appropriate cut-free fragment of GA with $(\wedge\Rightarrow)_1$ and $(\wedge\Rightarrow)_2$ replaced by $(\wedge\Rightarrow)$. We will establish cancellation elimination for GA* + (CAN). Since the rules for the extra connectives of GA are GA*-derivable, we also get cancellation elimination for GA.

Our proof proceeds in a similar vein to cut elimination, but with one premise rather than two. Just as we defined $\mathrm{CUT}(\mathcal{G},\mathcal{H})$ to capture the result of applications of (CUT) with a marked cut-formula A, so we define $\mathrm{CAN}(\mathcal{G},A)$ to capture the result of applications of (CAN) with a cancellation-formula A.

Definition 5.15. $\mathrm{CAN}(\mathcal{G},A)$ is the smallest set such that:

(1) $\mathcal{G} \in \mathrm{CAN}(\mathcal{G},A)$.
(2) $(\mathcal{H} \mid \Gamma \Rightarrow \Delta) \in \mathrm{CAN}(\mathcal{G},A)$ if $(\mathcal{H} \mid \Gamma, A \Rightarrow A, \Delta) \in \mathrm{CAN}(\mathcal{G},A)$.

The set $\mathrm{CAN}(\mathcal{G},A)$ is easy to visualize. Suppose that we have a hypersequent:

$$\mathcal{G} = (\Gamma_1, [A]^{\lambda_1} \Rightarrow [A]^{\lambda_1}, \Delta_1 \mid \ldots \mid \Gamma_n, [A]^{\lambda_n} \Rightarrow [A]^{\lambda_n}, \Delta_n)$$

where $A \notin \Gamma_i$ or $A \notin \Delta_i$ for $i = 1 \ldots n$. Then, letting $0 \leq \mu_i \leq \lambda_i$ for $i = 1 \ldots n$, $\mathrm{CAN}(\mathcal{G},A)$ consists of all hypersequents of the form:

$$\Gamma_1, [A]^{\lambda_1-\mu_1} \Rightarrow [A]^{\lambda_1-\mu_1}, \Delta_1 \mid \ldots \mid \Gamma_n, [A]^{\lambda_n-\mu_n} \Rightarrow [A]^{\lambda_n-\mu_n}, \Delta_n$$

We begin our proof with a kind of cut elimination for variables.

Lemma 5.16. *Suppose that:*

(1) $d \vdash_{\mathsf{GA}^*} \Gamma_1, [q]^{\lambda_1} \Rightarrow \Delta_1 \mid \ldots \mid \Gamma_n, [q]^{\lambda_n} \Rightarrow \Delta_n$.
(2) $\vdash_{\mathsf{GA}^*} \mathcal{G}_i \mid \Pi_i \Rightarrow [q]^{\lambda_i}, \Sigma_i$ *for* $i = 1 \ldots n$.

Then $\vdash_{\mathsf{GA}^*} \mathcal{G}_1 \mid \ldots \mid \mathcal{G}_n \mid \Gamma_1, \Pi_1 \Rightarrow \Sigma_1, \Delta_1 \mid \ldots \mid \Gamma_n, \Pi_n \Rightarrow \Sigma_n, \Delta_n$.

Proof. If $\lambda_1 = \ldots = \lambda_n = 0$, then the result follows easily using (MIX) and (EW), so assume without loss of generality that $\lambda_1 \geq 1$. We proceed by induction on $\mathrm{ht}(d)$. For $\mathrm{ht}(d) = 1$, the only possible case is $(\mathcal{G}' \mid q \Rightarrow q)$, an instance of (ID). So we obtain the result by applying (EW). For $\mathrm{ht}(d) > 1$, cases for the logical rules, (EW), and (EC) all involve straightforward applications of the induction hypothesis and the corresponding rule.

The remaining structural rules are another matter. Suppose that d ends with:

$$\dfrac{\dfrac{\vdots\, d_1}{[\Gamma_i, [q]^{\lambda_i} \Rightarrow \Delta_i]_{i=2}^n \mid \Gamma_1', [q]^{\lambda_1'} \Rightarrow \Delta_1'} \quad \dfrac{\vdots\, d_2}{[\Gamma_i, [q]^{\lambda_i} \Rightarrow \Delta_i]_{i=2}^n \mid \Gamma_1'', [q]^{\lambda_1''} \Rightarrow \Delta_1''}}{[\Gamma_i, [q]^{\lambda_i} \Rightarrow \Delta_i]_{i=2}^n \mid \Gamma_1, [q]^{\lambda_1} \Rightarrow \Delta_1}\ (\mathrm{MIX})$$

where $\Gamma_1 = \Gamma_1' \uplus \Gamma_1''$, $\Delta_1 = \Delta_1' \uplus \Delta_1''$, and $\lambda_1 = \lambda_1' + \lambda_1''$. Since $\mathrm{ht}(d_1) < \mathrm{ht}(d)$, by the induction hypothesis applied to the left premise:

$$\vdash_{\mathsf{GA}^*} \mathcal{G}_1 \mid \ldots \mid \mathcal{G}_n \mid [\Gamma_i, \Pi_i \Rightarrow \Sigma_i, \Delta_i]_{i=2}^n \mid \Gamma_1', \Pi_1 \Rightarrow [q]^{\lambda_1''}, \Sigma_1, \Delta_1'$$

But then since $\mathrm{ht}(d_2) < \mathrm{ht}(d)$, we can apply the induction hypothesis again (replacing $(\mathcal{G}_1 \mid \Pi_1 \Rightarrow [q]^{\lambda_1}, \Sigma_1)$ in (2) with the above hypersequent), and the result follows using (EC).

Now suppose that d ends with:

$$\dfrac{\dfrac{\vdots\, d'}{[\Gamma_i,[q]^{\lambda_i} \Rightarrow \Delta_i]_{i=3}^n \mid \Gamma_1,\Gamma_2,[q]^{\lambda_1+\lambda_2} \Rightarrow \Delta_1,\Delta_2}}{[\Gamma_i,[q]^{\lambda_i} \Rightarrow \Delta_i]_{i=3}^n \mid \Gamma_1,[q]^{\lambda_1} \Rightarrow \Delta_1 \mid \Gamma_2,[q]^{\lambda_2} \Rightarrow \Delta_2}\ (\textsc{split})$$

Since $\vdash_{\mathsf{GA}^*} \mathcal{G}_i \mid \Pi_i \Rightarrow [q]^{\lambda_i},\Sigma_i$ for $i = 1,2$, by (MIX) and (EW):

$$\vdash_{\mathsf{GA}^*} \mathcal{G}_1 \mid \mathcal{G}_2 \mid \Pi_1,\Pi_2 \Rightarrow [q]^{\lambda_1+\lambda_2},\Sigma_1,\Sigma_2$$

Hence, since $\mathrm{ht}(d') < \mathrm{ht}(d)$, by the induction hypothesis:

$$\vdash_{\mathsf{GA}^*} \mathcal{G}_1 \mid \ldots \mid \mathcal{G}_n \mid [\Gamma_i,\Pi_i \Rightarrow \Sigma_i,\Delta_i]_{i=3}^n \mid \Gamma_1,\Gamma_2,\Pi_1,\Pi_2 \Rightarrow \Sigma_1,\Sigma_2,\Delta_1,\Delta_2$$

and the result follows by (SPLIT). □

"Atomic cancellation elimination" can now be established as follows:

Lemma 5.17. *If $d \vdash_{\mathsf{GA}^*} \mathcal{G}$, then $\vdash_{\mathsf{GA}^*} \mathcal{H}$ for all $\mathcal{H} \in \mathrm{CAN}(\mathcal{G},q)$.*

Proof. We proceed by induction on $\mathrm{ht}(d)$. The base cases are easy, $\mathcal{G}$ is $(\mathcal{G}' \mid q \Rightarrow q)$, $(\mathcal{G}' \mid C \Rightarrow C)$, or $(\mathcal{G}' \mid\Rightarrow)$, and we can apply again (EMP) or (ID). For the cases where the last rule applied is (EC), (EW), (SPLIT), or one of the logical rules, the result follows by the induction hypothesis and an application of the corresponding rule (or if you like, the substitutivity of that rule).

The only tricky case is when d is of the form:

$$\dfrac{\dfrac{\vdots\, d_1}{\mathcal{H} \mid \Gamma_1,[q]^k \Rightarrow [q]^m,\Delta_1} \quad \dfrac{\vdots\, d_2}{\mathcal{H} \mid \Gamma_2,[q]^{n-k} \Rightarrow [q]^{n-m},\Delta_2}}{\mathcal{H} \mid \Gamma_1,\Gamma_2,[q]^n \Rightarrow [q]^n,\Delta_1,\Delta_2}\ (\textsc{mix})$$

and the member of $\mathrm{CAN}(\mathcal{G},q)$ is $(\mathcal{H}' \mid \Gamma_1,\Gamma_2 \Rightarrow \Delta_1,\Delta_2)$ for some $\mathcal{H}' \in \mathrm{CAN}(\mathcal{H},q)$. Without loss of generality, let $k \leq m$. Then by the induction hypothesis twice:

$$\vdash_{\mathsf{GA}^*} \mathcal{H}' \mid \Gamma_1 \Rightarrow [q]^{m-k},\Delta_1 \quad \text{and} \quad \vdash_{\mathsf{GA}^*} \mathcal{H}' \mid \Gamma_2,[q]^{m-k} \Rightarrow \Delta_2$$

So the result follows immediately by Lemma 5.16 and (EC). □

We now turn our attention to cancelling complex formulas. In GA* the logical rules are invertible on both sides of the sequent arrow. This means that we can push applications of (CAN) to complex formulas upwards in the derivation to applications on subformulas and ultimately to variables, where as we have just seen, they can be eliminated.

Lemma 5.18. *The rules $(\rightarrow\Rightarrow)_{\mathsf{A}}$, $(\Rightarrow\rightarrow)$, $(\wedge\Rightarrow)$, and $(\Rightarrow\wedge)$ are GA*-invertible.*

Proof. To cope with multiple occurrences of formulas, we will need to show the invertibility of more general rules. As it is the hardest case, let us just consider the following more general version of $(\wedge\Rightarrow)$:

$$\frac{[\Gamma_i, [A]^{\lambda_i} \Rightarrow \Delta_i \mid \Gamma_i, [B]^{\lambda_i} \Rightarrow \Delta_i]_{i=1}^n}{[\Gamma_i, [A \wedge B]^{\lambda_i} \Rightarrow \Delta_i]_{i=1}^n}$$

We prove that this rule is GA*-invertible, proceeding by induction on the height of a derivation d of the conclusion. If $\lambda_1 = \ldots = \lambda_n = 0$, then the result follows immediately using (EC), so let us assume without loss of generality that $\lambda_1 \geq 1$. Then for the base case, we have $\vdash_{\mathsf{GA}^*} \mathcal{G}' \mid A \wedge B \Rightarrow A \wedge B$. But $(A \Rightarrow A \wedge B \mid B \Rightarrow A \wedge B)$ is GA*-derivable (see Example 4.35) so the result holds using (EW). For the inductive step, the cases of (EW), (EC), (SPLIT), $(\rightarrow\Rightarrow)_{\mathsf{A}}$, and $(\Rightarrow\rightarrow)$ follow easily by an application of the induction hypothesis and the relevant rule. We consider the remaining tricky cases below.

Suppose first that d ends with:

$$\frac{\mathcal{G} \mid \Pi_1, [A \wedge B]^{\mu_1} \Rightarrow \Sigma_1 \quad \mathcal{G} \mid \Pi_2, [A \wedge B]^{\mu_2} \Rightarrow \Sigma_2}{\mathcal{G} \mid \Gamma_1, [A \wedge B]^{\lambda_1} \Rightarrow \Delta_1} \ (\text{MIX})$$

where $\mathcal{G} = [\Gamma_i, [A \wedge B]^{\lambda_i} \Rightarrow \Delta_i]_{i=2}^n$, $\lambda_1 = \mu_1 + \mu_2$, $\Gamma_1 = \Pi_1 \uplus \Pi_2$, and $\Delta_1 = \Sigma_1 \uplus \Sigma_2$. By the induction hypothesis twice:

$$\vdash_{\mathsf{GA}^*} \mathcal{G}' \mid \Pi_j, [A]^{\mu_j} \Rightarrow \Sigma_j \mid \Pi_j, [B]^{\mu_j} \Rightarrow \Sigma_j \quad \text{for } j = 1, 2$$

where $\mathcal{G}' = [\Gamma_i, [A]^{\lambda_i} \Rightarrow \Delta_i \mid \Gamma_i, [B]^{\lambda_i} \Rightarrow \Delta_i]_{i=2}^n$. So using (MIX), (SPLIT), and (EC):

$$\vdash_{\mathsf{GA}^*} \mathcal{G}' \mid \Pi_1, [A]^{\mu_1} \Rightarrow \Sigma_1 \mid \Pi_2, [B]^{\mu_2} \Rightarrow \Sigma_2$$

$$\vdash_{\mathsf{GA}^*} \mathcal{G}' \mid \Pi_2, [A]^{\mu_2} \Rightarrow \Sigma_2 \mid \Pi_1, [B]^{\mu_1} \Rightarrow \Sigma_1$$

But then using (MIX), (SPLIT), and (EC) again, as required:

$$\vdash_{\mathsf{GA}^*} \mathcal{G}' \mid \Pi_1, \Pi_2, [A]^{\mu_1+\mu_2} \Rightarrow \Sigma_1, \Sigma_2 \mid \Pi_1, \Pi_2, [B]^{\mu_1+\mu_2} \Rightarrow \Sigma_1, \Sigma_2$$

Suppose now that d ends with:

$$\frac{\mathcal{G} \mid \Gamma_1, [A \wedge B]^{\lambda_1} \Rightarrow C_1, \Delta_1' \quad \mathcal{G} \mid \Gamma_1, [A \wedge B]^{\lambda_1} \Rightarrow C_2, \Delta_1'}{\mathcal{G} \mid \Gamma_1, [A \wedge B]^{\lambda_1} \Rightarrow C_1 \wedge C_2, \Delta_1'} \ (\Rightarrow\wedge)$$

where $\Delta_1 = \Delta_1' \uplus [C_1 \wedge C_2]$ and $\mathcal{G} = [\Gamma_i, [A \wedge B]^{\lambda_i} \Rightarrow \Delta_i]_{i=2}^n$. By the induction hypothesis twice:

$$\vdash_{\mathsf{GA}^*} \mathcal{G}' \mid \Gamma_1, [A]^{\lambda_1} \Rightarrow C_j, \Delta_1' \mid \Gamma_1, [B]^{\lambda_1} \Rightarrow C_j, \Delta_1' \quad \text{for } j = 1, 2$$

where $\mathcal{G}' = [\Gamma_i, [A]^{\lambda_i} \Rightarrow \Delta_i \mid \Gamma_i, [B]^{\lambda_i} \Rightarrow \Delta_i]_{i=2}^n$. So using (MIX) and (SPLIT):

$$\vdash_{\mathsf{GA}^*} \mathcal{G}' \mid \Gamma_1, [A]^{\lambda_1} \Rightarrow C_1, \Delta_1' \mid \Gamma_1, [B]^{\lambda_1} \Rightarrow C_2, \Delta_1'$$

But then using $(\Rightarrow\wedge)$, we have as required:

$$\vdash_{\mathsf{GA}^*} \mathcal{G}' \mid \Gamma_1, [A]^{\lambda_1} \Rightarrow C_1 \wedge C_2, \Delta_1' \mid \Gamma_1, [B]^{\lambda_1} \Rightarrow C_1 \wedge C_2, \Delta_1'$$

Finally, suppose that d ends with an application of $(\wedge\Rightarrow)$. If the principal formula is not $A \wedge B$, then the result follows easily using the induction hypothesis. Otherwise, we have an application of the form:

$$\frac{\mathcal{G} \mid \Gamma_1, [A \wedge B]^{\lambda_1 - 1}, A \Rightarrow \Delta_1 \mid \Gamma_1, [A \wedge B]^{\lambda_1 - 1}, B \Rightarrow \Delta_1}{\mathcal{G} \mid \Gamma_1, [A \wedge B]^{\lambda_1} \Rightarrow \Delta_1} \; (\wedge\Rightarrow)$$

By the induction hypothesis:

$$\vdash_{\mathsf{GA}^*} \mathcal{G}' \mid \Gamma_1, [A]^{\lambda_1} \Rightarrow \Delta_1 \mid \Gamma_1, [B]^{\lambda_1} \Rightarrow \Delta_1 \mid \Gamma_1, [A]^{\lambda_1 - 1}, B \Rightarrow \Delta_1 \mid \Gamma_1, [B]^{\lambda_1 - 1}, A \Rightarrow \Delta_1$$

Then by multiple applications of (EC), (MIX), and (SPLIT) as required:

$$\vdash_{\mathsf{GA}^*} \mathcal{G}' \mid \Gamma_1, [A]^{\lambda_1} \Rightarrow \Delta_1 \mid \Gamma_1, [B]^{\lambda_1} \Rightarrow \Delta_1$$

□

Theorem 5.19. $\mathsf{GA}^* + $(CAN) *admits cancellation elimination.*

Proof. It is enough to show that an uppermost application of (CAN) in a derivation can be eliminated: then we can eliminate all applications one by one.

Claim. If $\vdash_{\mathsf{GA}^*} \mathcal{G} \mid \Gamma, A \Rightarrow A, \Delta$, then $\vdash_{\mathsf{GA}^*} \mathcal{G} \mid \Gamma \Rightarrow \Delta$.

Proof of claim. We proceed by induction on $\mathrm{cp}(A)$. The base case where A is a variable q follows immediately from Lemma 5.17. If A is $B \to C$, then by the invertibility of $(\to\Rightarrow)_{\mathsf{A}}$ and $(\Rightarrow\to)$ (Lemma 5.18):

$$\vdash_{\mathsf{GA}^*} \mathcal{G} \mid \Gamma, C, B \Rightarrow B, C, \Delta$$

So by the induction hypothesis twice, $\vdash_{\mathsf{GA}^*} \mathcal{G} \mid \Gamma \Rightarrow \Delta$ as required.

If A is $B \wedge C$, then by the invertibility of $(\wedge\Rightarrow)$ (Lemma 5.18):

$$\vdash_{\mathsf{GA}^*} \mathcal{G} \mid \Gamma, B \Rightarrow B \wedge C, \Delta \mid \Gamma, C \Rightarrow B \wedge C, \Delta$$

Using the invertibility of $(\Rightarrow\wedge)$ twice (Lemma 5.18):

$$\vdash_{\mathsf{GA}^*} \mathcal{G} \mid \Gamma, B \Rightarrow B, \Delta \mid \Gamma, C \Rightarrow C, \Delta$$

So by the induction hypothesis twice and (EC), $\vdash_{\mathsf{GA}^*} \mathcal{G} \mid \Gamma \Rightarrow \Delta$ as required. □

As mentioned above, cancellation elimination for $\mathsf{GA}^* + $(CAN) implies cut elimination for GA. Suppose that a hypersequent $\mathcal{G}$ is GA-derivable, assuming harmlessly that $\mathcal{G}$ contains only connectives of the restricted language $\mathcal{L}_{\mathsf{A}} = \{\to, \wedge\}$. Then as shown in Section 4.3.5, we can use (CAN) to remove all applications of (CUT) from the derivation to obtain a derivation in $\mathsf{GA}^* + (\text{CAN})$. But then by cancellation elimination, we obtain a cancellation-free derivation of $\mathcal{G}$ in GA^* and hence also a cut-free derivation in GA.

Corollary 5.20. GA *admits cut elimination.*

5.2.2 Abelian ℓ-Groups

Abelian Logic A is rather esoteric, claiming such bizarre theorems as $((A \to B) \to B) \to A$ that fail even in Classical Logic. From an algebraic perspective, however, A is pretty important. Each A-algebra is term equivalent to a *lattice-ordered abelian group*, or for short, *abelian ℓ-group*. That is, given any A-algebra $\mathbf{A} = \langle L, \wedge, \vee, \odot, \to, \mathrm{f}, \mathrm{e}\rangle$, recall that $\neg x =_{\mathrm{def}} x \to \mathrm{f}$, and consider:

$$\mathbf{G} = \langle L, \wedge, \vee, \odot, \neg, \mathrm{e}\rangle$$

$\langle L, \wedge, \vee\rangle$ is a lattice, $\odot$ is order preserving ($x \leq y$ implies $x \odot z \leq y \odot z$), and $\langle L, \odot, \neg, \mathrm{e}\rangle$ is a commutative (or abelian) group, so $\mathbf{G}$ is an abelian ℓ-group. Conversely, the extra operations of $\mathbf{A}$ can be expressed in abelian ℓ-groups via the identities $x \to y = \neg x \odot y$ and $\mathrm{f} = \mathrm{e}$.

We will show here that we can use our calculus GA^* to give "algorithmic" proofs of some fundamental results for the variety of A-algebras, and hence also for abelian ℓ-groups. First, we provide a completeness result for GA^* with an important algebraic corollary. Consider the following pcrls based on the integers and real numbers respectively:

$$\mathbf{Z} = \langle \mathbb{Z}, \min, \max, +, \to_+, 0, 0\rangle \quad \text{and} \quad \mathbf{R} = \langle \mathbb{R}, \min, \max, +, \to_+, 0, 0\rangle$$

where $x \to_+ y = y - x$. We will show that $\mathbf{Z}$ and hence also $\mathbf{R}$ (since any equation failing in the former, fails in the latter) generates the variety of A-algebras.

Proposition 5.21. *If* $\models_{\mathbf{Z}} \mathrm{I}(\mathcal{G})$, *then* $\vdash_{\mathsf{GA}^*} \mathcal{G}$.

Proof. It is easily checked using regular arithmetic that the logical rules of GA^* are all $\mathbf{Z}$-invertible. Also, the multiset complexity of each premise of these rules is strictly less (according to the multiset ordering) than that of the conclusion. Hence for any hypersequent $\mathcal{G}$, there exist strictly atomic hypersequents $\mathcal{G}_1 \ldots \mathcal{G}_n$ such that $\mathcal{G}_1, \ldots, \mathcal{G}_n \vdash_{\mathsf{GA}^*} \mathcal{G}$ and $\mathcal{G}$ is $\mathbf{Z}$-valid iff $\mathcal{G}_1, \ldots, \mathcal{G}_n$ are all $\mathbf{Z}$-valid.

Let us assume then that $\mathcal{G}$ is strictly atomic. We prove the proposition by induction on the number of different variables occurring in $\mathcal{G}$. For the base case, $\mathcal{G}$ consists only of empty sequents and is therefore derivable using (EW) and (EMP). For the induction step, fix a variable q, and observe that the following rules are both $\mathbf{Z}$-invertible and GA^*-derivable:

$$\frac{\mathcal{H} \mid \Gamma \Rightarrow \Delta}{\mathcal{H} \mid \Gamma, q \Rightarrow q, \Delta} \qquad \frac{\mathcal{H} \mid \Gamma^n \Rightarrow \Delta^n}{\mathcal{H} \mid \Gamma \Rightarrow \Delta}$$

So we can assume further that either q does not occur in a sequent of $\mathcal{G}$, or occurs exactly λ times only on either the left or the right; i.e. $\mathcal{G}$ is of the form:

$$\mathcal{G}' \mid [\Gamma_i, [q]^\lambda \Rightarrow \Delta_i]_{i=1}^n \mid [\Pi_j \Rightarrow [q]^\lambda, \Sigma_j]_{j=1}^m$$

where q does not occur in $\mathcal{G}'$ or Γ_i, Δ_i, Π_j, Σ_j for $i = 1 \ldots n$ and $j = 1 \ldots m$. We now define the hypersequent:

$$\mathcal{H} = \mathcal{G}' \mid [\Gamma_i, \Pi_j \Rightarrow \Sigma_j, \Delta_i]_{i=1\ldots n}^{j=1\ldots m}$$

It is not hard to see that $\mathcal{G}$ is GA*-derivable from $\mathcal{H}$ using the structural rules and initial sequents. Working backwards, we use (EC) and (SPLIT) to combine each pair $\Gamma_i, [q]^\lambda \Rightarrow \Delta_i$ and $\Pi_j \Rightarrow [q]^\lambda, \Sigma_j$ into one sequent, then remove the qs using (MIX) and (ID). Notice also that $\mathcal{H}$ contains fewer distinct variables than $\mathcal{G}$. But this means that it is enough to show $\models_{\mathbf{Z}} \mathrm{I}(\mathcal{H})$, since then by the induction hypothesis $\vdash_{\mathsf{GA}^*} \mathcal{H}$ and so $\vdash_{\mathsf{GA}^*} \mathcal{G}$.

Suppose for a contradiction that there exists a **Z**-valuation v such that $v(\mathrm{I}(\mathcal{H})) < 0$. Easily, $v'(\mathrm{I}(\mathcal{H})) < 0$ for any v' defined by $v'(p) = \mu v(p)$ for $\mu > 0$. So we can assume that every value $v(p)$ is divisible by 2λ. We let:

$$x = \max\{\Sigma_{A \in \Delta_i} v(A) - \Sigma_{B \in \Gamma_i} v(B) : 1 \leq i \leq n\}$$
$$y = \min\{\Sigma_{A \in \Pi_j} v(A) - \Sigma_{B \in \Sigma_j} v(B) : 1 \leq j \leq m\}$$

We claim that $x < y$. Otherwise for some i, j:

$$\sum_{B \in \Gamma_i} v(B) + \sum_{A \in \Pi_j} v(A) \leq \sum_{B \in \Sigma_j} v(B) + \sum_{A \in \Delta_i} v(A)$$

But this contradicts the fact that $v(\mathrm{I}(\mathcal{H})) < 0$. Hence we can alter v so that $x < \lambda v(q) < y$ (recalling that q does not occur in $\mathcal{H}$). It then follows that for $i = 1 \ldots n$ and $j = 1 \ldots m$:

$$\sum_{B \in \Gamma_i} v(B) + \lambda v(q) > \sum_{A \in \Delta_i} v(A) \quad \text{and} \quad \sum_{A \in \Pi_j} v(A) > \lambda v(q) + \sum_{B \in \Sigma_j} v(B)$$

So $v(\mathrm{I}(\mathcal{G})) < 0$ which contradicts $\models_{\mathbf{Z}} \mathrm{I}(\mathcal{G})$ as required. □

Combining this result and the soundness and completeness of GA* with respect to HA and **A**-algebras, we get the following completeness theorem.

Theorem 5.22. $\models_{\mathbf{R}} A$ *iff* $\models_{\mathbf{Z}} A$ *iff* $\models_{\mathsf{A}} A$ *iff* $\vdash_{\mathsf{HA}} A$.

Moreover, since an equation holds in all abelian ℓ-groups iff it holds in all A-algebras, we obtain also:

Corollary 5.23. *The variety of abelian ℓ-groups is generated by* **Z**.

Finally, notice that the procedure described in Proposition 5.21 provides an *algorithm* for deciding whether or not a formula is valid in A. We apply the invertible logical rules to reduce the problem to strictly atomic hypersequents, then eliminate occurrences of variables one by one.

Corollary 5.24. *The equational theory of abelian ℓ-groups is decidable.*

5.3 Density Elimination

In Chapter 3, we showed that adding a special density rule to any HUL$^-$-extension guarantees completeness with respect to dense chains. Here we will show that in many cases this rule is unnecessary. That is, we can eliminate applications of (DENSITY) from derivations in the corresponding hypersequent calculus.

Our method proceeds – like cut elimination – by removing applications which are uppermost in a derivation. Suppose that we have a derivation d ending:

$$\dfrac{\dfrac{\vdots}{\Gamma \Rightarrow p \mid \Pi, p \Rightarrow \Sigma}}{\Gamma, \Pi \Rightarrow \Sigma}\ (\text{DENSITY})$$

The idea of our proof is to replace occurrences of p in d in an "asymmetric" way: with Γ if p occurs on the left, and with Π on the left and Σ on the right, if p occurs on the right. What we get is not quite a derivation, but still a finite tree labelled with hypersequents, now ending:

$$\dfrac{\dfrac{\vdots}{\Gamma, \Pi \Rightarrow \Sigma \mid \Pi, \Gamma \Rightarrow \Sigma}}{\Gamma, \Pi \Rightarrow \Sigma}$$

The last step in this not-quite-a-derivation is an application of (EC). Moreover, the applications of logical rules and most structural rules in the original derivation are preserved by substitutivity. Where the derivation potentially breaks down is in rules like (COM) where ps can occur in premises on both the left and the right. For example, suppose that d ends with:

$$\dfrac{\dfrac{\dfrac{\dfrac{}{p \Rightarrow p}\ (\text{ID}) \quad \dfrac{\vdots}{\Gamma', \Pi \Rightarrow \Sigma}}{\Gamma' \Rightarrow p \mid \Pi, p \Rightarrow \Sigma}\ (\text{COM})}{\vdots}}{\dfrac{\Gamma \Rightarrow p \mid \Pi, p \Rightarrow \Sigma}{\Gamma, \Pi \Rightarrow \Sigma}\ (\text{DENSITY})}$$

If we replace ps as before, we get:

$$\dfrac{\dfrac{\Gamma,\Pi \Rightarrow \Sigma \quad \dfrac{\vdots}{\Gamma',\Pi \Rightarrow \Sigma}}{\Gamma',\Pi \Rightarrow \Sigma \mid \Gamma,\Pi \Rightarrow \Sigma}\ (\text{COM})}{\vdots}$$

$$\dfrac{\Gamma,\Pi \Rightarrow \Sigma \mid \Gamma,\Pi \Rightarrow \Sigma}{\Gamma,\Pi \Rightarrow \Sigma}\ (\text{EC})$$

But now we have a missing part of the derivation: the sub-derivation of $(\Gamma,\Pi \Rightarrow \Sigma)$, which was what we wanted to prove in the first place. However, notice that in this case, if we remove the branch ending with $(\Gamma,\Pi \Rightarrow \Sigma)$, then we can replace the application of (COM) with an application of (EW). More generally, we are able to use (CUT) and cut elimination to repair such derivations.

5.3.1 Calculi with Weakening

We begin our investigations with hypersequent calculi admitting the weakening rule (W). In fact, for single-conclusion regular calculi with weakening, we will obtain a very nice result: *all* such calculi admit density elimination. Let us assume then for now that we are dealing only with single-conclusion hypersequents. In this case, we deal in particular with hypersequents where the variable p occurs only in a limited fashion: not in complex formulas and not on both the left and right in the same sequent.

Definition 5.25. A hypersequent $\mathcal{G}$ is *p-regular* if it is of the form:

$$\Gamma_1 \Rightarrow p \mid \ldots \mid \Gamma_n \Rightarrow p \mid \Pi_1, [p]^{\lambda_1} \Rightarrow \Sigma_1 \mid \ldots \mid \Pi_m, [p]^{\lambda_m} \Rightarrow \Sigma_m$$

where p does not occur in $\Gamma_1,\ldots,\Gamma_n,\Pi_1,\ldots,\Pi_m,\Sigma_1,\ldots,\Sigma_m$.

We will also need a way of distinguishing the special occurrences of the variable p introduced by the density rule.

Definition 5.26. A *double-p-marked hypersequent* has two occurrences of p, both marked: one on the left in a sequent and one on the right in another sequent, written:

$$\mathcal{H} \mid \Gamma \Rightarrow \underline{p} \mid \Pi, \underline{p} \Rightarrow \Sigma$$

We will combine a p-regular hypersequent $\mathcal{G}$ with a double-p-marked hypersequent $(\mathcal{H} \mid \Gamma \Rightarrow \underline{p} \mid \Pi, \underline{p} \Rightarrow \Sigma)$ in a special way, essentially by applying (CUT) exhaustively to $(\mathcal{G} \mid \mathcal{H})$ and the sequents $(\Gamma \Rightarrow \underline{p})$ and $(\Pi, \underline{p} \Rightarrow \Sigma)$ with cut-formula p.

Definition 5.27. Suppose that:

1. $\mathcal{G} = [\Gamma_i \Rightarrow p]_{i=1}^{n} \mid [\Pi_j, [p]^{\lambda_j} \Rightarrow \Sigma_j]_{j=1}^{m}$ is a p-regular hypersequent.

2. $\mathcal{H}_p = (\mathcal{H} \mid \Gamma \Rightarrow \underline{p} \mid \Pi, \underline{p} \Rightarrow \Sigma)$ is a double-p-marked hypersequent.

Then $\mathrm{DEN}(\mathcal{G},\mathcal{H}_p) = (\mathcal{H} \mid [\Gamma_i, \Pi \Rightarrow \Sigma]_{i=1}^{n} \mid [\Pi_j, \Gamma^{\lambda_j} \Rightarrow \Sigma_j]_{j=1}^{m})$.

Example 5.28. Consider the p-regular and double-p-marked hypersequents:

$$\mathcal{G} = (q \to r \Rightarrow p \mid q \Rightarrow p \mid r \to q, p, p \Rightarrow r) \quad \text{and} \quad \mathcal{H}_p = (s \Rightarrow \underline{p} \mid s \to q, \underline{p} \Rightarrow r)$$

Then $\mathrm{DEN}(\mathcal{G},\mathcal{H}_p) = (q \to r, s \to q \Rightarrow r \mid q, s \to q \Rightarrow r \mid r \to q, s, s \Rightarrow r)$.

In particular, a double-p-marked hypersequent $\mathcal{H}_p = (\mathcal{H} \mid \Gamma \Rightarrow \underline{p} \mid \Pi, \underline{p} \Rightarrow \Sigma)$ is always p-regular and:

$$\mathrm{DEN}(\mathcal{H}_p, \mathcal{H}_p) = (\mathcal{H} \mid \mathcal{H} \mid \Gamma, \Pi \Rightarrow \Sigma \mid \Gamma, \Pi \Rightarrow \Sigma)$$

To preserve p-regularity in derivations, we need to ensure that rules do not allow context formulas to "jump" from one side of a sequent to another. We will also restrict the external structural rules allowed in the calculus.

Definition 5.29. A rule is called *local* if for each of its instances, any context formula occurring on the left (right) in a premise is a context formula on the left (right) in the conclusion. We call a hypersequent calculus *local* if all its rules are both local and – except for (EW), (EC), and (COM) – hypersequent versions of sequent rules.

Example 5.30. Both the following rules (and their hypersequent versions) are local:

$$\frac{\Gamma_1 \Rightarrow \Delta_1 \quad \Gamma_2 \Rightarrow \Delta_2}{\Gamma_1, \Gamma_2 \Rightarrow \Delta_1, \Delta_2}\ (\textsc{mix}) \qquad \frac{\Gamma_1 \Rightarrow A, \Delta_1 \quad \Gamma_2, B \Rightarrow \Delta_2}{\Gamma_1, \Gamma_2, A \to B \Rightarrow \Delta_1, \Delta_2}\ (\to\Rightarrow)$$

Clearly any formula on the left (right) of a premise of (MIX) is on the left (right) in the conclusion. This is also true of $(\to\Rightarrow)$ except for the formulas A and B, which are active. An example of a non-local rule is:

$$\frac{\Delta \Rightarrow \Gamma}{\Gamma \Rightarrow \Delta}$$

We now have enough tools to tackle our main theorem.

Theorem 5.31. *Let* GL *be a regular and local single-conclusion hypersequent calculus with weakening. Then density elimination holds for* GL^{D}.

Proof. For technical reasons, it will be useful to mimic the "," occurring in hypersequents and its unit by the connectives $\odot$ and e (or different symbols, if these are already taken). To this end, notice that we can assume that GL contains the logical rules $(\odot\Rightarrow)$, $(\Rightarrow\odot)$, $(\mathrm{e}\Rightarrow)$, and $(\Rightarrow\mathrm{e})$. If not, then suppose that the theorem holds for the calculus extended with these rules. Since cut-free derivations in this extended calculus have the subformula property, the theorem holds also for the original calculus.

As for cut elimination, it is sufficient to consider uppermost applications of (DENSITY) and remove these one by one. We prove the following:

Claim. Suppose that:

1. $\mathcal{H}_p = (\mathcal{H} \mid \Gamma \Rightarrow \underline{p} \mid \Pi, \underline{p} \Rightarrow \Sigma)$ is a double-p-marked hypersequent.
2. $\mathcal{G}$ is a p-regular hypersequent.

If $d \vdash_{\mathrm{GL}^\circ} \mathcal{G}$ and $d' \vdash_{\mathrm{GL}^\circ} \mathcal{H}_p$, then $\vdash_{\mathrm{GL}} \mathrm{DEN}(\mathcal{G}, \mathcal{H}_p)$.

To see that this suffices, we show that a single uppermost application of (DENSITY) can be eliminated. Let $\mathcal{G} = (\mathcal{G}' \mid \Gamma \Rightarrow p \mid \Pi, p \Rightarrow \Sigma)$ be the premise of such an application and suppose that $\vdash_{\mathrm{GL}} \mathcal{G}$. Then by cut elimination, $\vdash_{\mathrm{GL}^\circ} \mathcal{G}$ and it follows from the claim, with $\mathcal{H}_p = (\mathcal{G}' \mid \Gamma \Rightarrow \underline{p} \mid \Pi, \underline{p} \Rightarrow \Sigma)$, that $\vdash_{\mathrm{GL}} \mathcal{G}' \mid \mathcal{G}' \mid \Gamma, \Pi \Rightarrow \Sigma \mid \Gamma, \Pi \Rightarrow \Sigma$. So by (EC), $\vdash_{\mathrm{GL}} \mathcal{G}' \mid \Gamma, \Pi \Rightarrow \Sigma$ as required.

Proof of claim. We proceed by induction on $\mathrm{ht}(d)$. For the base case, suppose that $\mathcal{G} = (\mathcal{G}' \mid A \Rightarrow A)$ (where A cannot be p since $\mathcal{G}$ is p-regular). Then $\mathrm{DEN}(\mathcal{G}, \mathcal{H}_p) = (\mathcal{H}' \mid A \Rightarrow A)$ for some $\mathcal{H}'$ and is derivable by (ID). For all other cases, we consider the last rule (r) applied in d. If (r) is (EC) or (EW), then the claim follows by applying the induction hypothesis and (r). Otherwise:

- Suppose that (r) is a rule other than (EC), (EW), or (COM), and d ends with:

$$\frac{\mathcal{G}' \mid S_1 \quad \dots \quad \mathcal{G}' \mid S_n}{\mathcal{G}' \mid S}\ (\mathrm{r})$$

  $(\mathcal{G}' \mid S)$ is p-regular by assumption. Also GL is local so occurrences of p cannot "switch sides" in a sequent from the premises to the conclusion. Hence $(\mathcal{G}' \mid S_1), \dots, (\mathcal{G}' \mid S_n)$ are all p-regular, and by the induction hypothesis:

$$\vdash_{\mathrm{GL}} \mathrm{DEN}((\mathcal{G}' \mid S_i), \mathcal{H}_p) \quad \text{for } i = 1 \dots n.$$

  But $\mathrm{DEN}((\mathcal{G}' \mid S), \mathcal{H}_p)$ is the result of multiple applications of (CUT) between $(\mathcal{H} \mid \mathcal{G}' \mid S)$ and $(\Gamma \Rightarrow \underline{p})$ and $(\Pi, \underline{p} \Rightarrow \Sigma)$. Hence, using the substitutivity of (r) and the fact that (r) is local (which implies that p cannot occur in the premises of an instance of (r) with no p in the conclusion):

$$\frac{\mathrm{DEN}((\mathcal{G}' \mid S_1), \mathcal{H}_p) \quad \dots \quad \mathrm{DEN}((\mathcal{G}' \mid S_n), \mathcal{H}_p)}{\mathrm{DEN}((\mathcal{G}' \mid S), \mathcal{H}_p)}$$

  is an instance of (r). So $\vdash_{\mathrm{GL}} \mathrm{DEN}((\mathcal{G}' \mid S), \mathcal{H}_p)$ as required.

- Suppose now that (r) is (COM). If both premises are p-regular, then the claim follows by applying the induction hypothesis to the premises and using (COM). For example, suppose that d ends with:

$$\frac{\mathcal{G}' \mid \Gamma_1, \Pi_1 \Rightarrow p \quad \mathcal{G}' \mid \Gamma_2, \Pi_2, [p]^k \Rightarrow \Delta}{\mathcal{G}' \mid \Gamma_1, \Gamma_2, [p]^k \Rightarrow \Delta \mid \Pi_1, \Pi_2 \Rightarrow p}\ (\mathrm{COM})$$

  By the induction hypothesis twice:

$$\vdash_{\mathrm{GL}} \mathrm{DEN}(\mathcal{G}',\mathcal{H}_p) \mid \Gamma_1,\Pi_1,\Pi \Rightarrow \Sigma \quad \text{and} \quad \vdash_{\mathrm{GL}} \mathrm{DEN}(\mathcal{G}',\mathcal{H}_p) \mid \Gamma_2,\Pi_2,\Gamma^k \Rightarrow \Delta$$

Hence by (COM), as required:

$$\vdash_{\mathrm{GL}} \mathrm{DEN}(\mathcal{G}',\mathcal{H}_p) \mid \Gamma_1,\Gamma_2,\Gamma^k \Rightarrow \Delta \mid \Pi_1,\Pi_2,\Pi \Rightarrow \Sigma$$

Suppose then that one of the premises is not p-regular and d ends with:

$$\frac{\mathcal{G}' \mid \Gamma_1,\Pi_1,[p]^{m+1} \Rightarrow p \quad \mathcal{G}' \mid \Gamma_2,[p]^k,\Pi_2 \Rightarrow \Delta}{\mathcal{G}' \mid \Gamma_1,\Gamma_2,[p]^{k+m+1} \Rightarrow \Delta \mid \Pi_1,\Pi_2 \Rightarrow p}\ (\mathrm{COM})$$

Let $\mathcal{G}_1 = \mathrm{DEN}(\mathcal{G}',\mathcal{H}_p)$. Then by the induction hypothesis:

$$d_1 \vdash_{\mathrm{GL}} \mathcal{G}_1 \mid \Gamma_2,\Gamma^k,\Pi_2 \Rightarrow \Delta$$

Our aim is to find a derivation for:

$$\vdash_{\mathrm{GL}} \mathcal{G}_1 \mid \Gamma_1,\Gamma_2,\Gamma^{k+m+1} \Rightarrow \Delta \mid \Pi_1,\Pi_2,\Pi \Rightarrow \Sigma$$

Consider the GL°-derivation d' of $(\mathcal{H} \mid \Gamma \Rightarrow p \mid \Pi, p \Rightarrow \Sigma)$. By Lemma 4.13, we can substitute $\odot\Pi_2$ (recalling that $\odot[] = \mathrm{e}$) for p in this derivation to get:

$$d_2 \vdash_{\mathrm{GL}} \mathcal{H} \mid \Gamma \Rightarrow \odot\Pi_2 \mid \Pi, \odot\Pi_2 \Rightarrow \Sigma$$

Let d_3 be the (easy) derivation of $(\mathcal{H} \mid \Pi_2 \Rightarrow \odot\Pi_2)$ using $(\Rightarrow\odot)$, $(\Rightarrow\mathrm{e})$, and (ID), and let d_2' be the derivation:

$$\cfrac{\cfrac{\cfrac{\cfrac{\cfrac{\vdots\, d_2}{\mathcal{H} \mid \Gamma \Rightarrow \odot\Pi_2 \mid \Pi, \odot\Pi_2 \Rightarrow \Sigma}}{\mathcal{H} \mid \Gamma^{m+1} \Rightarrow \odot\Pi_2 \mid \Pi, \odot\Pi_2 \Rightarrow \Sigma}\ (\mathrm{W})}{\mathcal{H} \mid \Gamma^{m+1} \Rightarrow \odot\Pi_2 \mid \Pi_1, \odot\Pi_2, \Pi \Rightarrow \Sigma}\ (\mathrm{W}) \qquad \cfrac{\vdots\, d_3}{\mathcal{H} \mid \Pi_2 \Rightarrow \odot\Pi_2}}{\mathcal{H} \mid \Gamma^{m+1} \Rightarrow \odot\Pi_2 \mid \Pi_1,\Pi_2,\Pi \Rightarrow \Sigma}\ (\mathrm{CUT})}{\mathcal{G}_1 \mid \Gamma^{m+1} \Rightarrow \odot\Pi_2 \mid \Pi_1,\Pi_2,\Pi \Rightarrow \Sigma}\ (\mathrm{EW})$$

Also, let d_1' be the derivation:

$$\cfrac{\cfrac{\cfrac{\cfrac{\cfrac{\vdots\, d_1}{\mathcal{G}_1 \mid \Gamma_2,\Gamma^k,\Pi_2 \Rightarrow \Delta}}{\vdots}\ (\odot\Rightarrow) \text{ or } (\mathrm{e}\Rightarrow)}{\mathcal{G}_1 \mid \Gamma_2,\Gamma^k,\odot\Pi_2 \Rightarrow \Delta}\ (\odot\Rightarrow) \text{ or } (\mathrm{e}\Rightarrow)}{\mathcal{G}_1 \mid \Gamma_1,\Gamma_2,\Gamma^k,\odot\Pi_2 \Rightarrow \Delta}\ (\mathrm{W})}{}$$

Finally, putting these pieces together, we obtain the required derivation:

$$\dfrac{\dfrac{\vdots\, d_1'}{\mathcal{G}_1 \mid \Gamma_1, \Gamma_2, \Gamma^k, \odot\Pi_2 \Rightarrow \Delta} \quad \dfrac{\vdots\, d_2'}{\mathcal{G}_1 \mid \Gamma^{m+1} \Rightarrow \odot\Pi_2 \mid \Pi_1, \Pi_2, \Pi \Rightarrow \Sigma}}{\mathcal{G}_1 \mid \Gamma_1, \Gamma_2, \Gamma^{k+m+1} \Rightarrow \Delta \mid \Pi_1, \Pi_2, \Pi \Rightarrow \Sigma}\ (\text{CUT})^*$$

□

This result takes care of several concrete calculi that we have investigated.

Corollary 5.32. $\mathsf{GMTL^D}$, $\mathsf{GSMTL^D}$, $\mathsf{GMTL^D_n}$ *($n \geq 2$), and* $\mathsf{GG^D}$ *admit density elimination.*

What of other hypersequent calculi with weakening? Notice first that in such a calculus with (SPLIT), adding (DENSITY) allows us to prove anything:

$$\dfrac{\dfrac{\dfrac{\dfrac{}{p \Rightarrow p}\ (\text{ID})}{p \Rightarrow \mid \Rightarrow p}\ (\text{SPLIT})}{\Rightarrow}\ (\text{DENSITY})}{\Rightarrow A}\ (\text{W})$$

Let us consider the case of regular multiple-conclusion calculi with weakening. First, observe the following key fact. If such a calculus has weakening and communication, then any hypersequent $\mathcal{G}$ with two occurrences of p on the left of a sequent and two also on the right of another sequent, is derivable:

$$\dfrac{\dfrac{\dfrac{\dfrac{}{\mathcal{H} \mid p \Rightarrow p}\ (\text{ID}) \quad \dfrac{}{\mathcal{H} \mid p \Rightarrow p}\ (\text{ID})}{\mathcal{H} \mid p, p \Rightarrow \mid \Rightarrow p, p}\ (\text{COM})}{\mathcal{H} \mid p, p \Rightarrow \mid \Pi \Rightarrow p, p, \Sigma}\ (\text{W})}{\mathcal{H} \mid \Gamma, p, p \Rightarrow \Delta \mid \Pi \Rightarrow p, p, \Sigma}\ (\text{W})$$

In particular, adding (DENSITY) to a multiple-conclusion calculus with weakening and contraction allows us to derive any formula. Hence, either (as in the case of calculi for Classical Logic) (DENSITY) is not admissible for such a calculus, or the calculus is already trivial.

To deal with this feature of multiple-conclusion calculi with weakening, we generalize the notion of p-regularity for multiple-conclusion hypersequents to disallow such possibilities. We also update our notions of double-p-marked hypersequents and DEN$(\mathcal{G}, \mathcal{H})$.

Definition 5.33. A hypersequent $\mathcal{G}$ is *p-regular* if:

1. p does not occur in any complex formulas in $\mathcal{G}$.
2. p does not occur on both the left and right of any sequent in $\mathcal{G}$.
3. p does not occur twice on the left in one sequent and twice on the right in another sequent in $\mathcal{G}$.

A hypersequent $\mathcal{G}$ is *double-p-marked* if it is of the form $(\mathcal{H} \mid \Gamma \Rightarrow \underline{p}, \Delta \mid \Pi, \underline{p} \Rightarrow \Sigma)$ where p does not occur in $\mathcal{H}$, Γ, Δ, Π, or Σ. Suppose now that:

1. $\mathcal{G} = [\Gamma_i \Rightarrow [p]^{\mu_i}, \Delta_i]_{i=1}^n \mid [\Pi_j, [p]^{\lambda_j} \Rightarrow \Sigma_j]_{j=1}^m$ is p-regular.

2. $\mathcal{H}_p = (\mathcal{H} \mid \Gamma \Rightarrow \underline{p}, \Delta \mid \Pi, \underline{p} \Rightarrow \Sigma)$ is double-p-marked.

Then $\mathrm{DEN}(\mathcal{G}, \mathcal{H}_p) = (\mathcal{H} \mid [\Gamma_i, \Pi^{\mu_i} \Rightarrow \Sigma^{\mu_i}, \Delta_i]_{i=1}^n \mid [\Pi_j, \Gamma^{\lambda_j} \Rightarrow \Delta^{\lambda_j}, \Sigma_j]_{j=1}^m)$.

We can now adapt the previous density elimination proof to the particular case of IMTL, leaving the classification of a more general class of calculi for which this works to the ingenuity of the reader.

Theorem 5.34. *Density elimination holds for* GIMTL^{D}.

Proof. As in Theorem 5.31, it suffices to prove the following:

Claim. Suppose that:

1. $\mathcal{H}_p = (\mathcal{H} \mid \Gamma \Rightarrow \underline{p}, \Delta \mid \Pi, \underline{p} \Rightarrow \Sigma)$ is a double-p-marked hypersequent.
2. $\mathcal{G}$ is a p-regular hypersequent.

If $d \vdash_{\mathrm{GL}^\circ} \mathcal{G}$ and $d' \vdash_{\mathrm{GL}^\circ} \mathcal{H}_p$, then $\vdash_{\mathrm{GL}} \mathrm{DEN}(\mathcal{G}, \mathcal{H}_p)$.

We proceed again by induction on $\mathrm{ht}(d)$. For all the rules of GIMTL$^\circ$ except (COM), a p-regular conclusion means p-regular premises and we can follow the proof of Theorem 5.31. For (COM), we also follow the proof of Theorem 5.31, except when in one of the premises, p occurs at least twice on the left in one sequent, and at least twice on the right in another. Suppose for example (other cases are very similar) that d ends with an application of (COM):

$$\frac{\mathcal{G}' \mid \Gamma_1, \Pi_1 \Rightarrow p, p, \Sigma_1, \Delta_1 \quad \mathcal{G}' \mid \Gamma_2, \Pi_2 \Rightarrow \Sigma_2, \Delta_2}{\mathcal{G}' \mid \Gamma_1, \Gamma_2 \Rightarrow p, \Delta_1, \Delta_2 \mid \Pi_1, \Pi_2 \Rightarrow p, \Sigma_1, \Sigma_2}$$

where p occurs more than once on the left of some sequent in $\mathcal{G}'$.

Let $\mathcal{G}_1 = \mathrm{DEN}(\mathcal{G}', \mathcal{H}_p)$. Our aim is to show:

$$\vdash_{\mathrm{GL}} \mathcal{G}_1 \mid \Gamma_1, \Gamma_2, \Pi \Rightarrow \Sigma, \Delta_1, \Delta_2 \mid \Pi_1, \Pi_2, \Pi \Rightarrow \Sigma, \Sigma_1, \Sigma_2$$

Since p occurs at least twice on the left in some sequent of $\mathcal{G}'$, some sequent in $\mathcal{G}_1$ is of the form $(\Pi', \Gamma, \Gamma \Rightarrow \Delta, \Delta, \Sigma')$. So by (EC) and (W), it is enough to show:

$$\vdash_{\mathrm{GL}} \mathcal{G}_1 \mid \Gamma, \Gamma \Rightarrow \Delta, \Delta \mid \Gamma_2, \Pi \Rightarrow \Sigma, \Delta_2 \mid \Pi_2, \Pi \Rightarrow \Sigma, \Sigma_2$$

We now make use again of Lemma 4.13. Since $\vdash_{\mathrm{GL}^\circ} \mathcal{H} \mid \Gamma \Rightarrow p, \Delta \mid \Pi, p \Rightarrow \Sigma$:

$$\vdash_{\mathrm{GL}} \mathcal{H} \mid \Gamma \Rightarrow \mathrm{I}(\Gamma_2 \Rightarrow \Delta_2), \Delta \mid \Pi, \mathrm{I}(\Gamma_2 \Rightarrow \Delta_2) \Rightarrow \Sigma$$
$$\vdash_{\mathrm{GL}} \mathcal{H} \mid \Gamma \Rightarrow \mathrm{I}(\Gamma \Rightarrow \Delta), \Delta \mid \Pi, \mathrm{I}(\Gamma \Rightarrow \Delta) \Rightarrow \Sigma$$

Using the logical rules and some simple applications of (CUT):

$$\vdash_{\mathrm{GL}} \mathcal{H} \mid \Gamma_2 \Rightarrow \mathrm{I}(\Gamma \Rightarrow \Delta), \Delta_2 \mid \Pi, \mathrm{I}(\Gamma_2 \Rightarrow \Delta_2) \Rightarrow \Sigma$$
$$\vdash_{\mathrm{GL}} \mathcal{H} \mid \Gamma, \Gamma \Rightarrow \Delta, \Delta \mid \Pi, \mathrm{I}(\Gamma \Rightarrow \Delta) \Rightarrow \Sigma$$

So then by (EC) and (CUT) with cut-formula $\mathrm{I}(\Gamma \Rightarrow \Delta)$:

$$\vdash_{\mathrm{GL}} \mathcal{H} \mid \Gamma, \Gamma \Rightarrow \Delta, \Delta \mid \Pi, \Gamma_2 \Rightarrow \Delta_2, \Sigma \mid \Pi, \mathrm{I}(\Gamma_2 \Rightarrow \Delta_2) \Rightarrow \Sigma$$

But now, by the induction hypothesis applied to the right premise:

$$\vdash_{\text{GL}} \mathcal{G}_1 \mid \Gamma_2, \Pi_2 \Rightarrow \Delta_2, \Sigma_2$$

Using the logical rules, we have $\vdash_{\text{GL}} \mathcal{G}_1 \mid \Pi_2 \Rightarrow \text{I}(\Gamma_2 \Rightarrow \Delta_2), \Sigma_2$ and so by (CUT):

$$\vdash_{\text{GL}} \mathcal{G}_1 \mid \mathcal{H} \mid \Gamma, \Gamma \Rightarrow \Delta, \Delta \mid \Pi, \Gamma_2 \Rightarrow \Delta_2, \Sigma \mid \Pi, \Pi_2 \Rightarrow \Sigma_2, \Sigma$$

The result then follows by (EC). □

5.3.2 Calculi Without Weakening

All our density elimination proofs so far have relied quite considerably on the presence of the weakening rule (W). In particular, we have made essential use of the fact that hypersequents of the form $(\mathcal{G} \mid \Gamma, p \Rightarrow p, \Delta)$ and $(\mathcal{G} \mid \Gamma, p, p \Rightarrow \Delta \mid \Pi \Rightarrow p, p, \Sigma)$ are always derivable in such systems. To deal with hypersequent calculi lacking weakening, we must think again.

Let us begin once more with the single-conclusion case, and for clarity concentrate on GUL, the calculus for Uninorm Logic. The main difficulty in density elimination arises when the conclusion of a rule instance is p-regular (no ps on both sides of the same sequent), but one or more of the premises are not, as can happen with the communication rule. For calculi with weakening, non-p-regular hypersequents are derivable and, as we have seen, can be factored out of the derivation. Without weakening, sequents of the form $(\Gamma, p \Rightarrow p)$ may not be derivable and hence require more careful treatment. The key idea in our proof will be to perform a preliminary surgery on such sequents to make them p-regular.

Theorem 5.35. *Density elimination holds for* GUL^{D}.

Proof. The proof will be similar to those that we have already encountered, but with a preliminary replacement of sequents $(\Gamma, p \Rightarrow p)$ with $(\Gamma \Rightarrow \text{e})$. Suppose that $\mathcal{G}$ is p-regular. We define:

$$(\mathcal{G} \mid \Gamma_1, p \Rightarrow p \mid \ldots \mid \Gamma_n, p \Rightarrow p)^{\text{e}} = (\mathcal{G} \mid \Gamma_1 \Rightarrow \text{e} \mid \ldots \mid \Gamma_n \Rightarrow \text{e})$$

Then it is sufficient to establish the following:

Claim. Let $\mathcal{H}_p = (\mathcal{H} \mid \Gamma \Rightarrow \underline{p} \mid \Pi, \underline{p} \Rightarrow \Sigma)$ be a double-p-marked hypersequent:

$$\text{If } d \vdash_{\text{GUL}^\circ} \mathcal{G} \text{ and } d' \vdash_{\text{GUL}^\circ} \mathcal{H}_p, \text{ then } \vdash_{\text{GUL}} \text{DEN}(\mathcal{G}^{\text{e}}, \mathcal{H}_p) \mid \Gamma, \Pi \Rightarrow \Sigma.$$

Notice that we remove all the problematic occurrences of p on both sides of sequents from $\mathcal{G}$ so that $\mathcal{G}^{\text{e}}$ is p-regular.

We prove the claim as before by induction on $\text{ht}(d)$. If $\mathcal{G} = (\mathcal{G}' \mid p \Rightarrow p)$ or $\mathcal{G} = (\mathcal{G}' \mid C \Rightarrow C)$ for some other formula C, then the result follows by $(\Rightarrow \text{e})$ or (ID), respectively. Otherwise, let us consider the last rule (r) applied in d. The cases of (EC) and (EW) proceed easily using the induction hypothesis as in Theorem 5.31. For the logical rules, we have a rule instance of (r) of the form:

$$\frac{\mathcal{G}_1 \mid S_1 \ \ldots \ \mathcal{G}_1 \mid S_m}{\mathcal{G}_1 \mid S}$$

If S is p-regular, then we can proceed as in Theorem 5.31 since the premises must also (since the rule is local) be p-regular. Suppose then that S is of the form:

$$\Gamma', [p]^{k+1} \Rightarrow p$$

where p does not occur in Γ'. If one at least of $S_1 \ldots S_m$ is not p-regular, then the claim easily follows by the induction hypothesis and a subsequent application of (r). Hence assume that all of $S_1 \ldots S_m$ are p-regular, and so $S_i^{\mathrm{e}} = S_i$ for $i = 1 \ldots m$. By the induction hypothesis:

$$\vdash_{\mathsf{GUL}} \mathrm{DEN}((\mathcal{G}_1^{\mathrm{e}} \mid S_i), \mathcal{H}_p) \mid \Gamma, \Pi \Rightarrow \Sigma \quad \text{for } i = 1 \ldots m$$

But then using the substitutivity of (r):

$$\vdash_{\mathsf{GUL}} \mathrm{DEN}(\mathcal{G}_1^{\mathrm{e}}, \mathcal{H}_p) \mid \Gamma, \Pi \Rightarrow \Sigma \mid \Gamma', \Gamma^{k+1}, \Pi \Rightarrow \Sigma$$

So we can complete the required derivation as follows:

$$\dfrac{\dfrac{\mathrm{DEN}(\mathcal{G}_1^{\mathrm{e}}, \mathcal{H}_p) \mid \Gamma, \Pi \Rightarrow \Sigma \mid \Gamma', \Gamma^{k+1}, \Pi \Rightarrow \Sigma \quad \dfrac{}{\mathrm{DEN}(\mathcal{G}_1^{\mathrm{e}}, \mathcal{H}_p) \mid \Gamma, \Pi \Rightarrow \Sigma \mid \Rightarrow \mathrm{e}}\,(\Rightarrow \mathrm{e})}{\mathrm{DEN}(\mathcal{G}_1^{\mathrm{e}}, \mathcal{H}_p) \mid \Gamma, \Pi \Rightarrow \Sigma \mid \Gamma, \Pi \Rightarrow \Sigma \mid \Gamma', \Gamma^k \Rightarrow \mathrm{e}}\,(\textsc{com})}{\mathrm{DEN}(\mathcal{G}_1^{\mathrm{e}}, \mathcal{H}_p) \mid \Gamma, \Pi \Rightarrow \Sigma \mid \Gamma', \Gamma^k \Rightarrow \mathrm{e}}\,(\textsc{ec})$$

Suppose now that (r) is (COM), and let us first consider the case where both active sequents in the conclusion are p-regular. If all the active sequents in the premises are p-regular, then the claim follows by applying the induction hypothesis and (COM). Otherwise, we have a situation of the form:

$$\dfrac{\dfrac{\vdots\, d_1}{\mathcal{G}_1 \mid \Gamma_1, \Pi_1, [p]^k \Rightarrow \Delta_1} \quad \dfrac{\vdots\, d_2}{\mathcal{G}_1 \mid \Gamma_2, \Pi_2, [p]^{m+1} \Rightarrow p}}{\mathcal{G}_1 \mid \Gamma_1, \Gamma_2, [p]^{k+m+1} \Rightarrow \Delta_1 \mid \Pi_1, \Pi_2 \Rightarrow p}\,(\textsc{com})$$

Let $\mathcal{G}_2 = (\mathrm{DEN}(\mathcal{G}_1^{\mathrm{e}}, \mathcal{H}_p) \mid \Gamma, \Pi \Rightarrow \Sigma)$. We need to show that:

$$\vdash_{\mathsf{GUL}} \mathcal{G}_2 \mid \Gamma_1, \Gamma_2, \Gamma^{k+m+1} \Rightarrow \Delta_1 \mid \Pi_1, \Pi_2, \Pi \Rightarrow \Sigma$$

By the induction hypothesis:

$$d_3 \vdash_{\mathsf{GUL}} \mathcal{G}_2 \mid \Gamma_1, \Pi_1, \Gamma^k \Rightarrow \Delta_1 \quad \text{and} \quad d_4 \vdash_{\mathsf{GUL}} \mathcal{G}_2 \mid \Gamma_2, \Pi_2, \Gamma^m \Rightarrow \mathrm{e}$$

We first apply the rule $(\mathrm{e}\Rightarrow)$ to the end-hypersequent of d_3, obtaining a derivation of $(\mathcal{G}_2 \mid \Gamma_1, \Pi_1, \Gamma^k, \mathrm{e} \Rightarrow \Delta_1)$. Then by (CUT) with the end-hypersequent of d_4, we get:

$$d_5 \vdash_{\mathsf{GUL}} \mathcal{G}_2 \mid \Gamma_1, \Gamma_2, \Pi_1, \Pi_2, \Gamma^{k+m} \Rightarrow \Delta_1$$

Now let $A = \odot(\Pi_1 \uplus \Pi_2)$ and consider:

$$\frac{\mathcal{G}_2 \mid \Gamma \Rightarrow A \mid \Pi, A \Rightarrow \Sigma \quad \mathcal{G}_2 \mid \Gamma_1, \Gamma_2, \Gamma^{k+m}, A \Rightarrow \Delta_1 \mid \Pi, A \Rightarrow \Sigma}{\mathcal{G}_2 \mid \Gamma_1, \Gamma_2, \Gamma^{k+m+1} \Rightarrow \Delta_1 \mid \Pi, A \Rightarrow \Sigma}\ (\textsc{cut})$$

The left premise is derivable by (EW) and Lemma 4.13 applied to $\mathcal{H}_p$. The right premise is derivable by extending d_5 with (EW) and $(\mathrm{e}\Rightarrow)$ and $(\odot\Rightarrow)$ as necessary. The required derivation is then obtained by applying (CUT) to the conclusion with the easy derivation of $(\mathcal{G}_2 \mid \Gamma_1, \Gamma_2, \Gamma^k \Rightarrow \Delta_1 \mid \Pi_1, \Pi_2 \Rightarrow A)$.

Now consider the case for (COM) where exactly one active sequent in the conclusion of (COM) is p-regular. If one of the active sequents in the premises is not p-regular, then the claim easily follows by applying the induction hypothesis and (COM). Otherwise, d ends with:

$$\frac{\mathcal{G}_1 \mid \Gamma_1, \Pi_1 \Rightarrow p \quad \mathcal{G}_1 \mid \Gamma_2, \Pi_2, [p]^{k+m+1} \Rightarrow \Delta_1}{\mathcal{G}_1 \mid \Gamma_1, \Gamma_2, [p]^k \Rightarrow \Delta_1 \mid \Pi_1, \Pi_2, [p]^{m+1} \Rightarrow p}\ (\textsc{com})$$

Let $\mathcal{G}_2 = (\mathrm{DEN}(\mathcal{G}_1^{\mathrm{e}}, \mathcal{H}_p) \mid \Gamma, \Pi \Rightarrow \Sigma)$. By the induction hypothesis twice:

$$d_1 \vdash_{\mathsf{GUL}} \mathcal{G}_2 \mid \Gamma_1, \Pi_1, \Pi \Rightarrow \Sigma \quad \text{and} \quad d_2 \vdash_{\mathsf{GUL}} \mathcal{G}_2 \mid \Gamma_2, \Pi_2, \Gamma^{k+m+1} \Rightarrow \Delta_1$$

So we can construct the following derivation:

$$\dfrac{\dfrac{\dfrac{\dfrac{\vdots\, d_1}{\mathcal{G}_2 \mid \Gamma_1, \Pi_1, \Pi \Rightarrow \Sigma} \quad \dfrac{\vdots\, d_2}{\mathcal{G}_2 \mid \Gamma_2, \Pi_2, \Gamma^{k+m+1} \Rightarrow \Delta_1}}{\mathcal{G}_2 \mid \Gamma, \Pi \Rightarrow \Sigma \mid \Gamma_1, \Gamma_2, \Gamma^{k+m}, \Pi_1, \Pi_2 \Rightarrow \Delta_1}\ (\textsc{com})}{\mathcal{G}_2 \mid \Gamma_1, \Gamma_2, \Gamma^{k+m}, \Pi_1, \Pi_2 \Rightarrow \Delta_1}\ (\textsc{ec}) \quad \dfrac{}{\mathcal{G}_2 \mid \Rightarrow \mathrm{e}}\ (\Rightarrow\mathrm{e})}{\mathcal{G}_2 \mid \Gamma_1, \Gamma_2, \Gamma^k \Rightarrow \Delta_1 \mid \Pi_1, \Pi_2, \Gamma^m \Rightarrow \mathrm{e}}\ (\textsc{com})$$

Finally, suppose that neither active sequent in the conclusion of (COM) is p-regular. If the active sequents in both premises are not p-regular, then the claim follows by applying the induction hypothesis and (COM). Assume then that exactly one active sequent in the premises of (COM) is p-regular; i.e. d ends with:

$$\frac{\mathcal{G}_1 \mid \Gamma_1, \Pi_1, [p]^{k+m+2} \Rightarrow p \quad \mathcal{G}_1 \mid \Gamma_2, \Pi_2 \Rightarrow p}{\mathcal{G}_1 \mid \Gamma_1, \Gamma_2, [p]^{k+1} \Rightarrow p \mid \Pi_1, \Pi_2, [p]^{m+1} \Rightarrow p}\ (\textsc{com})$$

Let $\mathcal{G}_2 = (\mathrm{DEN}(\mathcal{G}_1^{\mathrm{e}}, \mathcal{H}_p) \mid \Gamma, \Pi \Rightarrow \Sigma)$. By the induction hypothesis twice:

$$d_1 \vdash_{\mathsf{GUL}} \mathcal{G}_2 \mid \Gamma_1, \Pi_1, \Gamma^{k+m+1} \Rightarrow \mathrm{e} \quad \text{and} \quad d_2 \vdash_{\mathsf{GUL}} \mathcal{G}_2 \mid \Gamma_2, \Pi_2, \Pi \Rightarrow \Sigma$$

We first apply the rule $(e\Rightarrow)$ to the end-hypersequent of d_2, obtaining a derivation of $(\mathcal{G}_2 \mid \Gamma_2,\Pi_2,\Pi,e \Rightarrow \Sigma)$. Then by (CUT) with the end-hypersequent of d_1, we obtain a derivation:

$$d_3 \vdash_{\mathsf{GUL}} \mathcal{G}_2 \mid \Gamma_1,\Gamma_2,\Pi_1,\Pi_2,\Gamma^{k+m+1},\Pi \Rightarrow \Sigma$$

The required derivation is then:

$$\dfrac{\dfrac{\dfrac{\dfrac{\vdots\, d_3}{\mathcal{G}_2 \mid \Gamma_1,\Gamma_2,\Pi_1,\Pi_2,\Gamma^{k+m+1},\Pi \Rightarrow \Sigma} \quad \dfrac{}{\mathcal{G}_2 \mid \Rightarrow e}\,(\Rightarrow e)}{\mathcal{G}_2 \mid \Gamma,\Pi \Rightarrow \Sigma \mid \Gamma_1,\Gamma_2,\Pi_1,\Pi_2,\Gamma^{k+m} \Rightarrow e}\,(\text{COM})}{\mathcal{G}_2 \mid \Gamma_1,\Gamma_2,\Pi_1,\Pi_2,\Gamma^{k+m} \Rightarrow e}\,(\text{EC}) \quad \dfrac{}{\mathcal{G}_2 \mid \Rightarrow e}\,(\Rightarrow e)}{\mathcal{G}_2 \mid \Gamma_1,\Gamma_2,\Gamma^{k} \Rightarrow e \mid \Pi_1,\Pi_2,\Gamma^{m} \Rightarrow e}\,(\text{COM})$$

□

This density elimination proof generalizes the proof for $\mathsf{GMTL}^{\mathsf{D}}$ and works also for calculi such as $\mathsf{GUL}^{\mathsf{D}}$ extended with rules (MIX), (EMP), or (SC_2). However, for calculi with contraction rules but lacking weakening, such as $\mathsf{GUML}^{\mathsf{D}}$ and $\mathsf{GIUML}^{\mathsf{D}}$, a slightly different strategy is required.

Theorem 5.36. *Density elimination holds for* $\mathsf{GUML}^{\mathsf{D}}$ *and* $\mathsf{GIUML}^{\mathsf{D}}$.

Proof. We will just sketch the proof for $\mathsf{GIUML}^{\mathsf{D}}$, the case of $\mathsf{GUML}^{\mathsf{D}}$ being very similar. Also, for convenience of exposition let us assume that derivations in this calculus use the derived rules (SPLIT) and (MIX), rather than (COM). We define:

$$(\Gamma,[p]^{\lambda} \Rightarrow [p]^{\mu},\Delta)^c = \begin{cases} (\Gamma,p \Rightarrow \Delta \mid \Gamma \Rightarrow p,\Delta \mid \Gamma \Rightarrow \Delta) & \text{if } \lambda,\mu \geq 1 \\ (\Gamma,p \Rightarrow \Delta) & \text{if } \lambda \geq 1, \mu = 0 \\ (\Gamma \Rightarrow p,\Delta) & \text{if } \lambda = 0, \mu \geq 1 \\ (\Gamma \Rightarrow \Delta) & \text{if } \lambda = \mu = 0 \end{cases}$$

$$(S_1 \mid \ldots \mid S_n)^c = (S_1)^c \mid \ldots \mid (S_n)^c$$

As in previous proofs, it is sufficient to establish the following:

Claim. Let $\mathcal{H}_p = (\mathcal{H} \mid \Gamma \Rightarrow \underline{p},\Delta \mid \Pi,\underline{p} \Rightarrow \Sigma)$ be a double-p-marked hypersequent. If $d \vdash_{\mathsf{GIMUL}^{\circ}} \mathcal{G}$ and $d' \vdash_{\mathsf{GIUML}^{\circ}} \mathcal{H}_p$, then $\vdash_{\mathsf{GIUML}} \mathrm{DEN}(\mathcal{G}^c,\mathcal{H}_p)$.

The proof is by induction on $\mathrm{ht}(d)$. We will just consider the trickiest case where d ends with an application of (SPLIT), leaving others for the reader's entertainment. Let us write $[p]^*$ to mean $[p]^k$ for some $k \geq 1$, and suppose that d ends with:

$$\dfrac{\mathcal{G}_1 \mid \Gamma_1,\Gamma_2,[p]^* \Rightarrow [p]^*,\Delta_1,\Delta_2}{\mathcal{G}_1 \mid \Gamma_1,[p]^* \Rightarrow \Delta_1 \mid \Gamma_2 \Rightarrow [p]^*,\Delta_2}\,(\text{SPLIT})$$

Let $\mathcal{G}_2 = \mathrm{DEN}(\mathcal{G}_1^c,\mathcal{H}_p)$. Then by the induction hypothesis:

$$\vdash_{\mathsf{GIUML}} \mathcal{G}_2 \mid \Gamma_1,\Gamma_2,\Gamma \Rightarrow \Delta,\Delta_1,\Delta_2 \mid \Gamma_1,\Gamma_2,\Pi \Rightarrow \Sigma,\Delta_1,\Delta_2 \mid \Gamma_1,\Gamma_2 \Rightarrow \Delta_1,\Delta_2$$

As in earlier proofs, by (EW), Lemma 4.13, and (CUT) with $\Gamma_1, \mathrm{I}(\Gamma_1 \Rightarrow \Delta_1) \Rightarrow \Delta_1$:

$$\vdash_{\mathsf{GIUML}} \mathcal{G}_2 \mid \Gamma, \Gamma_1 \Rightarrow \Delta_1, \Delta \mid \Pi, \mathrm{I}(\Gamma_1 \Rightarrow \Delta_1) \Rightarrow \Sigma$$

But then using the logical rules, (CUT), (EC), and (C):

$$\vdash_{\mathsf{GIUML}} \mathcal{G}_2 \mid \Gamma, \Gamma_1 \Rightarrow \Delta_1, \Delta \mid \Gamma_1, \Gamma_2, \Gamma, \Pi \Rightarrow \Sigma, \Delta, \Delta_1, \Delta_2 \mid \Gamma_2, \Pi \Rightarrow \Sigma, \Delta_2$$

So finally using (EC) and (SPLIT), as required:

$$\vdash_{\mathsf{GIUML}} \mathcal{G}_2 \mid \Gamma_1, \Gamma \Rightarrow \Delta, \Delta_1 \mid \Gamma_2, \Pi \Rightarrow \Sigma, \Delta_2$$ □

While for single-conclusion calculi with weakening, density elimination goes essentially hand-in-hand with cut elimination, calculi that lack weakening and/or are multiple-conclusion, are more problematic. In particular, no density elimination method is yet known for GIUL. The problem here, as may be gleaned from preceding proofs, is that a hypersequent may contain a sequent with more than one p on the left and another sequent with more than one p on the right. Such hypersequents are derivable with weakening and do not occur in single-conclusion calculi, but for GIUL, a multiple-conclusion calculus without weakening, a different strategy is required.

5.3.3 Standard Completeness

Perhaps the most interesting aspect of density elimination is that it fills a crucial gap in standard completeness proofs for a wide range of fuzzy logics. First notice that density elimination for a Gentzen system allows us to eliminate the density rule also from derivations in the corresponding Hilbert system.

Proposition 5.37. *For any* HUL^-*-extension* HL *and matching Gentzen system* GL *admitting density elimination:* $T \vdash_{\mathrm{HL}} A$ *iff* $T \vdash_{\mathrm{HL^D}} A$.

Proof. We have the following chain of reasoning:

$T \vdash_{\mathrm{HL}} A$	iff	$\vdash_{\mathrm{HL}} C \to A$ for some confusion C of T	Theorem 3.43
	iff	$\vdash_{\mathrm{GL}} C \Rightarrow A$ for some confusion C of T	Theorem 4.66
	iff	$\vdash_{\mathrm{GL^D}} C \Rightarrow A$ for some confusion C of T	Density Elimination
	iff	$\vdash_{\mathrm{HL^D}} C \to A$ for some confusion C of T	Theorem 4.66
	iff	$T \vdash_{\mathrm{HL^D}} A$	Theorem 3.61 □

But now we can appeal to Theorem 3.65 which tells us that certain Hilbert systems extended with (DENSITY) are complete with respect to standard algebras, to obtain:

Theorem 5.38. *For* $\mathrm{L} \in \{\mathsf{UL}, \mathsf{MTL}, \mathsf{SMTL}, \mathsf{IMTL}, \mathsf{G}, \mathsf{UML}, \mathsf{IUML}, \mathsf{MTL_n}\ (n \geq 2)\}$:

$$T \vdash_{\mathrm{HL}} A \quad \textit{iff} \quad T \models_{\mathsf{STAN(L)}} A$$

Moreover, the variety of L*-algebras is generated by* STAN(L).

Density elimination is a neat and uniform method for establishing standard completeness for many fuzzy logics. However, it does not always work. While adding (DENSITY) to *any* HUL^--extension HL gives completeness with respect to dense L-chains, in some cases such as Ł-algebras (MV-algebras) and BL-algebras, the Dedekind-MacNeille extension does not guarantee an algebra of the same class. The density elimination method also relies on the existence of a "suitable" hypersequent calculus. It is reasonable to suppose that the case of GIUL can be solved by choosing an appropriate induction hypothesis for the elimination proof, but for other logics the problems run deeper. In particular, although in the next chapter we define calculi for Ł, P, CRL, CHL, it is unclear whether density elimination can be obtained or is useful in these cases. Note nevertheless that with the exception of Ł, we can still use these calculi to obtain standard completeness. For Ł, as indeed for logics such as BL that (currently) lack a Gentzen system, algebraic techniques are required for proving standard completeness results.

5.4 Historical Remarks

Cut elimination was first established for the sequent calculi LK for Classical Logic and LJ for Intuitionistic Logic by Gentzen in the 1930s [93], and has remained a central topic of Proof Theory ever since (see e.g. the textbooks [204, 209]). Cut elimination proofs have also appeared regularly in the substructural logics literature, the most significant examples being the Lambek calculus [130], the contraction-free logics studied by Ono and Komori [177], and Girard's Linear Logic [97]. Conditions that guarantee cut elimination for a wide range of calculi for substructural logics – very similar to the notions of substitutivity and reductivity defined above – were provided by Restall in [186], inspired by Belnap's Display Logic framework [34]. Ciabattoni and Terui [57, 207] have also given sufficient (and given certain restrictions, necessary) conditions for cut elimination to hold for classes of single-conclusion sequent calculi. The semantic conditions used in this work derive from an algebraic approach to establishing cut admissibility pioneered by Okada in [169, 170].

The first cut elimination proofs at the hypersequent level were given by Avron in the late 1980s [9, 11] using the "history method", a rather complicated variant of Gentzen-style cut elimination. A simpler method, related to Schütte-Tait-style cut elimination [195, 202], that eliminates the largest cut in a derivation, was used by Metcalfe in [142] and Baaz, Ciabattoni, and Montagna in [17]. An alternative "cut elimination by substitutions" method was defined by Ciabattoni in 2004 [48] and used to obtain uniform proofs (similar to those given in this chapter) for single-conclusion hypersequent calculi. Finally, the algebraic method mentioned above for establishing cut admissibility has been extended to a broad class of single-conclusion hypersequent calculi by Ciabattoni, Galatos, and Terui in [53].

The cancellation elimination method described in this chapter was introduced by Ciabattoni and Metcalfe in 2004 [54] to establish cut elimination for the hypersequent calculus GŁ for Łukasiewicz Logic of Chapter 6. It was extended to Abelian Logic by Metcalfe in [143]. The generation of the variety of abelian ℓ-groups by $\mathbf{Z}$ was established by Weinberg in 1963 [218]. The algorithmic proof given here is new, but bears some similarity to Meyer and Slaney's 1989 completeness proof for Abelian Logic [149].

Finally, density elimination was established for a hypersequent calculus for first-order Gödel Logic by Baaz and Zach in 2000 [29], using a Gentzen-style proof that shifts applications of (DENSITY) upwards in derivations. This method was extended to a wide range of hypersequent calculi in [144] and used – as in this chapter – to establish standard completeness for corresponding fuzzy logics. The more elegant "density elimination by substitutions" method described above, where applications of (DENSITY) are removed by making suitable substitutions, was introduced by Ciabattoni and Metcalfe in [55].

Chapter 6
Fundamental Logics

The efforts of the last two chapters have given us elegant proof-theoretic presentations of a large family of fuzzy logics. We get different logics just by tinkering with the structural rules. The generality of the framework ensures uniform proofs of cut and density elimination, and, in many cases, properties such as decidability and standard completeness. Nevertheless, the reader will have noticed that several of the most important fuzzy logics mentioned in Chapter 2 are yet to be treated. Basic Logic remains an open problem (see the final chapter for some remarks). However, providing proof theory for Łukasiewicz Logic Ł and Product Logic P – together with Gödel Logic G, the *fundamental fuzzy logics* – will be the goal of this chapter.

6.1 Gödel Logic

Gödel Logic G is interesting not only from the fuzzy perspective but also as one of the main intermediate (between Intuitionistic and Classical) logics. It may be viewed as "the logic of linear order" since the multiplicative connectives $\odot$ and $\oplus$ collapse, respectively, to the additive connectives $\wedge$ and $\vee$, interpreted as min and max. For this reason, it is usual to base G on a reduced language $\mathcal{L}_{\mathrm{G}} = \{\wedge, \vee, \to, \bot, \top\}$, fixing as before a countably infinite set of variables X. We begin by refreshing our memory of the standard semantics of G based on the minimum t-norm and its residuum, for simplicity replacing references to the algebra $\mathbf{A}(*_{\mathrm{G}}, 0)$ with G:

A G-*valuation* is a function $v : \mathrm{Fm}_{\mathcal{L}_{\mathrm{G}}} \to [0,1]$ such that $v(\bot) = 0$, $v(\top) = 1$, and:

$$v(A \wedge B) = \min(v(A), v(B)) \qquad v(A \to B) = \begin{cases} v(B) & \text{if } v(A) > v(B) \\ 1 & \text{otherwise} \end{cases}$$
$$v(A \vee B) = \max(v(A), v(B))$$

An $\mathcal{L}_{\mathrm{G}}$-formula A is G-*valid*, written $\models_{\mathrm{G}} A$, iff $v(A) = 1$ for all G-valuations v. However, since we are dealing with just one logic in this section, we will drop the prefix G for valuations and validity.

Initial Hypersequents

$$\frac{}{\mathcal{G} \mid A \Rightarrow A}\ (\text{ID})$$

Structural Rules

$$\frac{\mathcal{G}}{\mathcal{G} \mid \mathcal{H}}\ (\text{EW}) \qquad \frac{\mathcal{G} \mid \mathcal{H} \mid \mathcal{H}}{\mathcal{G} \mid \mathcal{H}}\ (\text{EC}) \qquad \frac{\mathcal{G} \mid \Gamma_1, \Pi_1 \Rightarrow \Delta \quad \mathcal{G} \mid \Gamma_2, \Pi_2 \Rightarrow \Sigma}{\mathcal{G} \mid \Gamma_1, \Gamma_2 \Rightarrow \Delta \mid \Pi_1, \Pi_2 \Rightarrow \Sigma}\ (\text{COM})$$

$$\frac{\mathcal{G} \mid \Gamma, \Pi, \Pi \Rightarrow \Delta}{\mathcal{G} \mid \Gamma, \Pi \Rightarrow \Delta}\ (\text{C}) \qquad \frac{\mathcal{G} \mid \Gamma \Rightarrow \Delta}{\mathcal{G} \mid \Gamma, \Pi \Rightarrow \Delta}\ (\text{W})$$

Logical Rules

$$\frac{}{\mathcal{G} \mid \Gamma, \bot \Rightarrow \Delta}\ (\bot\Rightarrow) \qquad \frac{}{\mathcal{G} \mid \Gamma \Rightarrow \top}\ (\Rightarrow\top)$$

$$\frac{\mathcal{G} \mid \Gamma_1 \Rightarrow A \quad \mathcal{G} \mid \Gamma_2, B \Rightarrow \Delta}{\mathcal{G} \mid \Gamma_1, \Gamma_2, A \to B \Rightarrow \Delta}\ (\to\Rightarrow) \qquad \frac{\mathcal{G} \mid \Gamma, A \Rightarrow B}{\mathcal{G} \mid \Gamma \Rightarrow A \to B}\ (\Rightarrow\to)$$

$$\frac{\mathcal{G} \mid \Gamma, A \Rightarrow \Delta}{\mathcal{G} \mid \Gamma, A \wedge B \Rightarrow \Delta}\ (\wedge\Rightarrow)_1 \qquad \frac{\mathcal{G} \mid \Gamma, B \Rightarrow \Delta}{\mathcal{G} \mid \Gamma, A \wedge B \Rightarrow \Delta}\ (\wedge\Rightarrow)_2$$

$$\frac{\mathcal{G} \mid \Gamma, A \Rightarrow \Delta \quad \mathcal{G} \mid \Gamma, B \Rightarrow \Delta}{\mathcal{G} \mid \Gamma, A \vee B \Rightarrow \Delta}\ (\vee\Rightarrow) \qquad \frac{\mathcal{G} \mid \Gamma \Rightarrow A \quad \mathcal{G} \mid \Gamma \Rightarrow B}{\mathcal{G} \mid \Gamma \Rightarrow A \wedge B}\ (\Rightarrow\wedge)$$

$$\frac{\mathcal{G} \mid \Gamma \Rightarrow A}{\mathcal{G} \mid \Gamma \Rightarrow A \vee B}\ (\Rightarrow\vee)_1 \qquad \frac{\mathcal{G} \mid \Gamma \Rightarrow B}{\mathcal{G} \mid \Gamma \Rightarrow A \vee B}\ (\Rightarrow\vee)_2$$

Cut Rule

$$\frac{\mathcal{G} \mid \Gamma_1, A \Rightarrow \Delta \quad \mathcal{G} \mid \Gamma_2 \Rightarrow A}{\mathcal{G} \mid \Gamma_1, \Gamma_2 \Rightarrow \Delta}\ (\text{CUT})$$

Fig. 6.1 The hypersequent calculus GG

The axiomatization HG of Chapter 3 based on the language $\mathcal{L}_\text{B}$ is obtained by adding the contraction axioms $A \to (A \odot A)$ to HMTL. The corresponding G-algebras are idempotent prelinear integral bcrls (term equivalent to prelinear Heyting algebras). More commonly, Hilbert systems for G based on the language $\mathcal{L}_\text{G}$ are obtained by adding the prelinearity axioms $(A \to B) \vee (B \to A)$ to an axiomatization for Intuitionistic Logic.

6.1.1 The Hypersequent Calculus GG

The hypersequent calculus GG presented in Chapter 4 is an elegant and informative presentation of Gödel Logic, an extension both of the calculus GMTL for Monoidal t-norm Logic with contraction, and of a hypersequent version of Gentzen's calculus LJ for Intuitionistic Logic with communication. For convenience, we display the calculus GG in its more concise single-conclusion form in Fig. 6.1.

Recall from Chapter 4 that the invertible rules $(\wedge\Rightarrow)$ and $(\Rightarrow\vee)$ are derivable in this calculus. Also, note that since the connectives for G are both (as in Intuitionistic Logic and Classical Logic) additive and multiplicative, $(\wedge\Rightarrow)_1$ and $(\wedge\Rightarrow)_2$, and $(\to\Rightarrow)$ could be replaced with, respectively:

$$\frac{\mathcal{G} \mid \Gamma, A, B \Rightarrow \Delta}{\mathcal{G} \mid \Gamma, A \wedge B \Rightarrow \Delta}\ (\wedge\Rightarrow)_{\mathsf{G}} \qquad \frac{\mathcal{G} \mid \Gamma \Rightarrow A \quad \mathcal{G} \mid \Gamma, B \Rightarrow \Delta}{\mathcal{G} \mid \Gamma, A \to B \Rightarrow \Delta}\ (\to\Rightarrow)_{\mathsf{G}}$$

We could also use the availability of weakening and contraction to simplify the communication rule slightly to:

$$\frac{\mathcal{G} \mid \Gamma_1, \Gamma_2 \Rightarrow \Delta_1 \quad \mathcal{G} \mid \Gamma_1, \Gamma_2 \Rightarrow \Delta_2}{\mathcal{G} \mid \Gamma_1 \Rightarrow \Delta_1 \mid \Gamma_2 \Rightarrow \Delta_2}\ (\mathrm{COM})_{\mathsf{G}}$$

Example 6.1. Notice that both contraction and communication are required in the following derivation:

$$\dfrac{\dfrac{\dfrac{\dfrac{\dfrac{\dfrac{\dfrac{\dfrac{\overline{A \Rightarrow A} \quad \dfrac{\overline{B \Rightarrow B} \quad \overline{C \Rightarrow C}}{B \to C, B \Rightarrow C}\,(\to\Rightarrow)}{A \to (B \to C), A, B \Rightarrow C}\,(\to\Rightarrow) \quad \dfrac{\overline{A \Rightarrow A} \quad \dfrac{\overline{B \Rightarrow B} \quad \overline{C \Rightarrow C}}{B \to C, B \Rightarrow C}\,(\to\Rightarrow)}{A \to (B \to C), A, B \Rightarrow C}\,(\to\Rightarrow)}{A \to (B \to C), A, A \Rightarrow C \mid A \to (B \to C), B, B \Rightarrow C}\,(\mathrm{COM})}{A \to (B \to C), A, A \Rightarrow C \mid A \to (B \to C), B \Rightarrow C}\,(\mathrm{C})}{A \to (B \to C), A \Rightarrow C \mid A \to (B \to C), B \Rightarrow C}\,(\mathrm{C})}{A \to (B \to C), A \Rightarrow C \mid A \to (B \to C) \Rightarrow B \to C}\,(\Rightarrow\to)}{A \to (B \to C) \Rightarrow A \to C \mid A \to (B \to C) \Rightarrow B \to C}\,(\Rightarrow\to)}{A \to (B \to C) \Rightarrow (A \to C) \vee (B \to C)}\,(\Rightarrow\vee)}{\Rightarrow (A \to (B \to C)) \to ((A \to C) \vee (B \to C))}\,(\Rightarrow\to)$$

That is, $(A \to (B \to C)) \to ((A \to C) \vee (B \to C))$ is not a theorem of MTL or IL.

The crucial result, proved in the last chapter, is that cut elimination holds for GG. Among other things, this provides an easy proof of decidability, and, via density elimination, a more complicated proof of standard completeness. GG has its limitations, however. Even with loop-checking, which gives termination of the rules (read upwards), GG is not particularly efficient for theorem proving. One reason is the structural rules: external contraction can double the size of hypersequents. Another is that not all the logical rules of the calculus are invertible, so backtracking is required for proof search. Moreover, we may simply prefer to have a calculus for this logic in Gentzen's original formulation: that is, a sequent calculus. Such a calculus will also facilitate a more direct proof of standard completeness and help us to establish complexity bounds for G.

$$\frac{\Gamma, A, B \Rightarrow \Delta}{\Gamma, A \wedge B \Rightarrow \Delta}\ (\wedge\Rightarrow)_{\mathsf{G}} \qquad \frac{\Gamma, A \to C \Rightarrow \Delta \quad \Gamma, B \to C \Rightarrow \Delta}{\Gamma, A \to (B \to C) \Rightarrow \Delta}\ (\to(\to)\Rightarrow)$$

$$\frac{\Gamma \Rightarrow A, \Delta \quad \Gamma \Rightarrow B, \Delta}{\Gamma \Rightarrow A \wedge B, \Delta}\ (\Rightarrow\wedge) \qquad \frac{\Gamma, B \to C \Rightarrow A \to B, \Delta \quad \Gamma, C \Rightarrow \Delta}{\Gamma, (A \to B) \to C \Rightarrow \Delta}\ ((\to)\to\Rightarrow)$$

$$\frac{\Gamma, A \to C, B \to C \Rightarrow \Delta}{\Gamma, A \to (B \wedge C) \Rightarrow \Delta}\ (\to\wedge\Rightarrow) \qquad \frac{\Gamma, A \to C \Rightarrow \Delta \quad \Gamma, B \to C \Rightarrow \Delta}{\Gamma, (A \wedge B) \to C \Rightarrow \Delta}\ (\wedge\to\Rightarrow)$$

$$\frac{\Gamma \Rightarrow B \to C, A \to C, \Delta}{\Gamma \Rightarrow (A \wedge B) \to C, \Delta}\ (\Rightarrow\wedge\to) \qquad \frac{\Gamma \Rightarrow A \to B, \Delta \quad \Gamma \Rightarrow A \to C, \Delta}{\Gamma \Rightarrow A \to (B \wedge C), \Delta}\ (\Rightarrow\to\wedge)$$

$$\frac{\Gamma \Rightarrow A \to C, B \to C, \Delta}{\Gamma \Rightarrow A \to (B \to C), \Delta}\ (\Rightarrow\to(\to)) \qquad \frac{\Gamma \Rightarrow B \to C, \Delta \quad \Gamma, A \to B \Rightarrow C, \Delta}{\Gamma \Rightarrow (A \to B) \to C, \Delta}\ (\Rightarrow(\to)\to)$$

Fig. 6.2 Sequent decomposition rules for G

6.1.2 A Sequent Calculus

Sequents are not as flexible as hypersequents so we compensate in two ways. First, we use multiple-conclusion sequents where the right hand side is interpreted as an additive disjunction $\vee$ (recalling that $\bigwedge[] =_{\text{def}} \top$ and $\bigvee[] =_{\text{def}} \bot$).

Definition 6.2. $\mathrm{I}_{\mathsf{G}}(\Gamma \Rightarrow \Delta) =_{\text{def}} \bigwedge\Gamma \to \bigvee\Delta$.

In particular, we will call a hypersequent rule $\mathcal{G}_1, \ldots, \mathcal{G}_n \ / \ \mathcal{G}$ *sound* if whenever $\mathrm{I}_{\mathsf{G}}(\mathcal{G}_1), \ldots, \mathrm{I}_{\mathsf{G}}(\mathcal{G}_n)$ are all valid, then also $\mathrm{I}_{\mathsf{G}}(\mathcal{G})$ is valid, and *invertible* if the reverse implication holds.

Our second innovation is to use more complicated rules for logical connectives that "decompose" formulas into formulas with a smaller complexity. These are displayed in Fig. 6.2. Since the number of these rules increases exponentially with the number of connectives, we make use of a reduced language $\mathcal{L}_{\mathrm{I}} = \{\wedge, \to, \bot, \top\}$ with:[1]

$$A \vee B =_{\text{def}} ((A \to B) \to B) \wedge ((B \to A) \to A)$$

For the rest of this subsection we assume that all formulas, sequents, and so on are constructed using this language.

Example 6.3. Decomposition rules (working upwards) serve to reduce complicated formulas to less complicated formulas, e.g.

$$\dfrac{\dfrac{q \to r, q \to p \Rightarrow p \to q, r \to p}{q \to (r \wedge p) \Rightarrow p \to q, r \to p}\ (\to\wedge\Rightarrow) \qquad \dfrac{r, p \Rightarrow r \to p}{r \wedge p \Rightarrow r \to p}\ (\wedge\Rightarrow)_{\mathsf{G}}}{(p \to q) \to (r \wedge p) \Rightarrow r \to p}\ ((\to)\to\Rightarrow)$$

[1] Observe, that this definition of $A \vee B$ multiplies the number of occurrences of A and B by three, and should therefore be avoided for applications.

Unlike other logical rules we have seen, here there is a choice not only of which formula to decompose, but also of how to decompose each formula. We first decomposed $(p \to q) \to (r \wedge p)$ using the rule $((\to) \to \Rightarrow)$, but we could just as well have started with $(\to \wedge \Rightarrow)$. However, this choice does not affect the success of a derivation. The rules are all invertible – if the conclusion is valid, then so are the premises – so backtracking is unnecessary.

In fact, it is easy to see that the decomposition rules (read upwards) reduce all sequents to sequents containing only atoms and implications between atoms.

Definition 6.4. We call a sequent or hypersequent *atomic implicational* if it contain only atoms and implications of the form $a \to b$ where a and b are atoms.

Lemma 6.5. *Every sequent is derivable from atomic implicational sequents using the sequent decomposition rules for* G.

Proof. We prove the statement by induction on $\mathrm{cp}(S)$ for a sequent S. If S is atomic implicational, then we are done. Otherwise there is a rule instance $S_1, \ldots, S_n \,/\, S$ such that $\mathrm{cp}(S_i) <_{\mathsf{m}} \mathrm{cp}(S)$ for $i = 1 \ldots n$. The result then follows by the induction hypothesis applied to each premise. We consider the rule $((\to) \to \Rightarrow)$ as an example, leaving other cases to the reader's curiosity. Observe first that for the conclusion:

$$\mathrm{cp}(\Gamma, (A \to B) \to C \Rightarrow \Delta) = \mathrm{cp}(\Gamma \uplus \Delta) \uplus [\mathrm{cp}(A) + \mathrm{cp}(B) + \mathrm{cp}(C) + 2]$$

whereas the premises have strictly smaller (according to $<_{\mathsf{m}}$) complexities:

$$\begin{aligned}
\mathrm{cp}(\Gamma, C \Rightarrow \Delta) &= \mathrm{cp}(\Gamma \uplus \Delta) \uplus \mathrm{cp}(C) \\
\mathrm{cp}(\Gamma, B \to C \Rightarrow A \to B, \Delta) &= \mathrm{cp}(\Gamma \uplus [B \to C, A \to B] \uplus \Delta) \\
&= \mathrm{cp}(\Gamma \uplus \Delta) \uplus [\mathrm{cp}(A) + \mathrm{cp}(B) + 1, \mathrm{cp}(B) + \mathrm{cp}(C) + 1] \quad \square
\end{aligned}$$

The other crucial aspect of the decomposition rules is that they preserve validity in both directions, premises to conclusion and vice versa. That is, using simple arithmetic, we can show:

Lemma 6.6. *The decomposition rules for* G *are sound and invertible.*

Lemmas 6.5 and 6.6 imply that we can check if a sequent is valid by applying the decomposition rules (upwards) exhaustively, then checking the validity of the resulting atomic implicational sequents. For this latter step, it will be helpful to give a more immediately meaningful presentation of atomic implicational sequents.

Definition 6.7. A *set of inequalities* is a set α of ordered triples $a \lhd b$ where a and b are atoms and $\lhd \in \{<, \leq\}$. We say that α is *valid*, written $\models_{\mathsf{G}} \alpha$, iff for all valuations v, $v(a) \lhd v(b)$ for some $a \lhd b \in \alpha$.

Lemma 6.8. *Let* $\Gamma \Rightarrow \Delta$ *be an atomic implicational sequent and define the set* $\mathsf{Ineqs}(\Gamma \Rightarrow \Delta)$ *by the conditions:*

$$(a \leq b) \in \mathsf{Ineqs}(\Gamma \Rightarrow \Delta) \text{ if } (a \to b) \in \Delta \qquad (\top \leq b) \in \mathsf{Ineqs}(\Gamma \Rightarrow \Delta) \text{ if } b \in \Delta$$

$$(a < b) \in \mathsf{Ineqs}(\Gamma \Rightarrow \Delta) \text{ if } (b \to a) \in \Gamma \qquad (a < \top) \in \mathsf{Ineqs}(\Gamma \Rightarrow \Delta) \text{ if } a \in \Gamma$$

Then $\models_{\mathsf{G}} \mathrm{I}_{\mathsf{G}}(\Gamma \Rightarrow \Delta)$ *iff* $\models_{\mathsf{G}} \mathsf{Ineqs}(\Gamma \Rightarrow \Delta)$.

Proof. Observe that $\models_{\mathsf{G}} \mathrm{I}_{\mathsf{G}}(\Gamma \Rightarrow \Delta)$ iff, by the standard deduction theorem for G (Proposition 3.44), for every valuation v: either $v(A) < 1$ for some $A \in \Gamma$ or $v(B) = 1$ for some $B \in \Delta$. But since Γ and Δ contain only atoms and atomic implications, this means that either $v(a \to b) = 1$ for some $(a \to b) \in \Delta$, $v(b) = 1$ for some $b \in \Delta$, $v(b \to a) < 1$ for some $(b \to a) \in \Gamma$, or $v(a) < 1$ for some $a \in \Gamma$. But this holds iff $v(a) \lhd v(b)$ for some $(a \lhd b) \in \mathsf{Ineqs}(\Gamma \Rightarrow \Delta)$ as required. □

Definition 6.7 gives a criterion for validity of sets of inequalities, but what do valid sets actually look like? Consider the following examples:

$$\{(p \leq q), (q < r), (r \leq s), (s < p)\} \qquad \{(\bot < p), (p < q), (q \leq r)\}$$

Notice that in the first case we have a sequence of inequalities beginning and ending with p, and in the second we have a sequence beginning with $\bot$ and ending with r. This "chain-like" form is a common feature of all such sets.

Lemma 6.9. *A finite set of inequalities* α *is valid iff there exists* $(a_i \lhd_i a_{i+1}) \in \alpha$ *for* $i = 1 \ldots n$ *such that one of the following holds:*

(1) $a_1 = a_{n+1}$ *or* $a_1 = \bot$ *or* $a_{n+1} = \top$, *where* $\lhd_i$ *is* $\leq$ *for some* $i \in \{1, \ldots, n\}$.
(2) $a_1 = \bot$ *and* $a_{n+1} = \top$.

Proof. It is easy to check that α is valid if any of the above conditions are met. For the other direction, we proceed by induction on the number of different variables k occurring in α. Note first that if one of $a \leq a$, $a \leq \top$, $\bot \leq a$, or $\bot < \top$ occurs in α, then we are done. This takes care of the case where $k \leq 1$. Now for $k > 1$, we fix a variable q occurring in α, and define the following sets:

$$\begin{aligned}
\alpha_< &=_{\text{def}} \{a < b : \{a < q, q < b\} \subseteq \alpha\} \\
\alpha_\leq &=_{\text{def}} \{a \leq b : \{a \lhd_1 q, q \lhd_2 b\} \subseteq \alpha \text{ and } \leq \in \{\lhd_1, \lhd_2\}\} \\
\alpha' &=_{\text{def}} \{a \lhd b \in \alpha : a \neq q, b \neq q\} \cup \alpha_< \cup \alpha_\leq
\end{aligned}$$

α' has fewer variables than α. So if α' is valid, then applying the induction hypothesis to α', we have $(a_i \lhd_i a_{i+1}) \in \alpha'$ for $i = 1 \ldots n$, satisfying either (1) or (2) above. But then easily by replacing the inequalities $a_i \lhd_i a_{i+1}$ that occur in $\alpha_<$ or $\alpha_\leq$ appropriately by $a_i \lhd' q$ and $q \lhd'' a_{i+1}$, we get that (1) or (2) holds for α. Hence it is sufficient to show that α' is valid. Suppose otherwise, i.e. that there exists a valuation v such that $v(a) \lhd v(b)$ does not hold for any $a \lhd b \in \alpha'$. We show for a contradiction that α is not valid. Let:

$$x = \min\{v(a) : a \lhd q \in \alpha\} \quad \text{and} \quad y = \max\{v(b) : q \lhd b \in \alpha\}$$

Initial Sequents

$$\frac{}{\Gamma, A \Rightarrow A, \Delta}\ (\text{IDW}) \qquad \frac{}{\Gamma, \bot \Rightarrow \Delta}\ (\bot\Rightarrow) \qquad \frac{}{\Gamma \Rightarrow \top, \Delta}\ (\Rightarrow\top)$$

Decomposition Rules: as in Fig. 6.2.

Atomic Sequent Rules:

$$\frac{\Gamma, b \Rightarrow \Delta \quad \Gamma \Rightarrow a, \Delta}{\Gamma, a \to b \Rightarrow \Delta}\ (\to\Rightarrow)_a \qquad \frac{\Gamma, a \Rightarrow b}{\Gamma \Rightarrow a \to b, \Delta}\ (\Rightarrow\to)_a \qquad \frac{\Gamma, b \to a \Rightarrow a \to b, \Delta}{\Gamma \Rightarrow a \to b, \Delta}\ (\text{LIN})$$

Fig. 6.3 The sequent calculus GG_g

Note first that $x \geq y$. Otherwise it follows that for some a, b, we have $\{a \lhd_1 q, q \lhd_2 b\} \subseteq \alpha$ and $v(a) < v(b)$. But then $(a \lhd b) \in \alpha'$ so $v(a) \geq v(b)$, a contradiction. So there are two cases. If $x > y$, then we extend v such that $x > v(q) > y$. For any $(a \lhd q)$ or $(q \lhd b)$ in α, we have $v(a) \geq x > v(q) > y \geq v(b)$. Hence α is not valid, a contradiction. Now suppose that $x = y$ and extend v such that $v(q) = x$. We must have atoms a_0, b_0 such that $(a_0 < q)$ and $(q < b_0)$ are in α and $v(a_0) = v(b_0) = v(q)$. Now consider any $(a \lhd_1 q)$ or $(q \lhd_2 b)$ in α. Since $(a \lhd_1 b_0)$ and $(a_0 \lhd_2 b)$ are in α', $v(a) \lhd_1 v(q) = v(b_0)$ and $v(a_0) = v(q) \lhd_2 v(b)$ cannot hold. So α is again not valid, a contradiction. □

We get a calculus for G using just the decomposition rules and identifying initial sequents with valid atomic implicational sequents. As we will see in the next chapter, checking the validity of the latter can be done efficiently (in polynomial time, in fact). Here instead, however, we define a sequent calculus for G with very simple initial sequents by adding extra rules for atomic implicational sequents.

Take a look at Fig. 6.3. The rule $(\to\Rightarrow)_a$ is the usual classical implication left rule restricted to atoms, while $(\Rightarrow\to)_a$ combines applications of weakening (W) and implication right $(\Rightarrow\to)$ rules. The rule (LIN) is what really extends the calculus beyond Intuitionistic Logic and characterizes the linearity of the truth values. It captures the fact that $(a \to b) \vee (b \to a)$ is always a theorem of G.

Example 6.10. Consider the GG_g-derivation:

$$\frac{\dfrac{}{r \Rightarrow r, q \to p}\ (\text{IDW}) \qquad \dfrac{\dfrac{}{q \to r, p \to q \Rightarrow p \to q, r, q \to p}\ (\text{IDW})}{q \to r \Rightarrow p \to q, r, q \to p}\ (\text{LIN})}{(p \to q) \to r \Rightarrow r, q \to p}\ ((\to)\to\Rightarrow)$$

Theorem 6.11. $\vdash_{\text{GG}_\text{g}} S$ *iff* $\models_\text{G} \text{I}_\text{G}(S)$.

Proof. For soundness, we proceed by induction on the height of a proof of S in GG_g. Since the decomposition rules are sound and the initial sequents are obviously valid, it is sufficient to check the atomic sequent rules. $(\to\Rightarrow)_a$ and $(\Rightarrow\to)_a$ are already known to be sound, and for (LIN), it is enough to observe that for any valuation v, if $v(b \to a) < 1$, then $v(b) > v(a)$ and so $v(a \to b) = 1$.

For completeness, suppose that $\models_{\mathsf{G}} \mathrm{I}_{\mathsf{G}}(\Gamma \Rightarrow \Delta)$. Using Lemmas 6.5 and 6.6, we can assume that $\Gamma \Rightarrow \Delta$ is atomic implicational. Define:

$$\Gamma' = \Gamma \uplus [a \to b : b \to a \in \Delta]$$

Observe that $\Gamma \Rightarrow \Delta$ is derivable from $\Gamma' \Rightarrow \Delta$ using repeated applications of (LIN). Also, clearly $\models_{\mathsf{G}} \mathrm{I}_{\mathsf{G}}(\Gamma' \Rightarrow \Delta)$. Hence there exists a sequence of inequalities $(a_i \lhd_i a_{i+1}) \in \mathsf{Ineqs}(\Gamma' \Rightarrow \Delta)$ for $i = 1 \ldots n$ satisfying either (1) or (2) from Lemma 6.9. Moreover, we can assume that $\lhd_i$ is $\leq$ for at most one i. Otherwise, replacing any one of the two or more occurrences of $\leq$ with $<$ gives a sequence of inequalities that still satisfies either (1) or (2). If $a \leq b$ is replaced with $a < b$ where $(a \to b) \in \Delta$, then $(b \to a) \in \Gamma$. Also if $b \in \Delta$, then removing $\top \leq b$ still gives a sequence of inequalities satisfying (1) or (2).

We use the sequence to consider the form of the sequents. There are several cases. As a first example, suppose that Δ contains an atom a_1 and we have a sequence $\top \leq a_1 < \ldots < a_k < \top$. Then the sequent is of the form $(\Gamma'', a_k, a_k \to a_{k-1}, \ldots, a_2 \to a_1 \Rightarrow a_1, \Delta'')$ which is easily derived using $(\to\Rightarrow)_a$ and (IDW). Suppose now that Δ contains an implication $a_1 \to a_2$ and we have a sequence $a_1 \leq a_2 < \ldots < a_k < a_1$. Then the sequent is of the form $(\Gamma'', a_1 \to a_k, \ldots, a_3 \to a_2 \Rightarrow a_1 \to a_2, \Delta'')$ which is easily derived using $(\Rightarrow\to)_a$, $(\to\Rightarrow)_a$, and (IDW). Other cases are very similar. □

Notice that unlike the completeness proof for GG, the above proof is entirely semantic. This allows us to give a simple proof of standard completeness for the Hilbert system HG.

Theorem 6.12. *If* $\models_{\mathsf{G}} A$, *then* $\vdash_{\mathsf{HG}} A$.

Proof. Suppose that $\models_{\mathsf{G}} A$. Then by Theorem 6.11, $\vdash_{\mathsf{GG_g}} \Rightarrow A$. But $\mathsf{GG_g}$ is sound with respect to G-chains (the proof is the same as for the standard G-algebra). So A is valid in all G-chains, and hence by Theorem 3.56, $\vdash_{\mathsf{HG}} A$. □

6.1.3 Another Hypersequent Calculus

The sequent calculus $\mathsf{GG_g}$ has rather a lot of decomposition rules, even with a limited language (re-introducing $\vee$ as a primitive connective adds another four). However, if we deal with single-conclusion hypersequents, then we can have fewer rules and also make some modest improvements in efficiency. Proceeding rather informally to avoid repetition, we define a hypersequent calculus $\mathsf{GG_h}$ consisting of:

(1) all valid atomic implicational hypersequents.
(2) single-conclusion hypersequent versions of $(\wedge\Rightarrow)_{\mathsf{G}}$, $(\Rightarrow\wedge)$, $(\wedge\to\Rightarrow)$, $(\to\wedge\Rightarrow)$, $(\Rightarrow\to)$, $(\to(\to)\Rightarrow)$, plus:

$$\frac{\mathcal{G} \mid \Gamma, B \to C \Rightarrow \Delta \mid A \Rightarrow B \quad \mathcal{G} \mid \Gamma, C \Rightarrow \Delta}{\mathcal{G} \mid \Gamma, (A \to B) \to C \Rightarrow \Delta} \qquad \frac{\mathcal{G} \mid \Gamma \Rightarrow \Delta \mid p \to q \Rightarrow p \quad \mathcal{G} \mid \Gamma, q \Rightarrow \Delta}{\mathcal{G} \mid \Gamma, p \to q \Rightarrow \Delta}$$

Example 6.13. Consider the following application of the rule on the left in (2):

$$\frac{p \to r \Rightarrow r \mid p \to q \Rightarrow p \mid p \Rightarrow q \quad r \Rightarrow r \mid p \Rightarrow q}{((p \to q) \to p) \to r \Rightarrow r \mid p \Rightarrow q}$$

The right premise is valid because it contains $r \Rightarrow r$, and the left is valid because it contains $p \Rightarrow q$ (corresponding to $p \leq q$) and $p \to q \Rightarrow p$ (corresponding to $q < p$).

Soundness and completeness results proceed similarly to those above for the sequent calculus, and it is straightforward also to add rules for deriving atomic implicational hypersequents. We refer the interested reader to the historical remarks for further details.

6.1.4 A Sequent of Relations Calculus

There is also a way to have invertible logical rules that feature just one principal connective at a time: introduce more structure. In particular, we can allow two types of sequents, corresponding intuitively to $\leq$ and $<$, and treat sets of these sequents where exactly one formula occurs on each side.

Definition 6.14. A *sequent of relations* is a set of ordered triples:

$$\mathrm{S} = (\mathrm{A_1} \lhd_1 \mathrm{B_1} \mid \ldots \mid \mathrm{A_n} \lhd_n \mathrm{B_n})$$

where A_i and B_i are formulas and $\lhd_i \in \{<, \leq\}$ for $i = 1 \ldots n$. S is *valid*, written $\models_{\mathrm{G}} \mathrm{S}$, iff for all valuations v, $v(A_i) \lhd_i v(B_i)$ for some $i \in \{1, \ldots, n\}$.

By introducing more structure we get the best of both worlds: simple rules with the subformula property that are also invertible. Indeed, since there are not too many rules, we will return to using the full language $\mathcal{L}_{\mathrm{G}}$ with $\vee$ as a primitive connective. The sequent of relations calculus $\mathsf{GG_r}$ for this language is presented in Fig. 6.4.

Example 6.15. We illustrate $\mathsf{GG_r}$ with the following derivation:

$$\cfrac{\cfrac{\cfrac{\cfrac{p \leq q \mid \top \leq q \mid \top \leq p \mid q < p \quad p \leq q \mid \top \leq q \quad \top \leq p \mid q \leq p}{p \leq q \mid \top \leq q \mid p \to q \leq p \mid \top \leq p}\ (\to\leq)}{p \leq q \mid \top \leq q \mid \top \leq (p \to q) \to p}\ (\leq\to)}{\top \leq p \to q \mid \top \leq (p \to q) \to p}\ (\leq\to)}{\top \leq (p \to q) \vee ((p \to q) \to p)}\ (\leq\vee)$$

The uppermost sequents of relations of this derivation are valid since for any valuation v, always $v(p) \leq v(q)$ or $v(q) < v(p)$.

It is easy to show that the logical rules are sound and invertible, i.e. preserve validity in both directions. The cases of $\wedge$ and $\vee$ are almost immediate, and for the

Initial Sequents Of Relations: any valid atomic sequent of relations.

Logical Rules:

$$\frac{\mathcal{G} \mid A \lhd C \mid B \lhd C}{\mathcal{G} \mid A \wedge B \lhd C} \ (\wedge\lhd) \qquad \frac{\mathcal{G} \mid C \lhd A \quad \mathcal{G} \mid C \lhd B}{\mathcal{G} \mid C \lhd A \wedge B} \ (\lhd\wedge)$$

$$\frac{\mathcal{G} \mid A \lhd C \quad \mathcal{G} \mid B \lhd C}{\mathcal{G} \mid A \vee B \lhd C} \ (\vee\lhd) \qquad \frac{\mathcal{G} \mid C \lhd A \mid C \lhd B}{\mathcal{G} \mid C \lhd A \vee B} \ (\lhd\vee)$$

$$\frac{\mathcal{G} \mid B < A \quad \mathcal{G} \mid B < C}{\mathcal{G} \mid A \to B < C} \ (\to<) \qquad \frac{\mathcal{G} \mid A \leq B \mid C < B \quad \mathcal{G} \mid C < \top}{\mathcal{G} \mid C < A \to B} \ (<\to)$$

$$\frac{\mathcal{G} \mid \top \leq C \mid B < A \quad \mathcal{G} \mid B \leq C}{\mathcal{G} \mid A \to B \leq C} \ (\to\leq) \qquad \frac{\mathcal{G} \mid A \leq B \mid C \leq B}{\mathcal{G} \mid C \leq A \to B} \ (\leq\to)$$

Fig. 6.4 The sequent of relations calculus $\mathsf{GG_r}$

implication rules, just observe that $v(C) \lhd v(A \to B)$ iff either $v(C) \lhd v(B)$ or both $v(C) \lhd 1$ and $v(A) \leq v(B)$. So we can reduce the validity of a sequent of relations to the validity of atomic sequents of relations, and hence:

Theorem 6.16. $\vdash_{\mathsf{GG_r}} S$ *iff* $\models_{\mathsf{G}} S$.

6.2 Łukasiewicz Logic

Just as G may be considered the logic of order, so Łukasiewicz Logic Ł can be viewed as the logic of magnitude. In this logic, size matters. Whereas G is based on the (only) idempotent t-norm min, Łukasiewicz Logic is based on the nilpotent Archimedean t-norm $x *_{Ł} y = \min(1, 1-x+y)$. For simplicity, let us again use a restricted (but fully expressive) language, this time $\mathcal{L}_{\mathrm{L}} = \{\to, \bot\}$ with defined connectives:

$$\begin{array}{rlrl} \neg A &=_{\text{def}} A \to \bot & \top &=_{\text{def}} \neg\bot \\ A \odot B &=_{\text{def}} \neg(A \to \neg B) & A \oplus B &=_{\text{def}} \neg A \to B \\ A \wedge B &=_{\text{def}} A \odot (A \to B) & A \vee B &=_{\text{def}} (A \to B) \to B \end{array}$$

An Ł-*valuation* is a function $v : \mathrm{Fm}_{\mathcal{L}_{\mathrm{L}}} \to [0,1]$ such that $v(\bot) = 0$ and:

$$v(A \to B) = \min(1, 1 - v(A) + v(B))$$

where the valuations of the defined connectives emerge as expected as:

$$\begin{array}{rlrl} v(\neg A) &= 1 - v(A) & v(\top) &= 1 \\ v(A \odot B) &= \max(0, v(A) + v(B) - 1) & v(A \oplus B) &= \min(1, v(A) + v(B)) \\ v(A \wedge B) &= \min(v(A), v(B)) & v(A \vee B) &= \max(v(A), v(B)) \end{array}$$

A formula A is Ł-*valid*, written $\models_Ł A$, iff $v(A) = 1$ for all Ł-valuations v.

A Hilbert calculus for Ł in $\mathcal{L}_L$ consists of the rule (MP) and the axioms:

$$
\begin{aligned}
&(Ł1)\ A \to (B \to A)\\
&(Ł2)\ (A \to B) \to ((B \to C) \to (A \to C))\\
&(Ł3)\ ((A \to B) \to B) \to ((B \to A) \to A)\\
&(Ł4)\ ((A \to \bot) \to (B \to \bot)) \to (B \to A)
\end{aligned}
$$

It is straightforward to prove that this axiomatization derives the same theorems (using the defined connectives) as HŁ (HBL extended with the involution axioms). Hence it is complete with respect to Ł-chains (term equivalent to MV-chains). Moreover, using the methods of Chapter 3, these systems extended with the density rule are even complete with respect to dense Ł-chains. Unfortunately, however, a proof of the key standard completeness theorem is quite involved and beyond the proof-theoretic scope of this book. Instead, let us just state the theorem, referring to the historical remarks at the end of the chapter for references.

Theorem 6.17. $\models_Ł A$ *iff* $\vdash_{HŁ} A$.

For convenience, let us assume for the remainder of this section that all formulas are based on the language $\mathcal{L}_L$, and drop the prefix Ł for valuations and validity.

6.2.1 A Hypersequent Calculus

Unlike Gödel Logic, we have not yet encountered any sequent or hypersequent calculi for Łukasiewicz Logic. The natural and most desirable solution would be just to add some further structural rules to the calculus GIMTL: the closest we have come to a calculus for Ł so far. However, no such rules have yet been discovered, nor indeed do we expect them to be. Instead, we take a different approach: we interpret sequents outside of the language of Ł. Rather than give a direct interpretation of sequents and hypersequents as formulas, we define a semantic criterion for validity.

Definition 6.18. Let $\star_Ł^v(\Gamma) = 1 + \sum[v(A) - 1 : A \in \Gamma]$; we define:

$$\models_Ł \mathcal{G} \quad \text{iff} \quad \text{for all valuations } v\text{: } \star_Ł^v(\Gamma) \leq \star_Ł^v(\Delta) \text{ for some } (\Gamma \Rightarrow \Delta) \in \mathcal{G}.$$

A hypersequent rule $\mathcal{G}_1, \ldots, \mathcal{G}_n \ / \ \mathcal{G}$ is called *sound* if whenever $\models_Ł \mathcal{G}_i$ for $i = 1 \ldots n$, also $\models_Ł \mathcal{G}$, and *invertible* if the reverse implication holds.

This definition may seem a little strange. However, notice that in the single-conclusion case:

$$\models_Ł A_1, \ldots, A_n \Rightarrow B \quad \text{iff} \quad \models_Ł (A_1 \odot \ldots \odot A_n) \to B$$

Moreover, as we will see at the end of this section, there exists a method (based on McNaughton's theorem) for interpreting even multiple-conclusion sequents as

Initial Sequents

$$\frac{}{\mathcal{G} \mid A \Rightarrow A}\ (\text{ID}) \qquad \frac{}{\mathcal{G} \mid \Rightarrow}\ (\text{EMP}) \qquad \frac{}{\mathcal{G} \mid \Gamma, \bot \Rightarrow A}\ (\bot\Rightarrow)_{\text{Ł}}$$

Structural Rules:

$$\frac{\mathcal{G}}{\mathcal{G} \mid \mathcal{H}}\ (\text{EW}) \qquad \frac{\mathcal{G} \mid \mathcal{H} \mid \mathcal{H}}{\mathcal{G} \mid \mathcal{H}}\ (\text{EC}) \qquad \frac{\mathcal{G} \mid \Gamma \Rightarrow \Delta}{\mathcal{G} \mid \Gamma, \Pi \Rightarrow \Delta}\ (\text{W})$$

$$\frac{\mathcal{G} \mid \Gamma_1, \Gamma_2 \Rightarrow \Delta_1, \Delta_2}{\mathcal{G} \mid \Gamma_1 \Rightarrow \Delta_1 \mid \Gamma_2 \Rightarrow \Delta_2}\ (\text{SPLIT}) \qquad \frac{\mathcal{G} \mid \Gamma_1 \Rightarrow \Delta_1 \quad \mathcal{G} \mid \Gamma_2 \Rightarrow \Delta_2}{\mathcal{G} \mid \Gamma_1, \Gamma_2 \Rightarrow \Delta_1, \Delta_2}\ (\text{MIX})$$

Logical Rules

$$\frac{\mathcal{G} \mid \Gamma, B \Rightarrow A, \Delta}{\mathcal{G} \mid \Gamma, A \to B \Rightarrow \Delta}\ (\to\Rightarrow)_{\text{A}} \qquad \frac{\mathcal{G} \mid \Gamma \Rightarrow \Delta \quad \mathcal{G} \mid \Gamma, A \Rightarrow B, \Delta}{\mathcal{G} \mid \Gamma \Rightarrow A \to B, \Delta}\ (\Rightarrow\to)_{\text{Ł}}$$

Fig. 6.5 The hypersequent calculus GŁ

formulas of Ł. We will also describe alternative interpretations of hypersequents using Abelian Logic and a two-player dialogue game.

A hypersequent calculus GŁ based on Definition 6.18 is presented in Fig. 6.5. There are some key differences with calculi of previous chapters. First, but just for convenience, the calculus is cut-free. More importantly, the logical rules for $\to$ are non-standard. We use the Abelian Logic implication rule on the left, and a new rule on the right. Observe, however, that the standard implication rule $(\Rightarrow\to)$ is derivable for the single-conclusion case:

$$\frac{\dfrac{\dfrac{}{\mathcal{G} \mid \Rightarrow}\ (\text{EMP})}{\mathcal{G} \mid \Gamma \Rightarrow}\ (\text{W}) \qquad \mathcal{G} \mid \Gamma, A \Rightarrow B}{\mathcal{G} \mid \Gamma \Rightarrow A \to B}\ (\Rightarrow\to)_{\text{Ł}}$$

Note also that the standard rules for $\wedge$ and $\vee$ can be added, while the derived initial hypersequents for $\top$ are of the form $(\mathcal{G} \mid \Gamma \Rightarrow \top)$, i.e. $(\Rightarrow\top)$ restricted to the single-conclusion case like $(\bot\Rightarrow)_{\text{Ł}}$. On the other hand, the appropriate rules (derived and simplified) for the defined connective $\odot$ are non-standard, i.e.

$$\frac{\mathcal{G} \mid \Gamma, A, B \Rightarrow \Delta \quad \mathcal{G} \mid \Gamma, \bot \Rightarrow \Delta}{\mathcal{G} \mid \Gamma, A \odot B \Rightarrow \Delta}\ (\odot\Rightarrow)_{\text{Ł}} \qquad \frac{\mathcal{G} \mid \Gamma \Rightarrow A, B, \Delta \mid \Gamma \Rightarrow \bot, \Delta}{\mathcal{G} \mid \Gamma \Rightarrow A \odot B, \Delta}\ (\Rightarrow\odot)_{\text{Ł}}$$

Example 6.19. Consider the following proof of the axioms for (Ł3):

$$
\dfrac{\dfrac{\dfrac{\dfrac{\dfrac{\dfrac{}{B\Rightarrow B}\,(\text{ID}) \quad \dfrac{}{A\Rightarrow A}\,(\text{ID})}{B,A\Rightarrow A,B}\,(\text{MIX})}{B,B\to A\Rightarrow A}\,(\to\Rightarrow)_{\mathrm{A}} \quad \dfrac{\dfrac{\dfrac{}{B\Rightarrow B}\,(\text{ID}) \quad \dfrac{}{A\Rightarrow A}\,(\text{ID})}{B,A\Rightarrow A,B}\,(\text{MIX})}{B,B\to A,A\Rightarrow A,B}\,(\text{W})}{B,B\to A\Rightarrow A,A\to B}\,(\Rightarrow\to)_{\text{Ł}}}{(A\to B)\to B,B\to A\Rightarrow A}\,(\to\Rightarrow)_{\mathrm{A}}}{(A\to B)\to B\Rightarrow (B\to A)\to A}\,(\Rightarrow\to)}{\Rightarrow ((A\to B)\to B)\to((B\to A)\to A)}\,(\Rightarrow\to)
$$

In fact, all the axioms (Ł1) – (Ł4) can be proved using just sequents. Other theorems require the extra flexibility of hypersequents, however. Recall that (r)* denotes the combination of a rule (r) with (EW) and (EC):

$$
\dfrac{\dfrac{\dfrac{\dfrac{\dfrac{\dfrac{}{C\Rightarrow C}\,(\text{ID}) \quad \dfrac{\dfrac{\dfrac{\dfrac{\dfrac{}{C\Rightarrow C}\,(\text{ID}) \quad \dfrac{\dfrac{\dfrac{\dfrac{}{A\Rightarrow A}\,(\text{ID}) \quad \dfrac{}{B\Rightarrow B}\,(\text{ID})}{A,B\Rightarrow B,A}\,(\text{MIX})}{A\Rightarrow B\mid B\Rightarrow A}\,(\text{SPLIT})}{A\Rightarrow B\mid\Rightarrow B\to A}\,(\Rightarrow\to)}{A\Rightarrow B\mid C\Rightarrow C,B\to A}\,(\text{MIX})^*}{A\Rightarrow B\mid (B\to A)\to C\Rightarrow C}\,(\to\Rightarrow)_{\mathrm{A}}}{\Rightarrow A\to B\mid (B\to A)\to C\Rightarrow C}\,(\Rightarrow\to)}{C\Rightarrow C,A\to B\mid (B\to A)\to C\Rightarrow C}\,(\text{MIX})^*}{(A\to B)\to C\Rightarrow C\mid (B\to A)\to C\Rightarrow C}\,(\to\Rightarrow)_{\mathrm{A}}}{(A\to B)\to C,(B\to A)\to C\Rightarrow C\mid (B\to A)\to C\Rightarrow C}\,(\text{W})}{(A\to B)\to C,(B\to A)\to C\Rightarrow C\mid (A\to B)\to C,(B\to A)\to C\Rightarrow C}\,(\text{W})}{\dfrac{(A\to B)\to C,(B\to A)\to C\Rightarrow C}{\dfrac{(A\to B)\to C\Rightarrow ((B\to A)\to C)\to C}{\Rightarrow ((A\to B)\to C)\to((B\to A)\to C)\to C}\,(\Rightarrow\to)}\,(\Rightarrow\to)}\,(\text{EC})
$$

Soundness for GŁ is established with respect to Definition 6.18 as follows.

Theorem 6.20. *If $d \vdash_{\text{GŁ}} \mathcal{G}$, then $\models_{\text{Ł}} \mathcal{G}$.*

Proof. We proceed by induction on $\text{ht}(d)$. The base cases for the initial sequents are straightforward, as are the cases of (EC), (EW), and (W). For $(\to\Rightarrow)_{\mathrm{A}}$, consider a valuation v and suppose that $\star^v_{\text{Ł}}(\Gamma\uplus[B]) \leq \star^v_{\text{Ł}}(\Delta\uplus[A])$. Unravelling this inequation a bit we get:

$$\star^v_{\text{Ł}}(\Gamma)+(v(B)-1)\leq (v(A)-1)+\star^v_{\text{Ł}}(\Delta)$$

But then rearranging, $\star^v_{\text{Ł}}(\Gamma)+((1-v(A)+v(B))-1)\leq\star^v_{\text{Ł}}(\Delta)$ and hence also $\star^v_{\text{Ł}}(\Gamma)+(\min(1,1-v(A)+v(B))-1)\leq\star^v_{\text{Ł}}(\Delta)$; i.e. $\star^v_{\text{Ł}}(\Gamma\uplus[A\to B])\leq\star^v_{\text{Ł}}(\Delta)$.

For $(\Rightarrow\to)_{\text{Ł}}$, consider a valuation v. Then $\star^v_{\text{Ł}}(\Gamma)\leq\star^v_{\text{Ł}}(\Delta\uplus[A\to B])$ iff:

$$\star^v_{\text{Ł}}(\Gamma)\leq\star^v_{\text{Ł}}(\Delta)+(1-\min(1,1-v(A)+v(B)))$$

Rearranging a little, this holds iff:

$$\star_{\text{Ł}}^{v}(\Gamma) \leq \star_{\text{Ł}}^{v}(\Delta) \text{ and } \star_{\text{Ł}}^{v}(\Gamma) \leq \star_{\text{Ł}}^{v}(\Delta) + v(B) - v(A)$$

where the right inequality holds iff $\star_{\text{Ł}}^{v}(\Gamma \uplus [A]) \leq \star_{\text{Ł}}^{v}(\Delta \uplus [B])$. □

We begin our completeness proof by observing a useful equivalence.

Proposition 6.21. *For an atomic hypersequent* $\mathcal{G} = (\Gamma_1 \Rightarrow \Delta_1 \mid \ldots \mid \Gamma_n \Rightarrow \Delta_n)$, $\vdash_{\text{GŁ}} \mathcal{G}$ *iff there exists* $\lambda_1, \ldots, \lambda_n \in \mathbb{N}$ *with* $\lambda_i > 0$ *for some* $i \in \{1, \ldots, n\}$, *such that:*

$$\biguplus_{i=1}^{n} \Delta_i^{\lambda_i} \subseteq^* \biguplus_{i=1}^{n} \Gamma_i^{\lambda_i}$$

where $\Delta \subseteq^* \Gamma$ *if* $\Delta = \Pi \uplus [A_1, \ldots, A_n]$ *and* $\Gamma = \Pi \uplus [\bot]^n \uplus \Sigma$ *for some* Π *and* Σ.

Proof. For the right-to-left direction, we suppose that the required $\lambda_1, \ldots, \lambda_n$ exist and proceed backwards to obtain a GŁ-derivation of $\mathcal{G}$. First, apply (EC) and (EW) to arrive at a hypersequent:

$$\overbrace{\Gamma_1 \Rightarrow \Delta_1 \mid \ldots \mid \Gamma_1 \Rightarrow \Delta_1}^{\lambda_1} \mid \ldots \mid \overbrace{\Gamma_n \Rightarrow \Delta_n \mid \ldots \mid \Gamma_n \Rightarrow \Delta_n}^{\lambda_n}$$

Now apply (SPLIT) exhaustively to obtain:

$$\Gamma_1^{\lambda_1}, \ldots, \Gamma_n^{\lambda_n} \Rightarrow \Delta_1^{\lambda_1}, \ldots, \Delta_n^{\lambda_n}$$

But since $\biguplus_{i=1}^{n} \Delta_i^{\lambda_i} \subseteq^* \biguplus_{i=1}^{n} \Gamma_i^{\lambda_i}$, by an easy induction on sequent complexity, this is derivable using (W), (ID), (MIX), and $(\bot\Rightarrow)_{\text{Ł}}$.

For the other direction, we prove that if $d \vdash_{\text{GŁ}} \mathcal{G}$, then the condition holds for $\mathcal{G}$, proceeding by induction on $\text{ht}(d)$. The only case that does not follow easily from the induction hypothesis is (MIX). Let $\mathcal{G}' = (\Gamma_2 \Rightarrow \Delta_2 \mid \ldots \mid \Gamma_n \Rightarrow \Delta_n)$, and suppose that $\vdash_{\text{GŁ}} \mathcal{G}' \mid \Pi_j \Rightarrow \Sigma_j$ for $j = 1, 2$ where $\Gamma_1 = \Pi_1 \uplus \Pi_2$ and $\Delta_1 = \Sigma_1 \uplus \Sigma_2$. By the induction hypothesis, we have suitable $\lambda_{j1}, \ldots, \lambda_{jn}$ such that for $j = 1, 2$:

$$\Sigma_j^{\lambda_{j1}} \uplus \biguplus_{i=2}^{n} \Delta_i^{\lambda_{ji}} \subseteq^* \Pi_j^{\lambda_{j1}} \uplus \biguplus_{i=2}^{n} \Gamma_i^{\lambda_{ji}}$$

But then we take $\lambda_i = \lambda_{11}\lambda_{2i} + \lambda_{21}\lambda_{1i}$ for $i = 2 \ldots n$ and $\lambda_1 = \lambda_{11}\lambda_{21}$ and we are done. □

We now show that the derivability of hypersequents can be reduced using invertible rules to the derivability of atomic hypersequents.

Lemma 6.22. *The following rules are both invertible and* GŁ*-derivable:*

$$\frac{\mathcal{G} \mid \Gamma \Rightarrow \Delta \mid \Gamma, B \Rightarrow A, \Delta}{\mathcal{G} \mid \Gamma, A \to B \Rightarrow \Delta} \ (\to\Rightarrow)_{\text{Ł}} \qquad \frac{\mathcal{G} \mid \Gamma \Rightarrow \Delta \quad \mathcal{G} \mid \Gamma, A \Rightarrow B, \Delta}{\mathcal{G} \mid \Gamma \Rightarrow A \to B, \Delta} \ (\Rightarrow\to)_{\text{Ł}}$$

Proof. The rule $(\Rightarrow\rightarrow)_{Ł}$ is trivially GŁ-derivable (being a rule of the system), and Ł-invertibility was established implicitly in the soundness proof above. For $(\rightarrow\Rightarrow)_{Ł}$, we have the derivation:

$$\dfrac{\dfrac{\dfrac{\mathcal{G} \mid \Gamma \Rightarrow \Delta \mid \Gamma, B \Rightarrow A, \Delta}{\mathcal{G} \mid \Gamma \Rightarrow \Delta \mid \Gamma, A \rightarrow B \Rightarrow \Delta}\,(\rightarrow\Rightarrow)_{\mathsf{A}}}{\mathcal{G} \mid \Gamma, A \rightarrow B \Rightarrow \Delta \mid \Gamma, A \rightarrow B \Rightarrow \Delta}\,(\mathrm{W})}{\mathcal{G} \mid \Gamma, A \rightarrow B \Rightarrow \Delta}\,(\mathrm{EC})$$

For invertibility, consider a valuation v and suppose that:

$$\star_{Ł}^{v}(\Gamma) + (v(A \rightarrow B) - 1) \leq \star_{Ł}^{v}(\Delta)$$

If $v(A) \leq v(B)$, then $v(A \rightarrow B) = 1$ and we obtain $\star_{Ł}^{v}(\Gamma) \leq \star_{Ł}^{v}(\Delta)$. If $v(A) > v(B)$, then $v(A \rightarrow B) = 1 - v(A) + v(B)$ and we get $\star_{Ł}^{v}(\Gamma) + (v(B) - v(A)) \leq \star_{Ł}^{v}(\Delta)$. Adding $v(A) - 1$ to each side, $\star_{Ł}^{v}(\Gamma \uplus [A]) \leq \star_{Ł}^{v}(\Delta \uplus [B])$ as required. □

Lemma 6.23. *Every hypersequent $\mathcal{G}$ is derivable from atomic hypersequents using the rules $(\rightarrow\Rightarrow)_{Ł}$ and $(\Rightarrow\rightarrow)_{Ł}$.*

Proof. We proceed by induction on $\mathrm{cp}(\mathcal{G})$. If $\mathcal{G}$ is atomic, then we are done. Otherwise, we have an occurrence of $\rightarrow$ and $\mathcal{G}$ is derivable using $(\rightarrow\Rightarrow)_{Ł}$ or $(\Rightarrow\rightarrow)_{Ł}$ from hypersequents with lower complexity, and so by the induction hypothesis, also from atomic hypersequents. □

Theorem 6.24. *If $\models_{Ł} \mathcal{G}$, then $\vdash_{\mathrm{GŁ}} \mathcal{G}$.*

Proof. By the previous lemmas, we can just consider the case where $\mathcal{G}$ is atomic. We prove the theorem by induction on the number k of distinct propositional variables occurring on the left hand side of sequents in $\mathcal{G}$. Suppose that there are none. It follows that only $\bot$ occurs on the left of sequents. We claim that there must exist a sequent where the number of occurrences of $\bot$ on the left is greater than or equal to the number of formulas on the right. If not, then defining a valuation where all variables take the value 0, we obtain a contradiction. Hence we can easily derive $\mathcal{G}$ using (EW), (W), (MIX), and $(\bot\Rightarrow)_{Ł}$.

For $k > 0$, we pick a variable q occurring on the left of one of the sequents of $\mathcal{G}$. If q occurs on both sides in the same sequent, then we apply (MIX) and (ID) backwards to remove it, noting that the new hypersequent is also valid. Next, we use (EC) and (SPLIT) backwards to multiply sequents, giving (for some λ) a hypersequent:

$$\mathcal{G}' = (\mathcal{G}_0 \mid [\Gamma_i, [q]^{\lambda} \Rightarrow \Delta_i]_{i=1}^{n} \mid [\Pi_j \Rightarrow [q]^{\lambda}, \Sigma_j]_{j=1}^{m})$$

where q does not occur in $\mathcal{G}_0$, Γ_i, Δ_i, Π_j, or Σ_j for $i = 1 \ldots n$ and $j = 1 \ldots m$.

Observe that $\vdash_{\mathrm{GŁ}} \mathcal{G}$ if $\vdash_{\mathrm{GŁ}} \mathcal{G}'$. Also $\models_{Ł} \mathcal{G}'$. Now let:

$$\mathcal{H} = (\mathcal{G}_0 \mid [\Gamma_i, \Pi_j \Rightarrow \Sigma_j, \Delta_i]_{i=1\ldots n}^{j=1\ldots m} \mid [\Gamma_i \Rightarrow \Delta_i]_{i=1}^{n} \mid [\Pi_j \Rightarrow [q]^{\lambda}, \Sigma_j]_{j=1}^{m})$$

Clearly $\mathcal{H}$ contains fewer distinct variables occurring on the left of sequents. Also $\mathcal{G}'$ is derivable from $\mathcal{H}$. Reasoning backwards, we apply (EC) and (SPLIT) to $\mathcal{G}'$ to combine sequents of the form $(\Gamma_i, [q]^\lambda \Rightarrow \Delta_i)$ and $(\Pi_j \Rightarrow [q]^\lambda, \Sigma_j)$ into one: $(\Gamma_i, \Pi_j, [q]^\lambda \Rightarrow [q]^\lambda, \Delta_i, \Sigma_j)$. Then we apply (MIX) and (ID) backwards to remove the balanced occurrences of q, and (W) backwards to $(\Gamma_i, [q]^\lambda \Rightarrow \Delta_i)$ to get $(\Gamma_i \Rightarrow \Delta_i)$. Hence it is sufficient to show that $\mathcal{H}$ is valid, since then by the induction hypothesis $\vdash_{\text{GŁ}} \mathcal{H}$. Suppose otherwise for a contradiction, i.e. that there exists a valuation v such that $\star^v_{\text{Ł}}(\Gamma) > \star^v_{\text{Ł}}(\Delta)$ for all $(\Gamma \Rightarrow \Delta) \in \mathcal{H}$. Define:

$$x = \max(\{\star^v_{\text{Ł}}(\Delta_i) - \star^v_{\text{Ł}}(\Gamma_i) : 1 \leq i \leq n\} \cup \{-\lambda\})$$
$$y = \min(\{\star^v_{\text{Ł}}(\Pi_j) - \star^v_{\text{Ł}}(\Sigma_j) : 1 \leq j \leq m\} \cup \{0\})$$

We claim that $x < y$. Otherwise there exists i, j such that $\star^v_{\text{Ł}}(\Gamma_i) + \star^v_{\text{Ł}}(\Pi_j) \leq \star^v_{\text{Ł}}(\Sigma_j) + \star^v_{\text{Ł}}(\Delta_i)$, a contradiction. Now extend v such that $x < \lambda(v(q) - 1) < y$. Then for $i = 1 \ldots n$ and $j = 1 \ldots m$:

$$\star^v_{\text{Ł}}(\Delta_i) - \star^v_{\text{Ł}}(\Gamma_i) < \lambda(v(q) - 1) \quad \text{and} \quad \lambda(v(q) - 1) < \star^v_{\text{Ł}}(\Pi_j) - \star^v_{\text{Ł}}(\Sigma_j)$$

Hence $\star^v_{\text{Ł}}(\Gamma_i \uplus [q]^\lambda) > \star^v_{\text{Ł}}(\Delta_i)$ and $\star^v_{\text{Ł}}(\Pi_j) > \star^v_{\text{Ł}}(\Sigma_j \uplus [q]^\lambda)$. So $\mathcal{G}'$ is not valid, a contradiction. □

A nice by-product of this completeness proof is a syntactic proof of the *decidability* of Ł. A hypersequent is valid iff it is derivable by applying the invertible implication rules and then iteratively removing occurrences of each variable as described above. Hence we have a terminating procedure to decide whether any given formula A is or is not valid in this logic.

Theorem 6.25. *The validity problem for* Ł *is decidable.*

We have proved the completeness of GŁ semantically. However, it is also possible to give a syntactic proof of completeness using cut elimination, following the "cancellation elimination" procedure for Abelian Logic of the previous chapter. That is, we establish the GŁ-invertibility of $(\rightarrow\Rightarrow)_{\text{Ł}}$ and $(\Rightarrow\rightarrow)_{\text{Ł}}$. We then show that analogues of Lemmas 5.16 and 5.17 for GA also hold for GŁ. Cancellation elimination is established by an induction on the complexity of the cancelled formula A. If A is a variable, then we use these lemmas, otherwise we use the invertibility of the logical rules to decompose the formula.

6.2.2 A Sequent Calculus

Defining a hypersequent calculus for Ł requires ingenuity, but it does not come as a complete surprise. The existence of a sequent calculus for Ł on the other hand is quite unexpected. The key idea here is to represent differences between sequents occurring in a hypersequent using implicational formulas. Consider again the invertible rule $(\rightarrow\Rightarrow)_{\text{Ł}}$. Using the fact that $v(A) + v(A \rightarrow B) = v(B) + v(B \rightarrow A)$ for

Initial Sequents:

$$\frac{}{A \Rightarrow A}\ (\text{ID}) \qquad \frac{}{\Rightarrow}\ (\text{EMP}) \qquad \frac{}{\Gamma, \bot \Rightarrow A}\ (\bot\Rightarrow)_{Ł}$$

Structural Rules

$$\frac{\Gamma^n \Rightarrow \Delta^n}{\Gamma \Rightarrow \Delta}\ (\text{SC}_n)\ n \geq 2 \qquad \frac{\Gamma \Rightarrow \Delta}{\Gamma, \Pi \Rightarrow \Delta}\ (\text{W}) \qquad \frac{\Gamma_1 \Rightarrow \Delta_1 \quad \Gamma_2 \Rightarrow \Delta_2}{\Gamma_1, \Gamma_2 \Rightarrow \Delta_1, \Delta_2}\ (\text{MIX})$$

Logical Rules

$$\frac{\Gamma, B, B \to A \Rightarrow A, \Delta}{\Gamma, A \to B \Rightarrow \Delta}\ (\to\Rightarrow)^s_{Ł} \qquad \frac{\Gamma \Rightarrow \Delta \quad \Gamma, A \Rightarrow B, \Delta}{\Gamma \Rightarrow A \to B, \Delta}\ (\Rightarrow\to)_{Ł}$$

Fig. 6.6 The sequent calculus $GŁ_s$

all valuations v, we replace this rule at the sequent level with:

$$\frac{\Gamma, B, B \to A \Rightarrow A, \Delta}{\Gamma, A \to B \Rightarrow \Delta}\ (\to\Rightarrow)^s_{Ł}$$

Notice that the premise $(\Gamma, B, B \to A \Rightarrow A, \Delta)$ is derivable both from $(\Gamma, B \Rightarrow A, \Delta)$ using (W), and from $(\Gamma \Rightarrow \Delta)$ using $(\to\Rightarrow)^s_{Ł}$ again together with the sequent rules (MIX), (ID), and (W). Intuitively, we think of $(\Gamma, B, B \to A \Rightarrow A, \Delta)$ as representing both sequents. Moreover, the combination of such sequents – performed by the split rule in the hypersequent calculus – is achieved here by the global contraction rules (SC_n) for $n \geq 2$. The resulting calculus $GŁ_s$ is displayed in Fig. 6.6.

Example 6.26. Compare the following $GŁ_s$-derivation with the GŁ-derivation of the same sequent given in Example 6.19:

$$\dfrac{\dfrac{\dfrac{}{A \Rightarrow A}\ (\text{ID}) \quad \dfrac{\dfrac{\dfrac{\dfrac{}{B \Rightarrow B}\ (\text{ID}) \quad \dfrac{}{A \to B \Rightarrow A \to B}\ (\text{ID})}{B, A \to B \Rightarrow B, A \to B}\ (\text{MIX})}{B \to (A \to B), B, A \to B \Rightarrow B, A \to B}\ (\text{W})}{(A \to B) \to B, A \to B \Rightarrow B}\ (\to\Rightarrow)^s_{Ł}}{\dfrac{\dfrac{(A \to B) \to B, A \to B, A \Rightarrow A, B}{(A \to B) \to B, B \to A \Rightarrow A}\ (\to\Rightarrow)^s_{Ł}}{(A \to B) \to B \Rightarrow (B \to A) \to A}\ (\Rightarrow\to)}}{\Rightarrow ((A \to B) \to B) \to ((B \to A) \to A)}\ (\Rightarrow\to)$$

Example 6.27. Just as the axioms of Ł can be proved in GŁ without using hypersequents, so they can be proved in $GŁ_s$ without using (SC_n). Nevertheless these rules are necessary, as illustrated below:

$$\dfrac{\overline{A \Rightarrow A} \quad \dfrac{\dfrac{\overline{A \Rightarrow A} \quad \dfrac{\overline{C \Rightarrow C} \quad \dfrac{\dfrac{\overline{B \Rightarrow B} \quad \dfrac{\dfrac{\overline{B \Rightarrow B} \quad \overline{C \Rightarrow C}}{B,C \Rightarrow B,C}\,(\text{MIX})}{B,B \to C \Rightarrow C}\,(\to\Rightarrow)^s_{\text{Ł}} + (\text{W})}{B,B,B \to C \Rightarrow B,C}\,(\text{MIX})}{B,B,B \to (B \to C) \Rightarrow C}\,(\to\Rightarrow)^s_{\text{Ł}} + (\text{W})}{B,B,C,B \to (B \to C) \Rightarrow C,C}\,(\text{MIX})}{A,B,B,C,B \to (B \to C) \Rightarrow A,C,C}\,(\text{MIX})}{A,B,B,A \to C,B \to (B \to C) \Rightarrow C,C}\,(\to\Rightarrow)^s_{\text{Ł}} + (\text{W})}{\dfrac{\dfrac{A,A,B,B,A \to C,B \to (B \to C) \Rightarrow A,C,C}{A,A,B,B,A \to (A \to C),B \to (B \to C) \Rightarrow C,C}\,(\to\Rightarrow)^s_{\text{Ł}} + (\text{W})}{A,B,A \to (A \to C),B \to (B \to C) \Rightarrow C}\,(\text{sc}_2) + (\text{W})}\,(\text{MIX})$$

Using the definitions of $\wedge$, $\vee$, and $\odot$ we can derive and simplify (using admissibility in the calculus or the completeness results below) the following rules, noting that those for $\wedge$ and $\vee$ are standard on one side but not the other:

$$\dfrac{\Gamma, A, A \to B \Rightarrow \Delta}{\Gamma, A \wedge B \Rightarrow \Delta}\,(\wedge\Rightarrow)_{\text{Ł}} \qquad \dfrac{\Gamma \Rightarrow A, \Delta \quad \Gamma \Rightarrow B, \Delta}{\Gamma \Rightarrow A \wedge B, \Delta}\,(\Rightarrow\wedge)$$

$$\dfrac{\Gamma, A \Rightarrow \Delta \quad \Gamma, B \Rightarrow \Delta}{\Gamma, A \vee B \Rightarrow \Delta}\,(\vee\Rightarrow) \qquad \dfrac{\Gamma, B \to A \Rightarrow A, \Delta}{\Gamma \Rightarrow A \vee B, \Delta}\,(\Rightarrow\vee)_{\text{Ł}}$$

$$\dfrac{\Gamma, A, B \Rightarrow \Delta \quad \Gamma, \bot \Rightarrow \Delta}{\Gamma, A \odot B \Rightarrow \Delta}\,(\odot\Rightarrow)^s_{\text{Ł}} \qquad \dfrac{\Gamma, (A \to \bot) \to B \Rightarrow A, B, \Delta}{\Gamma \Rightarrow A \odot B, \Delta}\,(\Rightarrow\odot)^s_{\text{Ł}}$$

It is easy to see that GŁ_s is sound. We have already checked the soundness of most of the rules in our proofs for GŁ. The cases of (sc_n) and $(\to\Rightarrow)^s_{\text{Ł}}$ follow immediately using the definition of validity and the relationship $v(A) + v(A \to B) = v(B) + v(B \to A)$ for all valuations v. Completeness is much more complicated. Before getting started on the tricky proof, we first establish some useful invertibility properties for the implication rules.

Lemma 6.28. *$(\to\Rightarrow)^s_{\text{Ł}}$ and $(\Rightarrow\to)_{\text{Ł}}$ are GŁ_s-invertible.*

Proof. For $(\to\Rightarrow)^s_{\text{Ł}}$, we make use of the following derivation:

$$\dfrac{\dfrac{\overline{A \Rightarrow A}\,(\text{ID}) \quad \dfrac{\overline{B \Rightarrow B}\,(\text{ID}) \quad \Gamma, A \to B \Rightarrow \Delta}{\Gamma, B, A \to B \Rightarrow B, \Delta}\,(\text{MIX})}{\Gamma, B, A, A \to B \Rightarrow B, A, \Delta}\,(\text{MIX})}{\Gamma, B, B \to A \Rightarrow A, \Delta}\,(\to\Rightarrow)^s_{\text{Ł}}$$

For $(\Rightarrow\to)_{\text{Ł}}$, we prove the GŁ_s-invertibility of a more general rule.

Claim. If $d \vdash_{\text{GŁ}_\text{s}} \Gamma \Rightarrow [A \to B]^k, \Delta$, then $\vdash_{\text{GŁ}_\text{s}} \Gamma, [A]^i \Rightarrow [B]^i, \Delta$ for $i = 0 \ldots k$.

Proof of claim. We proceed by induction on $\text{ht}(d)$. Notice that when $k = 0$ the claim is immediate. For the base case, if d ends with $(A \to B \Rightarrow A \to B)$, then $(A \to B \Rightarrow)$

is derivable using (W) and (EMP). Also, $(A \to B, A \Rightarrow B)$ is derivable using $(\Rightarrow\to)_Ł$, (W), (MIX), and (ID). If d ends with $(\Gamma', \bot \Rightarrow A \to B)$, then the result follows using $(\bot\Rightarrow)_Ł$, (W), and (EMP).

For the inductive step, we consider the last rule applied. The cases of (W), (MIX), $(\to\Rightarrow)^s_Ł$, and $(\Rightarrow\to)_Ł$ with principal formula not $A \to B$, all involve straightforward applications of the induction hypothesis. For the remaining possibility for $(\Rightarrow\to)_Ł$, suppose that d ends with:

$$\frac{\Gamma, A \Rightarrow B, [A \to B]^{k-1}, \Delta \quad \Gamma \Rightarrow [A \to B]^{k-1}, \Delta}{\Gamma \Rightarrow [A \to B]^k, \Delta} \; (\Rightarrow\to)_Ł$$

We get that $\vdash_{GŁ_s} \Gamma, [A]^i \Rightarrow [B]^i, \Delta$ holds for $i = 1 \dots k$ and $i = 0$ as required, by applying the induction hypothesis to the left and right premises, respectively.

Finally for (SC$_n$), suppose that d ends with:

$$\frac{\Gamma^n \Rightarrow [A \to B]^{nk}, \Delta^n}{\Gamma \Rightarrow [A \to B]^k, \Delta} \; (\text{SC}_n)$$

By the induction hypothesis $(\Gamma^n, [A]^{ni} \Rightarrow [B]^{ni}, \Delta^n)$ is derivable for $i = 0 \dots k$. Hence by (SC$_n$), $(\Gamma, [A]^i \Rightarrow [B]^i, \Delta)$ is derivable for $i = 0 \dots k$. □

The basic idea of the completeness proof is to show that $GŁ_s$ simulates $GŁ$, making use of labels to track different sequents in hypersequent derivations.

Definition 6.29. The set of *labels* Lab is built from atomic labels $\{a_i\}_{i \in \mathbb{N}}$ as follows:

(1) $1 \in$ Lab and $a_i \in$ Lab for all $i \in \mathbb{N}$.

(2) If $x \in$ Lab and $y \in$ Lab, then $xy \in$ Lab.

A *labelled formula* is of the form $x{:}A$ where $x \in$ Lab and A is a formula. Also:

- $\Gamma^L = [(1{:}A) : A \in \Gamma]$ for a multiset of formulas Γ.
- $\Gamma^U = [A : (x{:}A) \in \Gamma]$ for a multiset of labelled formulas Γ.

A *labelled sequent* is a structure $[\Gamma \,;\, \Sigma] \Rightarrow \Delta$ where Γ and Δ are multisets of labelled formulas, and Σ is a multiset of formulas.

We map from labelled to unlabelled sequents using *labelling functions* that map the set of labels Lab into the set $\{0, 1\}$, removing formulas labelled with 0 from a sequent and leaving behind those labelled with a 1. Unlabelled formulas are ignored altogether. A labelled sequent S is interpreted by taking the hypersequent formed by applying all possible labelling functions to S. More formally:

Definition 6.30. A *labelling function* is a function $f : \text{Lab} \to \{0, 1\}$ where:

(1) $f(1) = 1$ and $f(a_i) \in \{0, 1\}$ for all $i \in \mathbb{N}$.

(2) $f(xy) = f(x)f(y)$.

Definition 6.31. For a labelled sequent $[\Gamma \,;\, \Sigma] \Rightarrow \Delta$:

$$\models_{\text{Ł}} [\Gamma\ ;\ \Sigma] \Rightarrow \Delta \quad \text{iff} \quad \models_{\text{Ł}} [f(\Gamma) \Rightarrow f(\Delta) : f \text{ a labelling function}]$$

where $f(\Gamma) = [A : (x{:}A) \in \Gamma \text{ and } f(x) = 1]$.

Example 6.32. Consider the labelled sequent:

$$[x{:}p, 1{:}q\ ;\ p \to q] \Rightarrow 1{:}p, x{:}q$$

mapped by a labelling function f to $(q \Rightarrow p)$ if $f(x) = 0$, and to $(p, q \Rightarrow p, q)$ if $f(x) = 1$. Hence, the corresponding hypersequent is $(q \Rightarrow p \mid p, q \Rightarrow p, q)$.

We now introduce rules for labelled sequents that record both sequents for a hypersequent derivation in GŁ and formulas added by the $(\to\Rightarrow)^s_{\text{Ł}}$ rule for GŁ_s. We establish two useful properties for labelled sequents obtained by applying these rules upwards to a labelled sequent of the form $([\Gamma^{\text{L}}\ ;\ ] \Rightarrow \Delta^{\text{L}})$. In particular, we show that if one of the obtained labelled sequents is valid, then removing the labels, it is GŁ_s-derivable.

Lemma 6.33. *Consider the following* intermediate rules for Ł*:*

$$\frac{[\Gamma\ ;\ \Sigma] \Rightarrow \Delta \quad [\Gamma, x{:}A\ ;\ \Sigma] \Rightarrow x{:}B, \Delta}{[\Gamma\ ;\ \Sigma] \Rightarrow x{:}A \to B, \Delta}\ (\Rightarrow\to)^l_{\text{Ł}}$$

$$\frac{[\Gamma, xy{:}B\ ;\ \Sigma, B \to A] \Rightarrow xy{:}A, \Delta}{[\Gamma, x{:}A \to B\ ;\ \Sigma] \Rightarrow \Delta}\ (\to\Rightarrow)^l_{\text{Ł}}$$

where y is a new atomic label not occurring in the conclusion of $(\to\Rightarrow)^l_{\text{Ł}}$.

Let d be a derivation using the intermediate rules for Ł of $[\Gamma_0^{\text{L}}\ ;\ ] \Rightarrow \Delta_0^{\text{L}}$ from a set of labelled sequents X where the set of labelled sequents at the leaves of d is X. Then:

(a) $\vdash_{\text{GŁ}_\text{s}} \Gamma^{\text{U}} \ominus f(\Gamma), \Sigma \Rightarrow \Delta^{\text{U}} \ominus f(\Delta)$ *for all labelling functions f.*
(b) $\vdash_{\text{GŁ}_\text{s}} \Gamma^{\text{U}}, \Sigma \Rightarrow \Delta^{\text{U}}$ *for all $[\Gamma\ ;\ \Sigma] \Rightarrow \Delta \in X$ such that $\models_{\text{Ł}} [\Gamma\ ;\ \Sigma] \Rightarrow \Delta$.*

Proof. (a) We proceed by induction on $\text{ht}(d)$. For the base case, $X = \{[\Gamma_0^{\text{L}}\ ;\ ] \Rightarrow \Delta_0^{\text{L}}\}$. The only label is 1 and for every labelling function f, $f(\Gamma_0^{\text{L}}) = \Gamma_0$ and $f(\Delta_0^{\text{L}}) = \Delta_0$. But $\Rightarrow$ is derivable in GŁ_s by (EMP) so we are done. For the inductive step, we consider the beginning of the derivation and a rule application to members of X.

Suppose first that the rule application is $(\to\Rightarrow)^l_{\text{Ł}}$ and the member of X is the premise $([\Gamma, xy : B\ ;\ \Sigma, B \to A] \Rightarrow xy : A, \Delta)$. We obtain a shorter derivation by removing this first rule application and starting the branch from the conclusion $([\Gamma, x{:}A \to B\ ;\ \Sigma] \Rightarrow \Delta)$. By the induction hypothesis, the lemma then applies to all members of X except possibly the premise. However, also by the induction hypothesis for any labelling function f:

$$\vdash_{\text{GŁ}_\text{s}} (\Gamma^{\text{U}} \uplus [A \to B]) \ominus f(\Gamma \uplus [x{:}A \to B]), \Sigma \Rightarrow \Delta^{\text{U}} \ominus f(\Delta) \qquad (6.1)$$

We want to show that the following sequent is GŁ_s-derivable:

$$(\Gamma^{U} \uplus [B]) \ominus f(\Gamma \uplus [xy:B]), \Sigma, B \to A \Rightarrow (\Delta^{U} \uplus [A]) \ominus f(\Delta \uplus [xy:A])$$

There are two cases:

1. If $f(x) = f(y) = 1$, then $\vdash_{\mathrm{G}Ł_s} \Gamma^{U} \ominus f(\Gamma), \Sigma, B \to A \Rightarrow \Delta^{U} \ominus f(\Delta)$ directly by 6.1.
2. If $f(x) = 0$ or $f(y) = 0$, then since $\vdash_{\mathrm{G}Ł_s} \Gamma^{U} \ominus f(\Gamma), A \to B, \Sigma \Rightarrow \Delta^{U} \ominus f(\Delta)$, by Lemma 6.28 as required:

$$\vdash_{\mathrm{G}Ł_s} \Gamma^{U} \ominus f(\Gamma), B, \Sigma, B \to A \Rightarrow \Delta^{U} \ominus f(\Delta), A$$

Suppose now that the rule application is $(\Rightarrow\to)^{l}_{Ł}$ and the members of X are the premises $([\Gamma\ ;\ \Sigma] \Rightarrow \Delta)$ and $([\Gamma, x:A\ ;\ \Sigma] \Rightarrow x:B, \Delta)$. We obtain a shorter derivation by removing this first rule application and starting the branch from the conclusion $([\Gamma\ ;\ \Sigma] \Rightarrow x:A \to B, \Delta)$. By the induction hypothesis, the lemma then applies to all members of X except possibly the premises. However, also by the induction hypothesis for any labelling function f:

$$\vdash_{\mathrm{G}Ł_s} \Gamma^{U} \ominus f(\Gamma), \Sigma \Rightarrow \Delta^{U} \uplus [A \to B] \ominus f(\Delta \uplus [x:A \to B])$$

Again, there are two cases:

1. If $f(x) = 1$, then $\vdash_{\mathrm{G}Ł_s} \Gamma^{U} \ominus f(\Gamma), \Sigma \Rightarrow \Delta^{U} \ominus f(\Delta)$ as required for both premises.
2. If $f(x) = 0$, then $\vdash_{\mathrm{G}Ł_s} \Gamma^{U} \ominus f(\Gamma), \Sigma \Rightarrow \Delta^{U} \ominus f(\Delta), A \to B$ and the result follows by Lemma 6.28.

For (b), suppose that $\models_{Ł} [\Gamma\ ;\ \Sigma] \Rightarrow \Delta$. We proceed by induction on $\mathrm{cp}(\Gamma^{U} \uplus \Delta^{U})$. If Γ^{U} and Δ^{U} are atomic, then by the completeness of GŁ (Theorem 6.24), and Proposition 6.21 there exist labelling functions $f_1, \ldots, f_n$ such that $\biguplus_{i=1}^{n} f_i(\Delta) \subseteq^{*} \biguplus_{i=1}^{n} f_i(\Gamma)$. Consider the following $\mathrm{G}Ł_s$-derivation for the sequent $\Gamma^{U}, \Sigma \Rightarrow \Delta^{U}$:

$$\dfrac{\dfrac{[f_i(\Gamma)]_{i=1}^{n} \Rightarrow [f_i(\Delta)]_{i=1}^{n} \qquad [\Gamma^{U} \ominus f_i(\Gamma)]_{i=1}^{n}, \Sigma^{n} \Rightarrow [\Delta^{U} \ominus f_i(\Delta)]_{i=1}^{n}}{(\Gamma^{U})^{n}, \Sigma^{n} \Rightarrow (\Delta^{U})^{n}}\ (\text{MIX})}{\Gamma^{U}, \Sigma \Rightarrow \Delta^{U}}\ (\text{SC}_n)$$

The left premise is easily derived using (MIX) and the initial sequents. The right premise is derived by repeated applications of (MIX) and part (a).

If Γ^{U} and Δ^{U} are not atomic, then we have two cases:

- If $\models_{Ł} [\Gamma', x:A \to B\ ;\ \Sigma] \Rightarrow \Delta$, then $\models_{Ł} [\Gamma', xy:B\ ;\ \Sigma, B \to A] \Rightarrow xy:A, \Delta$. Also, by the induction hypothesis:

$$\vdash_{\mathrm{G}Ł_s} \Gamma', B, \Sigma, B \to A \Rightarrow A, \Delta$$

  So by an application of $(\to\Rightarrow)^{s}_{Ł}$, $\vdash_{\mathrm{G}Ł_s} \Gamma', A \to B, \Sigma \Rightarrow \Delta$ as required.
- If $\models_{Ł} [\Gamma; \Sigma] \Rightarrow x:A \to B, \Delta'$, then $\models_{Ł} [\Gamma; \Sigma] \Rightarrow \Delta'$ and $\models_{Ł} [\Gamma, x:A\ ;\ \Sigma] \Rightarrow x:B, \Delta'$. By the induction hypothesis twice:

$$\vdash_{\mathrm{G}Ł_s} \Gamma, \Sigma \Rightarrow \Delta \quad \text{and} \quad \vdash_{\mathrm{G}Ł_s} \Gamma, A, \Sigma \Rightarrow B, \Delta$$

So by an application of $(\Rightarrow\rightarrow)_{\text{Ł}}$, $\vdash_{\text{GŁ}_\text{s}} \Gamma, \Sigma \Rightarrow A \rightarrow B, \Delta$ as required. □

Finally, we can get the following completeness theorem for GŁ_s. Just observe that if $\models_{\text{Ł}} \Gamma \Rightarrow \Delta$, then easily also $\models_{\text{Ł}} [\Gamma^{\text{L}}\ ;\ ] \Rightarrow \Delta^{\text{L}}$, and hence by part (b) of the previous lemma, $\vdash_{\text{GŁ}_\text{s}} \Gamma \Rightarrow \Delta$.

Theorem 6.34. $\models_{\text{Ł}} S$ *iff* $\vdash_{\text{GŁ}_\text{s}} S$.

6.2.3 An Embedding into Abelian Logic

While we have given sequent and hypersequent calculi that are sound and complete for Ł, it is perhaps something of a mystery as to what they actually *mean*. Can we interpret sequents and hypersequents in a more concrete fashion than simply stating when they are valid? We will give three (different) positive answers to this question: one making use of McNaughton's theorem, another a dialogue game, and first, here, by giving an "embedding" of Ł into Abelian Logic A.

The idea for this embedding is to translate implicational formulas $B \rightarrow C$ of Ł into formulas of A by bounding the antecedent B from above by e, and the succedent C from below by a new variable $q_\bot$ playing the role of falsity.

Theorem 6.35. $\models_{\text{Ł}} A$ *iff* $\models_{\text{A}} A^*$ *where for a fixed variable* $q_\bot$ *not occurring in* A*:*

$$\begin{aligned} p^* &= p \\ \bot^* &= q_\bot \\ (B \rightarrow C)^* &= (B^* \wedge \text{e}) \rightarrow (C^* \vee q_\bot) \end{aligned}$$

Proof. Recall that by Theorem 5.22, $\models_{\text{A}} C$ iff $\models_{\mathbf{R}} C$ for any $C \in \text{Fm}_{\mathcal{L}_\text{F}}$ where $\mathbf{R} = \langle \mathbb{R}, \min, \max, +, \rightarrow_+, 0, 0\rangle$ with $x \rightarrow_+ y = y - x$. For the left-to-right direction, we first check (an easy exercise) that the translated versions of the axioms (Ł1) – (Ł4) are A-tautologies. We then show that the translated version of (MP) is A-admissible; i.e. we suppose that $\models_{\text{A}} A^*$ and $\models_{\text{A}} (A \rightarrow B)^*$, and show that $\models_{\text{A}} B^*$. By the first assumption, $v(A^*) \geq 0$ for all $\mathbf{R}$-valuations v. Hence using the second assumption, $0 \leq v((A \rightarrow B)^*) = v((A^* \wedge \text{e}) \rightarrow (B^* \vee q_\bot)) = v(B^* \vee q_\bot)$. If B is atomic, then for $v(B) = v(q_\bot) = -1$ in $\mathbf{R}$ we have $v((A^* \wedge \text{e}) \rightarrow (B^* \vee q_\bot)) < 0$, a contradiction. So $B = C \rightarrow D$ and $B^* = (C^* \wedge \text{e}) \rightarrow (D^* \vee q_\bot)$. But then $v(B^*) \geq v(q_\bot)$ and $v(B^*) = v(B^* \vee q_\bot) \geq 0$ as required.

For the right-to-left direction, suppose that $\not\models_{\text{Ł}} A$ where $q_\bot$ does not occur in A. So $v(A) < 1$ for some Ł-valuation v. We define an $\mathbf{R}$-valuation v' as follows:

$$v'(p) = \begin{cases} -1 & \text{if } p \text{ is } q_\bot \\ v(p) - 1 & \text{otherwise.} \end{cases}$$

Claim. $v'(B^* \wedge \text{e}) = v(B) - 1$ for all $B \in \text{Fm}_{\mathcal{L}_\text{L}}$.

Note that if the claim holds, then $v'(A^*) = v(A) - 1 < 0$ so $\not\models_{\text{A}} A^*$ as required.

Proof of claim. We proceed by induction on $\mathrm{cp}(B)$. The base cases hold by definition. If $B = C \to D$ then (using the induction hypothesis for the last step):

$$\begin{aligned} v'(B^* \wedge \mathrm{e}) &= v'(((C^* \wedge \mathrm{e}) \to (D^* \vee q_\perp)) \wedge \mathrm{e}) \\ &= \min(0, v'((C^* \wedge \mathrm{e}) \to (D^* \vee q_\perp))) \\ &= \min(0, v'(D^* \vee q_\perp) - v'(C^* \wedge \mathrm{e})) \\ &= \min(0, \max(v'(D^*), -1) - v(C) + 1) \end{aligned}$$

Using the induction hypothesis again, $v'(D^* \wedge \mathrm{e}) = v(D) - 1$. Hence, if $v'(D^*) \leq 0$, then $v'(B^* \wedge \mathrm{e}) = \min(0, (v(D) - 1) - v(C) + 1) = \min(1, 1 - v(C) + v(D)) - 1 = v(B) - 1$. Also, if $v'(D^*) > 0$, then $v(D) = 1$. So $v'(B^* \wedge \mathrm{e}) = \min(0, v'(D^*) - v(C) + 1) = 0 = v(B) - 1$. □

This result can be understood as showing that Łukasiewicz Logic Ł is a (quite natural) *fragment* of Abelian Logic A. Moreover, it can be used to give an interpretation of hypersequents for Ł as formulas of A. Let $\Gamma^* = [A^* : A \in \Gamma]$ for any multiset Γ of formulas from $\mathrm{Fm}_{\mathcal{L}_\mathrm{L}}$. Then using Theorem 6.35, it is not hard to show that:

$$\models_{\text{Ł}} \Gamma_1 \Rightarrow \Delta_1 \mid \ldots \mid \Gamma_n \Rightarrow \Delta_n \quad \text{iff} \quad \models_{\mathsf{A}} \mathrm{I}(\Gamma_1^* \Rightarrow \Delta_1^* \mid \ldots \mid \Gamma_n^* \Rightarrow \Delta_n^*)$$

6.2.4 McNaughton Functions

Let us now consider a different approach. Rather than interpret sequents outside of Ł (i.e. in A), we can construct a (rather complicated) formula interpretation within the logic itself. To achieve this, a special class of functions is introduced. For convenience, we use vector notation, writing $\bar{x}$ and $\bar{a}$ for an n-tuple of variables $x_1, \ldots, x_n$ or numbers $a_1, \ldots, a_n$, respectively, and denoting by $\bar{a}\bar{x}$ the standard scalar product $\sum_{i=1}^n a_i x_i$ of $\bar{a}$ and $\bar{x}$.

Definition 6.36. $f : [0,1]^n \to [0,1]$ is a *McNaughton function* (of n variables) if:

1. f is continuous.
2. f is piecewise linear; i.e. there exist $g_i(\bar{x}) = \bar{a}_i\bar{x} + b_i$ with $\bar{a}_i, b_i \in \mathbb{Z}$ for $i = 1 \ldots k$, and for any $\bar{x} \in [0,1]^n$, $f(\bar{x}) = g_j(\bar{x})$ for some $j \in \{1, \ldots, k\}$.

A McNaughton function f is called *simple* if there exists $g(\bar{x}) = \bar{a}\bar{x} + b$ with $\bar{a}, b \in \mathbb{Z}$ such that $f(\bar{x}) = g^\#(\bar{x})$ where $g^\#(\bar{x}) =_{\mathrm{def}} \min(1, \max(0, g(\bar{x})))$.

It is easy to see that the interpretation of any formula A in the standard algebra for Ł determines a McNaughton function.

Definition 6.37. A formula A containing variables $p_1, \ldots, p_n$ uniquely determines a McNaughton function $m_A : [0,1]^n \to [0,1]$ as follows:

(1) If $A = p_i$, then $m_A(\bar{x}) = x_i$.
(2) If $A = \perp$, then $m_A(\bar{x}) = 0$.
(3) If $A = B \to C$, then $m_A(\bar{x}) = \min(1, 1 - m_B(\bar{x}) + m_C(\bar{x}))$.

The tricky part is of course to go the other way. Here we restrict our attention to certain combinations of simple McNaughton functions, starting with a constructive procedure that for any simple McNaughton function $f(x_1,\dots,x_n)$, outputs a formula $A_f(p_1,\dots,p_n)$ such that $m_{A_f} = f$.

Proposition 6.38. *Let $f = \bar{a}\bar{x} + b$ with $\bar{a}, b \in \mathbb{Z}$. The formula B_f is defined by induction on $\sigma(f) = \sum_i |a_i|$ as follows:*

1. If $\sigma(f) = 0$, then $f = b$ and let:

$$B_f = \begin{cases} \top & \text{if } b \geq 1 \\ \bot & \text{if } b \leq 0 \end{cases}$$

2. For $\sigma(f) > 0$, let $j = \min\{j : a_j \neq 0\}$ and let:

$$B_f = \begin{cases} (B_{f-x_j} \oplus p_j) \odot B_{f-x_j+1} & \text{if } a_j > 0 \\ B_f = \neg((B_{1-f-x_j} \oplus p_j) \odot B_{2-f-x_j}) & \text{if } a_j < 0 \end{cases}$$

Let A_f be B_f with occurrences of $C \odot \top$ and $C \oplus \bot$ replaced by C. Then $m_{A_f} = f^\#$.

Proof. Clearly it is enough to show that $m_{B_f} = f^\#$. We proceed by induction on $\sigma(f)$. The base case is immediate. For $\sigma(f) > 0$, assume without loss of generality that the first $a_j \neq 0$ is a_1. For $a_1 > 0$, let $h = f - x_1$. Then from the above definition:

$$B_f = (B_h \oplus p_1) \odot B_{h+1}$$

By the induction hypothesis, $m_{B_h} = h^\#$ and $m_{B_{h+1}} = (h+1)^\#$. We need to show that $m_{B_f} = (h + x_1)^\# = f^\#$. We have different cases depending on $\bar{x}$. The cases where $h(\bar{x}) > 1$ or $h(\bar{x}) < -1$ are immediate. If $h(\bar{x}) \in [0,1]$, then $h^\#(\bar{x}) = h(\bar{x})$ and $(h(\bar{x})+1)^\# = 1$. Hence $(h(\bar{x})+x_1)^\# = \min(1, h(\bar{x})+x_1) = m_{B_f}(\bar{x})$. If $h(\bar{x}) \in [-1,0]$, then $h^\# = 0$ and $(h(\bar{x})+1)^\# = h(\bar{x})+1$. So $(h(\bar{x})+x_1)^\# = \max(0, h(\bar{x})+x_1) = \max(0, (x_1 + (h(\bar{x})+1) - 1)$ as required. If $a_1 < 0$, then consider B_{1-f}. By the above reasoning, we obtain $m_{B_{1-f}} = 1 - f^\#$. So since $B_f = \neg B_{1-f}$, we obtain $m_{B_f} = f^\#$ as required. □

Example 6.39. Consider these formulas for some basic linear functions:

$$\begin{aligned} A_x &= p & A_{-x} &= \bot \\ A_{1-x} &= \neg p & A_{x-1} &= \bot \\ A_{x+y} &= p \oplus q & A_{x+y-1} &= p \odot q \\ A_{x-y} &= p \odot \neg q & A_{x-y+1} &= p \oplus \neg q \end{aligned}$$

Now observe that using the usual distributivity laws for min, max, and +, we have a way of writing the non-standard interpretation of sequents in Definition 6.18 as the minimum of maximums of linear functions. A sequent interpretation is then obtained by taking the conjunction of the disjunction of the formulas associated with these functions by the previous proposition.

More formally, let:

$$\begin{aligned} m_{(\Gamma\Rightarrow\Delta)} &= 1 - \Sigma_{C\in\Gamma}(m_C - 1) + \Sigma_{D\in\Delta}(m_D - 1) \\ &= \min_{i=1\dots n} \max_{j=1\dots m_i} g_{ij} \end{aligned}$$

where g_{ij} is linear for $i = 1\dots n$ and $j = 1\dots m_i$, and define:

$$\mathrm{I}_{\text{Ł}}(\Gamma \Rightarrow \Delta) =_{\text{def}} \bigwedge_{i=1}^{n} \bigvee_{j=1}^{m_i} A_{g_{ij}} \quad \text{and} \quad \mathrm{I}_{\text{Ł}}(S_1 \mid \dots \mid S_n) =_{\text{def}} \mathrm{I}_{\text{Ł}}(S_1) \vee \dots \vee \mathrm{I}_{\mathrm{P}}(S_n).$$

Then putting everything together, for any hypersequent $\mathcal{G}$:

$$\models_{\text{Ł}} \mathcal{G} \quad \text{iff} \quad \models_{\text{Ł}} \mathrm{I}_{\text{Ł}}(\mathcal{G})$$

Example 6.40. Consider the sequent $(p \to q \Rightarrow r)$. First we calculate:

$$\begin{aligned} m_{(p\to q\Rightarrow r)} &= 1 - (m_{p\to q} - 1) + (m_r - 1) \\ &= 1 - (\min(1, 1 - x + y) - 1) + (z - 1) \\ &= \max(z, x - y + z) \end{aligned}$$

and then the corresponding formulas:

$$\begin{aligned} A_z &= r \\ A_{x-y+z} &= (A_{x-y} \oplus r) \odot A_{x-y-1} \\ &= ((p \odot \neg q) \oplus r) \odot (p \oplus \neg q) \\ \mathrm{I}_{\text{Ł}}(p \to q \Rightarrow r) &= r \vee (((p \odot \neg q) \oplus r) \odot (p \oplus \neg q)) \end{aligned}$$

Note that this last formula is equivalent in Ł to $(p \to q) \to r$.

Now consider the sequent $(\Rightarrow p, p)$:

$$\begin{aligned} m_{(\Rightarrow p,p)} &= 1 + (m_p - 1) + (m_p - 1) \\ &= 1 + (x - 1) + (x - 1) \\ &= 2x - 1 \end{aligned}$$

Since $A_{2x-1} = p \odot p$, we get $\mathrm{I}_{\text{Ł}}(\Rightarrow p, p) = p \odot p$. Hence also:

$$\mathrm{I}_{\text{Ł}}(p \to q \Rightarrow r \mid \Rightarrow p, p) = r \vee (((p \odot \neg q) \oplus r) \odot (p \oplus \neg q)) \vee (p \odot p)$$

6.2.5 Giles's Game

We conclude our discussion of interpretations for Ł with a game introduced by Robin Giles in the 1970s. The situation is as follows: two players – me and you, say – assert a number of statements and agree to pay \$1 to their opponent for every false statement made. To focus on the connections with proof theory, let us represent this state of affairs by a sequent built from formulas (representing the statements) in the language $\mathcal{L}_{\text{L}} = \{\to, \bot\}$:

$$A_1, \ldots, A_n \Rightarrow B_1, \ldots, B_m$$

where $A_1, \ldots, A_n$ are asserted by you and $B_1, \ldots, B_m$ are asserted by me.

The two players make moves in the game in any order[2] according to the following rule, reflecting the idea that if I (or you) state "A implies B", then I (or you) should be prepared to state B if you (or I) state A:

> If I assert $A \to B$, then you can choose *either* to attack this statement by asserting A, in which case I have to assert also B, *or* not to attack this statement in which case it is removed from the statements that I assert. (And *vice versa*, i.e. for the roles of me and you switched.)

The choices for me (the left two) and you (the right two) can be written as:

$$\frac{\Gamma, B \Rightarrow A, \Delta}{\Gamma, A \to B \Rightarrow \Delta} \qquad \frac{\Gamma \Rightarrow \Delta}{\Gamma, A \to B \Rightarrow \Delta} \qquad \frac{\Gamma, A \Rightarrow B, \Delta}{\Gamma \Rightarrow A \to B, \Delta} \qquad \frac{\Gamma \Rightarrow \Delta}{\Gamma \Rightarrow A \to B, \Delta}$$

Each premise here represents the sequent resulting from you (for the first two) or me (for the second two) either attacking or not attacking $A \to B$ in the conclusion.

What then does it mean to win such a game? For Ł, one answer is to regard each occurrence of a variable q as stating that "a certain *elementary experiment* E_q yields a positive result" where for each run of the game the outcome of E_q has a fixed probability $v(q)$ of being positive. For the atomic statement $\bot$ we let $v(\bot) = 0$; i.e. the experiment $E_\bot$ always yields a negative answer. To illustrate this idea, consider the sequent $(p \Rightarrow q, q)$: the experiment E_p has to be performed once and the experiment E_q has to be performed twice. If, for example, all three results are negative, then I owe you \$2 and you owe me \$1. If the results of E_p and the first trial of E_q are positive, and the second trial of E_q is negative, then I owe you \$1 and you owe me nothing.

A *run* of the game for a sequent $\Gamma \Rightarrow \Delta$ consists of:

1. a sequence of sequents obtained one from the other by the above rules, starting with $\Gamma \Rightarrow \Delta$ and ending in an atomic sequent.
2. an assignment v of risk values from $[0,1]$ to variables occurring in the game.

I *win* a run of the game if I do not expect any loss of money resulting from betting on results of the elementary experiments in the final atomic sequent $(\Pi \Rightarrow \Sigma)$. This means that for every p that I (or you) assert, i.e. occurring in Σ (or Π), my (or your) expected loss is $1 - v(p)$. Hence easily, I win iff $\star_{\text{Ł}}^{v}(\Pi) \leq \star_{\text{Ł}}^{v}(\Sigma)$. Note that this winning condition refers to *average* pay-offs and not to a certain instance of evaluating the final elementary sequent by performing corresponding experiments.

[2] Note that an order can be imposed on the moves in the game without affecting the reasoning here, but we choose to follow Giles' original description.

E.g. $p \Rightarrow p$ is a winning sequent for me, although it may happen that the result of performing experiment E_p is positive for your assertion of p, while it is negative for my assertion of p, implying that I owe you \$1.

Example 6.41. Consider the game where I initially assert $p \rightarrow q$; i.e. the game starts with the sequent $(\Rightarrow p \rightarrow q)$. For your turn (there is nothing I could do if it were my turn), you can either assert p in order to force me to assert q, or explicitly refuse to attack $p \rightarrow q$. In the first case, the game ends with $(p \Rightarrow q)$, in the second, with $(\Rightarrow)$. For a given assignment v of risk values where $v(p) \geq v(q)$, I win the game in both cases. In other words: I have a *winning strategy* for $(\Rightarrow p \rightarrow q)$ for all assignments v satisfying $v(p) \geq v(q)$. In particular, I have a winning strategy for $(\Rightarrow p \rightarrow p)$ independently from the risk value assigned to p.

A *winning strategy* for me for a sequent $(\Gamma \Rightarrow \Delta)$ with associated risk function v can be represented as a finite tree labelled with sequents such that:

1. $(\Gamma \Rightarrow \Delta)$ labels the root.
2. Every leaf is labelled with an atomic sequent that I win.
3. Every node labelled with a non-atomic sequent has one child node labelled with a sequent resulting from a move by me (if one exists) and a child node for every sequent resulting from a move by you.

We will say that there is a winning strategy for me for $(\Gamma \Rightarrow \Delta)$ if for *all* assignments of risk values, there exists a winning strategy for $(\Gamma \Rightarrow \Delta)$. Generalizing even further, we will say that there is a winning strategy for a hypersequent $\mathcal{G}$ if for all assignments of risk values, there is a winning strategy for me for some $S \in \mathcal{G}$.

We can represent winning strategies for hypersequents (and sequents) by derivations in a suitable calculus:

Definition 6.42. GŁ$_{\text{giles}}$ consists of all atomic hypersequents for which I have a winning strategy, plus:

$$\frac{\begin{array}{c}\mathcal{G} \mid [(\Gamma \ominus [A \rightarrow B] \Rightarrow \Delta), (\Gamma \ominus [A \rightarrow B], B \Rightarrow A, \Delta) : A \rightarrow B \in \Gamma] \\ \{(\mathcal{G} \mid \Gamma \Rightarrow \Delta \ominus [A \rightarrow B]), (\mathcal{G} \mid \Gamma, A \Rightarrow B, \Delta \ominus [A \rightarrow B]) : A \rightarrow B \in \Delta\}\end{array}}{\mathcal{G} \mid \Gamma \Rightarrow \Delta}\ (\rightarrow)_{\text{Ł}}$$

The first premise corresponds to *my* choice of moves in the game: two sequents for each implicational formula asserted by you that I can either attack or disregard. The remaining premises correspond to the different possible moves made by *you*: two for each implicational formula asserted by me that you can attack or disregard. So there is a winning strategy for the conclusion hypersequent iff there is a winning strategy for all the premise hypersequents. So it follows by a simple induction on hypersequent complexity that:

Theorem 6.43. $\vdash_{\text{GŁ}_{\text{giles}}} \mathcal{G}$ *iff there exists a winning strategy for* $\mathcal{G}$.

A hypersequent derivation of a sequent may also be interpreted as a non-deterministic meta-winning strategy for that sequent: for any choices made by you, there is a set of choices listed for me, one of which will win. Also, since $(\rightarrow)_{Ł}$ is both derivable in GŁ-derivable and GŁ-invertible, there exists a winning strategy for a hypersequent $\mathcal{G}$ iff $\vdash_{GŁ_{giles}} \mathcal{G}$ iff $\vdash_{GŁ} \mathcal{G}$. Hence, using the soundness and completeness of GŁ:

Corollary 6.44. *$\models_{Ł} A$ iff there exists a winning strategy for $(\Rightarrow A)$.*

6.3 Product Logic

Product Logic P, the third fundamental fuzzy logic, possesses features of both of its more famous siblings. Roughly speaking, it behaves like Ł on the interval $(0,1]$, and like G at 0. Let us work again with a more restricted language $\mathcal{L}_T = \{\odot, \rightarrow, \bot\}$ with defined connectives:

$$\neg A =_{\text{def}} A \rightarrow \bot \qquad A \wedge B =_{\text{def}} A \odot (A \rightarrow B)$$
$$\top =_{\text{def}} \neg\bot \qquad A \vee B =_{\text{def}} ((A \rightarrow B) \rightarrow B) \wedge ((B \rightarrow A) \rightarrow A)$$

P-valuations are then functions $v : \text{Fm}_{\mathcal{L}_T} \rightarrow [0,1]$ such that $v(\bot) = 0$ and:

$$v(A \odot B) = v(A) \cdot v(B) \qquad v(A \rightarrow B) = \begin{cases} v(B)/v(A) & \text{if } v(A) > v(B) \\ 1 & \text{otherwise} \end{cases}$$

A formula A is P-valid, written $\models_P A$, iff $v(A) = 1$ for all P-valuations v. As before, we will drop the prefix P for the rest of the section, and speak just of valuations and validity.

A Hilbert system for Product Logic in the language $\text{Fm}_{\mathcal{L}_T}$ consists of the axiomatization of BL mentioned in Chapter 3 extended with:

$$(\Pi 1) \quad \neg\neg A \rightarrow ((A \rightarrow (A \odot B)) \rightarrow B)$$
$$(\Pi 2) \quad (A \odot (A \rightarrow \bot)) \rightarrow \bot$$

In developing Gentzen systems for P, the problems are similar to those encountered for Ł. It does not seem to be possible to obtain a calculus by adding structural rules to a standard calculus like GMTL. Rather, as for Ł, we make use of a non-standard interpretation for sequents and then develop tailored logical rules for that interpretation.

Definition 6.45 (Product Interpretation).

$$\text{I}_P(\Gamma \Rightarrow \Delta) =_{\text{def}} \odot\Gamma \rightarrow \odot\Delta$$
$$\text{I}_P(S_1 \mid \ldots \mid S_n) =_{\text{def}} \text{I}_P(S_1) \vee \ldots \vee \text{I}_P(S_n)$$

Hence, letting $\star_P^v(\Gamma) =_{\text{def}} \prod_{A \in \Gamma} v(A)$ where $\star_P^v([\,]) = 1$, we have $\models_P \text{I}_P(\mathcal{G})$ iff for all valuations v:

$$\star_P^v(\Gamma) \leq \star_P^v(\Delta) \text{ for some } (Ga \Rightarrow \Delta) \in \mathcal{G}$$

Initial Sequents

$$\frac{}{\mathcal{G} \mid A \Rightarrow A}\ (\text{ID}) \qquad \frac{}{\mathcal{G} \mid \Rightarrow}\ (\text{EMP}) \qquad \frac{}{\mathcal{G} \mid \Gamma, \bot \Rightarrow \Delta}\ (\bot\Rightarrow)$$

Structural Rules:

$$\frac{\mathcal{G}}{\mathcal{G} \mid \mathcal{H}}\ (\text{EW}) \qquad \frac{\mathcal{G} \mid \mathcal{H} \mid \mathcal{H}}{\mathcal{G} \mid \mathcal{H}}\ (\text{EC}) \qquad \frac{\mathcal{G} \mid \Gamma \Rightarrow \Delta}{\mathcal{G} \mid \Gamma, \Pi \Rightarrow \Delta}\ (\text{W})$$

$$\frac{\mathcal{G} \mid \Gamma_1, \Gamma_2 \Rightarrow \Delta_1, \Delta_2}{\mathcal{G} \mid \Gamma_1 \Rightarrow \Delta_1 \mid \Gamma_2 \Rightarrow \Delta_2}\ (\text{SPLIT}) \qquad \frac{\mathcal{G} \mid \Gamma_1 \Rightarrow \Delta_1 \quad \mathcal{G} \mid \Gamma_2 \Rightarrow \Delta_2}{\mathcal{G} \mid \Gamma_1, \Gamma_2 \Rightarrow \Delta_1, \Delta_2}\ (\text{MIX})$$

Logical Rules

$$\frac{\mathcal{G} \mid \Gamma, A, B \Rightarrow \Delta}{\mathcal{G} \mid \Gamma, A \odot B \Rightarrow \Delta}\ (\odot\Rightarrow) \qquad \frac{\mathcal{G} \mid \Gamma \Rightarrow A, B, \Delta}{\mathcal{G} \mid \Gamma \Rightarrow A \odot B, \Delta}\ (\Rightarrow\odot)_{\mathsf{A}}$$

$$\frac{\mathcal{G} \mid \Gamma \Rightarrow A}{\mathcal{G} \mid \Gamma, \neg A \Rightarrow \Delta}\ (\neg\Rightarrow)_{\mathsf{P}} \qquad \frac{\mathcal{G} \mid \Gamma \Rightarrow \Delta \quad \mathcal{G} \mid \Gamma, A \Rightarrow B, \Delta}{\mathcal{G} \mid \Gamma \Rightarrow A \to B, \Delta}\ (\Rightarrow\to)_{\text{Ł}}$$

$$\frac{\mathcal{G} \mid \Gamma, \neg A \Rightarrow \Delta \quad \mathcal{G} \mid \Gamma, B \Rightarrow A, \Delta}{\mathcal{G} \mid \Gamma, A \to B \Rightarrow \Delta}\ (\to\Rightarrow)_{\mathsf{P}}$$

Fig. 6.7 The hypersequent calculus GP

As before, a hypersequent rule $\mathcal{G}_1, \ldots, \mathcal{G}_n \ / \ \mathcal{G}$ is *sound* if whenever $\mathrm{I}_{\mathsf{P}}(\mathcal{G}_1), \ldots, \mathrm{I}_{\mathsf{P}}(\mathcal{G}_n)$ are all valid, then also $\mathrm{I}_{\mathsf{P}}(\mathcal{G})$ is valid, and *invertible* if the reverse implication holds.

Moreover, we can connect validity in Ł and P in the following useful way:

Lemma 6.46. *Let $\mathcal{G}$ be a strictly atomic hypersequent. Then $\models_{\text{Ł}} \mathcal{G}$ iff $\models_{\mathsf{P}} \mathrm{I}_{\mathsf{P}}(\mathcal{G})$.*

Proof. Let $\mathcal{G} = (\Gamma_1 \Rightarrow \Delta_1 \mid \ldots \mid \Gamma_n \Rightarrow \Delta_n)$ be a strictly atomic hypersequent. Then $\not\models_{\mathsf{P}} \mathrm{I}_{\mathsf{P}}(\mathcal{G})$ iff there exists a valuation v such that $\star_{\mathsf{P}}^{v}(\Gamma_i) > \star_{\mathsf{P}}^{v}(\Delta_i)$ for $i = 1 \ldots n$. Moreover, we can assume that $v(p) \neq 0$ for each variable p, since it is not possible that $v(p) = 0$ for $p \in \bigcup_{i=1}^{n} \Gamma_i$, and if $v(q) = 0$ for some $q \in \bigcup_{i=1}^{n} \Delta_i$, then we can easily take $v(q) > 0$ small enough so that the inequalities still hold. But now observe that $\star_{\mathsf{P}}^{v}(\Gamma_i) > \star_{\mathsf{P}}^{v}(\Delta_i)$ iff $\log \star_{\mathsf{P}}^{v}(\Gamma_i) > \log \star_{\mathsf{P}}^{v}(\Delta_i)$ iff $\sum_{p \in \Gamma_i} \log(v(p)) > \sum_{q \in \Delta_i} \log(v(q))$. Let w be the least value $\log(v(p))$ for p occurring in $\mathcal{G}$. Then $\sum_{p \in \Gamma_i} \log(v(p)) > \sum_{q \in \Delta_i} \log(v(q))$ for $i = 1 \ldots n$ iff $\sum_{p \in \Gamma_i} \log(v(p))/w > \sum_{q \in \Delta_i} \log(v(q))/w$. Also, $-1 \leq \log(v(p))/w \leq 0$ for every p occurring in $\mathcal{G}$. So consider the Ł-valuation $v'(p) = 1 + \log(v(p))/w$. Then $\sum_{p \in \Gamma_i} \log(v(p)) > \sum_{q \in \Delta_i} \log(v(q))$ for $i = 1 \ldots n$ iff $\star_{\text{Ł}}^{v'}(\Gamma_i) > \star_{\text{Ł}}^{v'}(\Delta_i)$ for $i = 1 \ldots n$. So $\not\models_{\mathsf{P}} \mathrm{I}_{\mathsf{P}}(\mathcal{G})$ iff $\not\models_{\text{Ł}} \mathcal{G}$ as required. □

Our hypersequent calculus for P based on this interpretation is displayed in Fig. 6.7, where $\neg$ is used here just as a convenient abbreviation for $A \to \bot$ and does not require special rules. Should we choose to expand the language, rules for $\wedge$ and $\vee$ are just the standard ones introduced in Chapter 4.

Example 6.47. We illustrate the calculus with a derivation of the axioms of $(\Pi 1)$, noting that the single-conclusion rule $(\Rightarrow\to)$ is derivable in GP exactly as in GŁ:

$$
\begin{prooftree}
\AxiomC{}
\RightLabel{(ID)}
\UnaryInfC{$A \Rightarrow A$}
\AxiomC{}
\RightLabel{(ID)}
\UnaryInfC{$B \Rightarrow B$}
\RightLabel{(MIX)}
\BinaryInfC{$A, B \Rightarrow A, B$}
\RightLabel{$(\odot\Rightarrow)$}
\UnaryInfC{$A \odot B \Rightarrow A, B$}
\RightLabel{(W)}
\UnaryInfC{$\neg\neg A, A \odot B \Rightarrow A, B$}
\AxiomC{}
\RightLabel{(ID)}
\UnaryInfC{$\neg A \Rightarrow \neg A$}
\RightLabel{$(\neg\Rightarrow)_{\mathsf{P}}$}
\UnaryInfC{$\neg\neg A, \neg A \Rightarrow B$}
\RightLabel{$(\rightarrow\Rightarrow)_{\mathsf{P}}$}
\BinaryInfC{$\neg\neg A, A \rightarrow (A \odot B) \Rightarrow B$}
\RightLabel{$(\Rightarrow\rightarrow)$}
\UnaryInfC{$\neg\neg A \Rightarrow (A \rightarrow (A \odot B)) \rightarrow B$}
\RightLabel{$(\Rightarrow\rightarrow)$}
\UnaryInfC{$\Rightarrow \neg\neg A \rightarrow ((A \rightarrow (A \odot B)) \rightarrow B)$}
\end{prooftree}
$$

We also point out the following useful derivation:

$$
\begin{prooftree}
\AxiomC{}
\RightLabel{(ID)}
\UnaryInfC{$A \Rightarrow A$}
\AxiomC{}
\RightLabel{(ID)}
\UnaryInfC{$B \Rightarrow B$}
\RightLabel{(MIX)}
\BinaryInfC{$A, B \Rightarrow A, B$}
\RightLabel{$(\Rightarrow\odot)$}
\UnaryInfC{$A, B \Rightarrow A \odot B$}
\RightLabel{$(\neg\Rightarrow)_{\mathsf{P}}$}
\UnaryInfC{$\neg(A \odot B), A, B \Rightarrow \bot, \bot$}
\RightLabel{(W)}
\UnaryInfC{$\neg(A \odot B), A, \neg(A \odot B), B \Rightarrow \bot, \bot$}
\RightLabel{(SPLIT)}
\UnaryInfC{$\neg(A \odot B), A \Rightarrow \bot \mid \neg(A \odot B), B \Rightarrow \bot$}
\RightLabel{$(\Rightarrow\rightarrow)$}
\UnaryInfC{$\neg(A \odot B), A \Rightarrow \bot \mid \neg(A \odot B) \Rightarrow \neg B$}
\RightLabel{$(\Rightarrow\rightarrow)$}
\UnaryInfC{$\neg(A \odot B) \Rightarrow \neg A \mid \neg(A \odot B) \Rightarrow \neg B$}
\end{prooftree}
$$

Soundness for GP is proved using the familiar induction on the height of a derivation. We have to check that for each rule, the conclusion is valid whenever the premises are valid. Let us take a look at a couple of the more different cases, noting that we can easily ignore the side-hypersequent $\mathcal{G}$ as in previous proofs:

- $(\rightarrow\Rightarrow)_{\mathsf{P}}$. Suppose that $(\Gamma, \neg A \Rightarrow \Delta)$ and $(\Gamma, B \Rightarrow A, \Delta)$ are valid, and let v be a valuation. If $v(A) = 0$, then $v(\neg A) = 1$ and $v(A \rightarrow B) = 1$, so $\star_{\mathsf{P}}^{v}(\Gamma \uplus [A \rightarrow B]) = \star_{\mathsf{P}}^{v}(\Gamma) \leq \star_{\mathsf{P}}^{v}(\Delta)$. Otherwise $v(A \rightarrow B) \leq v(B)/v(A)$ and $\star_{\mathsf{P}}^{v}(\Gamma \uplus [A \rightarrow B]) \leq (\star_{\mathsf{P}}^{v}\Gamma) \cdot (v(B)/v(A)) \leq \star_{\mathsf{P}}^{v}(\Delta)$. So $(\Gamma, A \rightarrow B \Rightarrow \Delta)$ is valid.
- $(\neg\Rightarrow)_{\mathsf{P}}$. Suppose that $(\Gamma \Rightarrow A)$ is valid. If $v(A) = 0$, then $\star_{\mathsf{P}}^{v}(\Gamma) \leq 0$ and hence also $\star_{\mathsf{P}}^{v}(\Gamma) \cdot v(\neg A) \leq \star_{\mathsf{P}}^{v}(\Delta)$. If $v(A) > 0$, then $v(\neg A) = 0$ and $\star_{\mathsf{P}}^{v}(\Gamma) \cdot v(\neg A) \leq \star_{\mathsf{P}}^{v}(\Delta)$. So $(\Gamma, \neg A \Rightarrow \Delta)$ is valid.

Theorem 6.48. *If* $\vdash_{\mathsf{GP}} \mathcal{G}$, *then* $\models_{\mathsf{P}} \mathrm{I}_{\mathsf{P}}(\mathcal{G})$.

We turn our attention now to completeness. Eliminating a cut rule for GP looks tricky so we will proceed semantically. First note that the following rule is easily seen to be GP-derivable using (EC), (W), $(\rightarrow\Rightarrow)_{\mathsf{P}}$, and (EW):

$$
\frac{\mathcal{G} \mid \Gamma, \neg A \Rightarrow \Delta \quad \mathcal{G} \mid \Gamma \Rightarrow \Delta \mid \Gamma, B \Rightarrow A, \Delta}{\mathcal{G} \mid \Gamma, A \rightarrow B \Rightarrow \Delta}\ (\rightarrow\Rightarrow)_{\mathsf{P}}^{i}
$$

We will now extend GP with extra invertible "decomposition rules" for negated formulas $\neg A$ and demonstrate that this extended system is complete. We can then show that GP is complete by showing that the extra rules can be eliminated.

Definition 6.49. GP^+ is GP extended with the following rules:

$$\frac{\mathcal{G} \mid \Gamma, \neg B \Rightarrow \Delta \mid A \Rightarrow B}{\mathcal{G} \mid \Gamma, \neg(A \to B) \Rightarrow \Delta}\ (\neg\!\to\Rightarrow) \qquad \frac{\mathcal{G} \mid \Gamma, \neg A \Rightarrow \Delta \quad \mathcal{G} \mid \Gamma, \neg B \Rightarrow \Delta}{\mathcal{G} \mid \Gamma, \neg(A \odot B) \Rightarrow \Delta}\ (\neg\odot\Rightarrow)$$

The nice thing about these rules is that, along with some but not all of the logical rules of GP, they are invertible. This allows us to reduce the validity of a hypersequent to the validity of hypersequents containing just atoms and negated atoms.

Lemma 6.50. *$(\odot\Rightarrow)$, $(\Rightarrow\odot)_{\mathsf{A}}$, $(\Rightarrow\to)_{\mathsf{Ł}}$, $(\to\Rightarrow)^i_{\mathsf{P}}$, $(\neg\to\Rightarrow)$, and $(\neg\odot\Rightarrow)$ are invertible.*

Proof. Let us just consider some of the less obvious cases, again ignoring side-hypersequents:

- $(\to\Rightarrow)^i_{\mathsf{P}}$. Suppose that $(\Gamma, A \to B \Rightarrow \Delta)$ is valid. Fix a valuation v. For the left premise, if $v(A) > 0$, then $v(\neg A) = 0$ and $\star^v_{\mathsf{P}}(\Gamma).v(\neg A) \leq \star^v_{\mathsf{P}}(\Delta)$, and if $v(A) = 0$, then $v(\neg A) = v(A \to B) = 1$ and $\star^v_{\mathsf{P}}(\Gamma) \leq \star^v_{\mathsf{P}}(\Delta)$. For the right premise, if $v(A) \leq v(B)$, then $v(A \to B) = 1$ and $\star^v_{\mathsf{P}}(\Gamma) \leq \star^v_{\mathsf{P}}(\Delta)$, and if $v(A) > v(B)$, then $v(A \to B) = v(B)/v(A)$ and $\star^v_{\mathsf{P}}(\Gamma) \cdot v(B) \leq \star^v_{\mathsf{P}}(\Delta) \cdot v(A)$. So both $(\Gamma, \neg A \Rightarrow \Delta)$ and $(\Gamma \Rightarrow \Delta \mid \Gamma, B \Rightarrow A, \Delta)$ are valid.
- $(\neg\to\Rightarrow)$. Suppose that $(\Gamma, \neg(A \to B) \Rightarrow \Delta)$ is valid. If $v(B) > 0$, then $v(\neg B) = 0$ and $\star^v_{\mathsf{P}}(\Gamma).v(\neg B) \leq \star^v_{\mathsf{P}}(\Delta)$. If $v(A) = 0$, then $v(A) \leq v(B)$. If $v(B) = 0$ and $v(A) > 0$, then $v(\neg B) = v(\neg(A \to B)) = 1$ and $\star^v_{\mathsf{P}}(\Gamma) \leq \star^v_{\mathsf{P}}(\Delta)$. So $(\Gamma, \neg B \Rightarrow \Delta \mid A \Rightarrow B)$ is valid. □

Theorem 6.51. *If $\models_{\mathsf{P}} \mathrm{I}_{\mathsf{P}}(\mathcal{G})$, then $\vdash_{\mathsf{GP}^+} \mathcal{G}$.*

Proof. We prove the claim by induction on $\mathrm{cp}(\mathcal{G})$. If $\mathcal{G}$ is strictly atomic, then by Lemma 6.46, $\models_{\mathsf{Ł}} \mathcal{G}$. Hence $\vdash_{\mathsf{GŁ}} \mathcal{G}$ and, since they share the same structural rules, also $\vdash_{\mathsf{GP}^+} \mathcal{G}$. If $\mathcal{G}$ has either a formula on the left of a sequent which is not an atom or negated atom, or a formula on the right which is not an atom, then in both cases we can apply an appropriate invertible rule. Moreover this process is terminating since all the invertible rules strictly reduce the complexity of the hypersequent with respect to the multiset order. Three cases remain:

- If there is an occurrence of $\bot$ on the left of a sequent, then $\mathcal{G}$ is an instance of $(\bot\Rightarrow)$ and we are done.
- Suppose that we have a hypersequent:

$$\mathcal{G} = (\mathcal{G}' \mid \Gamma \Rightarrow \bot, \Delta)$$

where $\bot \notin \Gamma$. If $\mathcal{G}'$ is valid, then by the induction hypothesis $\mathcal{G}'$ is derivable and by (EW) so is $\mathcal{G}$. Suppose then that $\mathcal{G}'$ is not valid, i.e. there is a valuation v such that $\star^v_{\mathsf{P}}(\Delta') < \star^v_{\mathsf{P}}(\Gamma')$ for all $(\Gamma' \Rightarrow \Delta') \in \mathcal{G}'$. We define a valuation v_ε for $0 < \varepsilon < 1$ as follows:

$$v_\varepsilon(p) = \begin{cases} \varepsilon & \text{if } v(p) = 0 \\ v(p) & \text{otherwise} \end{cases}$$

But now for each $(\Gamma' \Rightarrow \Delta') \in \mathcal{G}'$, $\star_{\mathsf{P}}^{v_\varepsilon}(\Gamma') = \star_{\mathsf{P}}^{v}(\Gamma')$ and we can take ε small enough so that $\star_{\mathsf{P}}^{v_\varepsilon}(\Delta') < \star_{\mathsf{P}}^{v_\varepsilon}(\Gamma')$. Since also $\star_{\mathsf{P}}^{v_\varepsilon}(\Gamma) > 0$, we get that $\mathcal{G}$ is not valid, a contradiction.

- Suppose that for an atom ($\bot$ or variable) a:

$$\mathcal{G} = (\Gamma_1, \neg a \Rightarrow \Delta_1 \mid \Gamma_2 \Rightarrow \Delta_2 \mid \ldots \mid \Gamma_m \Rightarrow \Delta_m)$$

If $a \in \Gamma_i$ for some $i \in \{1, \ldots, m\}$, then $\mathcal{G}$ is derivable (working backwards) by applying (SPLIT) (if needed) to obtain a sequent containing $\neg a$ and a, then $(\neg\Rightarrow)$, (W), and (ID). For $a \notin \Gamma_i$ for $i = 1 \ldots m$, consider:

$$\mathcal{G}' = (\Gamma_1 \Rightarrow \Delta_1 \mid \ldots \mid \Gamma_m \Rightarrow \Delta_m)$$

If $\mathcal{G}'$ is valid, then by the induction hypothesis $\mathcal{G}'$ is derivable. Hence, by (W), so is $\mathcal{G}$. Suppose for a contradiction then that $\mathcal{G}'$ is not valid; i.e. there is a valuation v such that $\star_{\mathsf{P}}^{v}(\Gamma_i) > \star_{\mathsf{P}}^{v}(\Delta_i)$ for $i = 1 \ldots m$. Define a new valuation v' such that $v'(a) = 0$ and $v'(q) = v(q)$ for $q \neq a$. Then $v'(\neg a) = 1$ and $\star_{\mathsf{P}}^{v'}(\Gamma_i) = \star_{\mathsf{P}}^{v}(\Gamma_i) > \star_{\mathsf{P}}^{v}(\Delta_i) \geq \star_{\mathsf{P}}^{v'}(\Delta_i)$. So $\mathcal{G}$ is not valid, a contradiction. □

Lemma 6.52. *$(\neg\!\rightarrow\Rightarrow)$ and $(\neg\odot\Rightarrow)$ can be eliminated from* GP^+.

Proof. For $(\neg\!\rightarrow\Rightarrow)$, first consider the derivation:

$$\dfrac{\dfrac{\dfrac{\dfrac{\mathcal{G} \mid \Gamma, \neg(A \to B) \Rightarrow \Delta \mid A \Rightarrow B}{\mathcal{G} \mid \Gamma, \neg(A \to B) \Rightarrow \Delta \mid \Gamma, A \Rightarrow B}\,(\mathrm{W})}{\mathcal{G} \mid \Gamma, \neg(A \to B) \Rightarrow \Delta \mid \Gamma \Rightarrow A \to B}\,(\Rightarrow\to)}{\mathcal{G} \mid \Gamma, \neg(A \to B) \Rightarrow \Delta \mid \Gamma, \neg(A \to B) \Rightarrow \Delta}\,(\neg\Rightarrow)_{\mathsf{P}}}{\mathcal{G} \mid \Gamma, \neg(A \to B) \Rightarrow \Delta}\,(\mathrm{EC})$$

This means that to show that $\vdash_{\mathsf{GP}} \mathcal{G} \mid \Gamma, \neg(A \to B) \Rightarrow \Delta \mid A \Rightarrow B$ follows from $\vdash_{\mathsf{GP}} \mathcal{G} \mid \Gamma, \neg B \Rightarrow \Delta \mid A \Rightarrow B$, it is enough to establish (by a simple induction):

Claim. If $\vdash_{\mathsf{GP}} [\Gamma_i, [\neg B]^{\lambda_i} \Rightarrow \Delta_i]_{i=1}^n$, then $\vdash_{\mathsf{GP}} [\Gamma_i, [\neg(A \to B)]^{\lambda_i} \Rightarrow \Delta_i]_{i=1}^n$.

For $(\neg\odot\Rightarrow)$, it is enough to show the following:

Claim. Suppose that:

1. $\mathcal{H}_1 = [\Gamma_i, [\neg A]^{\lambda_i} \Rightarrow \Delta_i]_{i=1}^k$.
2. $\mathcal{H}_2 = [\Gamma_i, [\neg B]^{\lambda_i} \Rightarrow \Delta_i]_{i=k+1}^n$.
3. $\mathcal{H} = [\Gamma_i, [\neg(A \odot B)]^{\lambda_i} \Rightarrow \Delta_i]_{i=1}^n$.

If $d_1 \vdash_{\mathsf{GP}} \mathcal{H}_1$ and $d_2 \vdash_{\mathsf{GP}} \mathcal{H}_2$, then $\vdash_{\mathsf{GP}} \mathcal{H}$.

Proof of claim. We proceed by induction on $\mathrm{ht}(d_1)+\mathrm{ht}(d_2)$. If $\mathcal{H}_1$ or $\mathcal{H}_2$ has no occurrences of $\neg A$ or $\neg B$ respectively, then clearly $\mathcal{H}$ is derivable so we assume that each has at least one such occurrence. If $\bot$ occurs on the left of a sequent of $\mathcal{H}_1$ or $\mathcal{H}_2$, then we are done so also assume no such occurrences. For the base case we have $\mathcal{H}_1 = (\neg A \Rightarrow \neg A)$ and $\mathcal{H}_2 = (\neg B \Rightarrow \neg B)$. Then $\mathcal{H} = (\neg(A \odot B) \Rightarrow \neg A \mid \neg(A \odot B) \Rightarrow \neg B)$ is derived as in Example 6.47. For the inductive step, we consider the last rule applications in d_1 or d_2. If d_1 or d_2 ends with a structural rule or a logical rule on a side-formula (not a distinguished occurrence of $\neg A$ or $\neg B$), then we can just use the induction hypothesis. There are two other possibilities.

Suppose first that d_1 and d_2 end with, respectively:

$$\frac{\mathcal{G}_1 \mid \Pi_1 \Rightarrow A}{\mathcal{G}_1 \mid \Pi_1, \neg A \Rightarrow \Sigma_1}\ (\neg\Rightarrow) \qquad \frac{\mathcal{G}_2 \mid \Pi_2 \Rightarrow B}{\mathcal{G}_2 \mid \Pi_2, \neg B \Rightarrow \Sigma_2}\ (\neg\Rightarrow)$$

By the induction hypothesis applied to the premise of one of the above with the conclusion of the other, we obtain GP-derivations of:

$$\mathcal{G}' \mid \Pi'_2, \neg(A \odot B) \Rightarrow \Sigma_2 \mid \Pi'_1 \Rightarrow A \quad \text{and} \quad \mathcal{G}' \mid \Pi'_1, \neg(A \odot B) \Rightarrow \Sigma_1 \mid \Pi'_2 \Rightarrow B$$

where $\mathcal{G}' = (\mathcal{G}'_1 \mid \mathcal{G}'_2)$ and $\mathcal{H} = (\mathcal{G}' \mid \Pi'_1, \neg(A \odot B) \Rightarrow \Sigma_1 \mid \Pi'_2, \neg(A \odot B) \Rightarrow \Sigma_2)$.

We give the following GP-derivation of $\mathcal{H}$:

$$\dfrac{\dfrac{\dfrac{\dfrac{\dfrac{\dfrac{\dfrac{\mathcal{G}' \mid \Pi'_2, \neg(A \odot B) \Rightarrow \Sigma_2 \mid \Pi'_1 \Rightarrow A}{\mathcal{H} \mid \Pi'_1 \Rightarrow A}\,(\text{EW}) \qquad \dfrac{\mathcal{G}' \mid \Pi'_1, \neg(A \odot B) \Rightarrow \Sigma_1 \mid \Pi'_2 \Rightarrow B}{\mathcal{H} \mid \Pi'_2 \Rightarrow B}\,(\text{EW})}{\mathcal{H} \mid \Pi'_1, \Pi'_2 \Rightarrow A, B}\,(\text{MIX})}{\mathcal{H} \mid \Pi'_1, \Pi'_2 \Rightarrow A \odot B}\,(\Rightarrow\odot)}{\mathcal{H} \mid \Pi'_1, \Pi'_2, \neg(A \odot B) \Rightarrow \Sigma_1, \Sigma_2}\,(\neg\Rightarrow)}{\mathcal{H} \mid \Pi'_1, \Pi'_2, \neg(A \odot B), \neg(A \odot B) \Rightarrow \Sigma_1, \Sigma_2}\,(\text{W})}{\mathcal{H} \mid \Pi'_1, \neg(A \odot B) \Rightarrow \Sigma_1 \mid \Pi'_2, \neg(A \odot B) \Rightarrow \Sigma_2}\,(\text{SPLIT})}{\mathcal{G}' \mid \Pi'_1, \neg(A \odot B) \Rightarrow \Sigma_1 \mid \Pi'_2, \neg(A \odot B) \Rightarrow \Sigma_2}\,(\text{EC})$$

For the second possibility, suppose that d_1 ends with (ID) where $\mathcal{H}_1 = (\neg A \Rightarrow \neg A)$ and d_2 ends with $(\neg\Rightarrow)$ applied to an occurrence of $\neg B$ where $\mathcal{H}_2 = (\mathcal{G}_2 \mid \Pi_2, \neg B \Rightarrow \Sigma_2)$. $\mathcal{H}$ is of the form $(\mathcal{G}'_2 \mid \Pi'_2, \neg(A \odot B) \Rightarrow \Sigma_2 \mid \neg(A \odot B) \Rightarrow \neg A)$. Since $(\mathcal{G}_2 \mid \Pi_2 \Rightarrow B)$ is derivable, by the induction hypothesis, $(\mathcal{G}'_2 \mid \Pi'_2 \Rightarrow B \mid \neg(A \odot B) \Rightarrow \neg A)$ is derivable. So $\mathcal{H}$ is derivable as follows:

$$
\dfrac{\dfrac{\dfrac{\dfrac{\dfrac{\dfrac{\dfrac{\dfrac{}{\mathcal{G}_2' \mid \neg(A \odot B) \Rightarrow \neg A \mid A \Rightarrow A}\,(\text{ID}) \qquad \mathcal{G}_2' \mid \Pi_2' \Rightarrow B \mid \neg(A \odot B) \Rightarrow \neg A}{\mathcal{G}_2' \mid \neg(A \odot B) \Rightarrow \neg A \mid \Pi_2', A \Rightarrow A, B}\,(\text{MIX})}{\mathcal{G}_2' \mid \neg(A \odot B) \Rightarrow \neg A \mid \Pi_2', A \Rightarrow A \odot B}\,(\Rightarrow\odot)}{\mathcal{G}_2' \mid \neg(A \odot B) \Rightarrow \neg A \mid \Pi_2', \neg(A \odot B), A \Rightarrow \Sigma_2, \bot}\,(\neg\Rightarrow)}{\mathcal{G}_2' \mid \neg(A \odot B) \Rightarrow \neg A \mid \Pi_2', \neg(A \odot B), \neg(A \odot B), A \Rightarrow \Sigma_2, \bot}\,(\text{W})}{\mathcal{G}_2' \mid \Pi_2', \neg(A \odot B) \Rightarrow \Sigma_2 \mid \neg(A \odot B) \Rightarrow \neg A \mid \neg(A \odot B), A \Rightarrow \bot}\,(\text{SPLIT})}{\mathcal{G}_2' \mid \Pi_2', \neg(A \odot B) \Rightarrow \Sigma_2 \mid \neg(A \odot B) \Rightarrow \neg A \mid \neg(A \odot B) \Rightarrow \neg A}\,(\Rightarrow\to)}{\mathcal{G}_2' \mid \Pi_2', \neg(A \odot B) \Rightarrow \Sigma_2 \mid \neg(A \odot B) \Rightarrow \neg A}\,(\text{EC}) \qquad \square
$$

Hence, combining the preceding lemma with Theorems 6.48 and 6.51, we have:

Theorem 6.53. $\models_{\mathsf{P}} I_{\mathsf{P}}(\mathcal{G})$ *iff* $\vdash_{\mathsf{GP}} \mathcal{G}$.

We can even use this calculus to establish standard completeness for HP. That is, we can show that a formula A is valid in the algebra $\mathbf{A}(*_{\mathsf{P}}, 0)$, i.e. $\models_{\mathsf{P}} A$, iff A is derivable in the Hilbert system HP (or equivalently, valid in all P-algebras).

Theorem 6.54. *If* $\models_{\mathsf{P}} A$*, then* $\vdash_{\mathsf{HP}} A$

Proof. If $\models_{\mathsf{P}} A$, then by the completeness of GP, $\vdash_{\mathsf{GP}} (\Rightarrow A)$. But GP is sound with respect to P-chains (the proof is the same as for the standard algebra). Hence A is valid in all P-chains, and so by Theorem 3.56, $\vdash_{\mathsf{HP}} A$. □

Moreover, we can extract a decidability result from our completeness proof for GP^+. A formula A is valid iff applying the logical rules of GP^+ backwards to $(\Rightarrow A)$ (a terminating procedure) ends with valid hypersequents containing only atoms and negated atoms. But then we can follow the proof of Theorem 6.51 to deal with constants and negated atoms to arrive at strictly atomic hypersequents. Checking the validity of such hypersequents is the same for P as for Ł, and therefore decidable.

Theorem 6.55. *The validity problem for* P *is decidable.*

We can also develop a sequent calculus $\mathsf{GP_s}$ for Product Logic, displayed in Fig. 6.8, proceeding along the same lines as for Ł.

Example 6.56. Below we give a $\mathsf{GP_s}$-derivation of the $(\Pi 2)$ axioms:

$$
\dfrac{\dfrac{}{\Rightarrow}\,(\text{EMP}) \qquad \dfrac{\dfrac{\dfrac{\dfrac{\dfrac{}{\neg A \Rightarrow \neg A}\,(\text{ID})}{\neg A, \neg\neg A \Rightarrow \bot}\,(\neg\Rightarrow)_{\mathsf{P}} \qquad \dfrac{\dfrac{}{A \Rightarrow A}\,(\text{ID})}{\neg A, A \Rightarrow \neg A, \bot}\,(\neg\Rightarrow)_{\mathsf{P}}}{\neg A, \neg A \to A \Rightarrow \bot}\,(\to\Rightarrow)^{s}_{\mathsf{P}} + (\mathsf{w})}{\neg A \odot (\neg A \to A) \Rightarrow \bot}\,(\odot\Rightarrow)}{\Rightarrow \neg(\neg A \odot (\neg A \to A))}\,(\Rightarrow\to)_{Ł}
$$

Soundness is proved for $\mathsf{GP_s}$ in the usual way. Proving completeness is much more complicated, but relies essentially on the same techniques used in the completeness proof for $\mathsf{GŁ_s}$. Here, we just state the theorem without proof:

Theorem 6.57. $\vdash_{\mathsf{GP_s}} \Gamma \Rightarrow \Delta$ *iff* $\models_{\mathsf{P}} I_{\mathsf{P}}(\Gamma \Rightarrow \Delta)$.

Initial Sequents:

$$\frac{}{A \Rightarrow A}\ (\text{ID}) \qquad \frac{}{\Rightarrow}\ (\text{EMP}) \qquad \frac{}{\Gamma, \bot \Rightarrow \Delta}\ (\bot\Rightarrow)$$

Structural Rules:

$$\frac{\Gamma^n \Rightarrow \Delta^n}{\Gamma \Rightarrow \Delta}\ (\text{SC}_\text{n})\ n \geq 2 \qquad \frac{\Gamma \Rightarrow \Delta}{\Gamma, \Pi \Rightarrow \Delta}\ (\text{W}) \qquad \frac{\Gamma_1 \Rightarrow \Delta_1 \quad \Gamma_2 \Rightarrow \Delta_2}{\Gamma_1, \Gamma_2 \Rightarrow \Delta_1, \Delta_2}\ (\text{MIX})$$

Logical Rules:

$$\frac{\Gamma, A, B \Rightarrow \Delta}{\Gamma, A \odot B \Rightarrow \Delta}\ (\odot\Rightarrow) \qquad \frac{\Gamma \Rightarrow A, B, \Delta}{\Gamma \Rightarrow A \odot B, \Delta}\ (\Rightarrow\odot)_\text{A} \qquad \frac{\Gamma \Rightarrow A}{\Gamma, \neg A \Rightarrow \Delta}\ (\neg\Rightarrow)_\text{P}$$

$$\frac{\Gamma \Rightarrow \Delta \quad \Gamma, A \Rightarrow B, \Delta}{\Gamma \Rightarrow A \to B, \Delta}\ (\Rightarrow\to)_\text{Ł} \qquad \frac{\Gamma, \neg A \Rightarrow \Delta \quad \Gamma, B, B \to A \Rightarrow A, \Delta}{\Gamma, A \to B \Rightarrow \Delta}\ (\to\Rightarrow)^s_\text{P}$$

Fig. 6.8 The sequent calculus $\mathsf{GP_s}$

6.4 Related Logics

Some related logics, not as important as the three just covered, but interesting nonetheless, can be tackled using similar techniques. Recall first that *Cancellative Hoop Logic* CHL – the logic of cancellative hoops – emerges by removing $\bot$ from the language of Product Logic P and restricting P-valuations to the half-open interval $(0,1]$. More precisely, CHL is based on the language $\mathcal{L}_\text{C} = \{\odot, \to\}$ and CHL-valuations are functions $v : \text{Fm}_{\mathcal{L}_\text{C}} \to (0,1]$ where $v(A \odot B) = v(A) \cdot v(B)$ and $v(A \to B) = v(B)/v(A)$ if $v(A) > v(B)$; 1 otherwise.

A Hilbert system for CHL in this language consists of the axiomatization for BL given in Chapter 3 with (A7) removed, extended with the cancellation axiom:

$$(\text{CAN})\ (A \to (A \odot B)) \to B$$

A hypersequent calculus GCHL, displayed in Fig. 6.9, is obtained by interpreting hypersequents as in P, removing $(\bot\Rightarrow)_\text{Ł}$ from GŁ and adding the rules $(\odot\Rightarrow)$ and $(\Rightarrow\odot)_\text{P}$ of GP.

Example 6.58. The cancellation axiom is derived in GCHL as follows, noting that again the single-conclusion version of $(\Rightarrow\to)$ is derivable in this calculus:

$$\dfrac{\dfrac{\dfrac{\dfrac{\dfrac{}{A \Rightarrow A}\ (\text{ID}) \quad \dfrac{}{B \Rightarrow B}\ (\text{ID})}{A, B \Rightarrow A, B}\ (\text{MIX})}{A \odot B \Rightarrow A, B}\ (\odot\Rightarrow)}{A \to (A \odot B) \Rightarrow B}\ (\to\Rightarrow)_\text{A}}{\Rightarrow (A \to (A \odot B)) \to B}\ (\Rightarrow\to)$$

Proofs of soundness and completeness for this calculus follow exactly the same pattern as those for GŁ and GP. Namely we show that all the rules of GCHL are sound

Initial Sequents

$$\frac{}{\mathcal{G} \mid A \Rightarrow A} \ (\text{ID}) \qquad \frac{}{\mathcal{G} \mid \Rightarrow} \ (\text{EMP})$$

Structural Rules:

$$\frac{\mathcal{G}}{\mathcal{G} \mid \mathcal{H}} \ (\text{EW}) \qquad \frac{\mathcal{G} \mid \mathcal{H} \mid \mathcal{H}}{\mathcal{G} \mid \mathcal{H}} \ (\text{EC}) \qquad \frac{\mathcal{G} \mid \Gamma \Rightarrow \Delta}{\mathcal{G} \mid \Gamma, \Pi \Rightarrow \Delta} \ (\text{W})$$

$$\frac{\mathcal{G} \mid \Gamma_1, \Gamma_2 \Rightarrow \Delta_1, \Delta_2}{\mathcal{G} \mid \Gamma_1 \Rightarrow \Delta_1 \mid \Gamma_2 \Rightarrow \Delta_2} \ (\text{SPLIT}) \qquad \frac{\mathcal{G} \mid \Gamma_1 \Rightarrow \Delta_1 \quad \mathcal{G} \mid \Gamma_2 \Rightarrow \Delta_2}{\mathcal{G} \mid \Gamma_1, \Gamma_2 \Rightarrow \Delta_1, \Delta_2} \ (\text{MIX})$$

Logical Rules

$$\frac{\mathcal{G} \mid \Gamma, A, B \Rightarrow \Delta}{\mathcal{G} \mid \Gamma, A \odot B \Rightarrow \Delta} \ (\odot\Rightarrow) \qquad \frac{\mathcal{G} \mid \Gamma \Rightarrow A, B, \Delta}{\mathcal{G} \mid \Gamma \Rightarrow A \odot B, \Delta} \ (\Rightarrow\odot)_{\mathsf{A}}$$

$$\frac{\mathcal{G} \mid \Gamma, B \Rightarrow A, \Delta}{\mathcal{G} \mid \Gamma, A \rightarrow B \Rightarrow \Delta} \ (\rightarrow\Rightarrow)_{\mathsf{A}} \qquad \frac{\mathcal{G} \mid \Gamma \Rightarrow \Delta \quad \mathcal{G} \mid \Gamma, A \Rightarrow B, \Delta}{\mathcal{G} \mid \Gamma \Rightarrow A \rightarrow B, \Delta} \ (\Rightarrow\rightarrow)_{\text{Ł}}$$

Fig. 6.9 The hypersequent calculus GCHL

and that suitable versions of the logical rules are also invertible. It then follows that any valid hypersequent is derivable from valid atomic hypersequents, which are themselves shown to be derivable as in e.g. Theorem 6.24 for GŁ.

Theorem 6.59. $\vdash_{\mathsf{GCHL}} \mathcal{G}$ *iff* $\models_{\mathsf{CHL}} \mathrm{I_P}(\mathcal{G})$.

A sequent calculus is obtained for CHL by removing $(\bot\Rightarrow)_{\text{Ł}}$ from GŁ_{s} and adding the rules $(\odot\Rightarrow)$ and $(\Rightarrow\odot)_{\mathsf{A}}$ of $\mathsf{GP_s}$.

As a second example, recall that *Cross Ratio Logic* CRL is the logic based on the algebra $\mathbf{A}(*_{\mathsf{CR}}, \frac{1}{2})$ where $*_{\mathsf{CR}}$ is the cross-ratio uninorm. An axiomatization in the standard language $\mathcal{L}_{\mathrm{B}} = \{\wedge, \vee, \odot, \rightarrow, \mathrm{f}, \mathrm{e}, \bot, \top\}$ is obtained by adding to HMALL, the axioms (e) $\mathrm{e} \rightarrow \mathrm{f}$ and (f) $\mathrm{f} \rightarrow \mathrm{e}$, and the restricted cancellation axiom:

$$(\text{RCAN}) \quad (\top \rightarrow A) \vee (A \rightarrow \bot) \vee ((A \rightarrow (A \odot B)) \rightarrow B)$$

To avoid repetition, however, it will be convenient to define a calculus for CRL based on a language $\mathcal{L}_{\mathrm{E}}$ with primitive connectives $\rightarrow$, $\wedge$, e, $\Box$, and $\diamond$, where:

$$\begin{array}{rlrl} \top & =_{\text{def}} \diamond \mathrm{e} & \bot & =_{\text{def}} \Box \mathrm{e} \\ \neg A & =_{\text{def}} A \rightarrow \mathrm{e} & A \vee B & =_{\text{def}} \neg(\neg A \wedge \neg B) \\ A \odot B & =_{\text{def}} \neg(A \rightarrow \neg B) & A \oplus B & =_{\text{def}} \neg A \rightarrow B \end{array}$$

CRL-valuations are then functions $v : \mathrm{Fm}_{\mathcal{L}_{\mathrm{E}}} \rightarrow [0,1]$ satisfying:

$$v(A \rightarrow B) = \begin{cases} \dfrac{(1 - v(A))v(B)}{v(A)(1 - v(B)) - (1 - v(A))v(B)} & \text{if } \{v(A), v(B)\} \notin \{\{0\}, \{1\}\} \\ 1 & \text{otherwise} \end{cases}$$

Initial Sequents

$$\frac{}{\mathcal{G} \mid A \Rightarrow A}\ (\text{ID}) \qquad \frac{}{\mathcal{G} \mid \Rightarrow}\ (\text{EMP}) \qquad \frac{}{\mathcal{G} \mid \Gamma, \Box A \Rightarrow A, \Delta}\ (\Box)$$

$$\frac{}{\mathcal{G} \mid \Gamma, A \Rightarrow \diamond A, \Delta}\ (\diamond) \qquad \frac{}{\mathcal{G} \mid \Gamma, \Box \mathrm{e} \Rightarrow \Delta}\ (\bot) \qquad \frac{}{\mathcal{G} \mid \Gamma \Rightarrow \diamond \mathrm{e}, \Delta}\ (\top)$$

Structural Rules

$$\frac{\mathcal{G}}{\mathcal{G} \mid \mathcal{H}}\ (\text{EW}) \qquad \frac{\mathcal{G} \mid \mathcal{H} \mid \mathcal{H}}{\mathcal{G} \mid \mathcal{H}}\ (\text{EC})$$

$$\frac{\mathcal{G} \mid \Gamma_1, \Gamma_2 \Rightarrow \Delta_1, \Delta_2}{\mathcal{G} \mid \Gamma_1 \Rightarrow \Delta_1 \mid \Gamma_2 \Rightarrow \Delta_2}\ (\text{SPLIT}) \qquad \frac{\mathcal{G} \mid \Gamma_1 \Rightarrow \Delta_1 \quad \mathcal{G} \mid \Gamma_2 \Rightarrow \Delta_2}{\mathcal{G} \mid \Gamma_1, \Gamma_2 \Rightarrow \Delta_1, \Delta_2}\ (\text{MIX})$$

Logical Rules

$$\frac{\mathcal{G} \mid \Gamma \Rightarrow \Delta}{\mathcal{G} \mid \Gamma, \mathrm{e} \Rightarrow \Delta}\ (\mathrm{e}\Rightarrow) \qquad \frac{\mathcal{G} \mid \Gamma \Rightarrow \Delta}{\mathcal{G} \mid \Gamma \Rightarrow \mathrm{e}, \Delta}\ (\Rightarrow \mathrm{e})$$

$$\frac{\mathcal{G} \mid \Gamma, B \Rightarrow A, \Delta \quad \mathcal{G} \mid \Gamma, \Box(A \to B) \Rightarrow \Delta}{\mathcal{G} \mid \Gamma, A \to B \Rightarrow \Delta}\ (\to\Rightarrow)_c \qquad \frac{\mathcal{G} \mid \Gamma, A \Rightarrow B, \Delta}{\mathcal{G} \mid \Gamma \Rightarrow A \to B, \Delta}\ (\Rightarrow\to)$$

$$\frac{\mathcal{G} \mid \Gamma, A \Rightarrow \Delta \mid \Gamma, B \Rightarrow \Delta}{\mathcal{G} \mid \Gamma, A \wedge B \Rightarrow \Delta}\ (\wedge\Rightarrow) \qquad \frac{\mathcal{G} \mid \Gamma \Rightarrow A, \Delta \quad \mathcal{G} \mid \Gamma \Rightarrow B, \Delta}{\mathcal{G} \mid \Gamma \Rightarrow A \wedge B, \Delta}\ (\Rightarrow\wedge)$$

$$\frac{\mathcal{G} \mid \Gamma, A \Rightarrow \Box \mathrm{e}, \Delta}{\mathcal{G} \mid \Gamma, \diamond A \Rightarrow \Delta}\ (\diamond\Rightarrow) \qquad \frac{\mathcal{G} \mid \Gamma, \diamond \mathrm{e} \Rightarrow A, \Delta}{\mathcal{G} \mid \Gamma \Rightarrow \Box A, \Delta}\ (\Rightarrow\Box)$$

$$\frac{\mathcal{G} \mid \Gamma, \Box A, \Box B \Rightarrow \Delta}{\mathcal{G} \mid \Gamma, \Box(A \wedge B) \Rightarrow \Delta}\ (\Box\wedge\Rightarrow) \qquad \frac{\mathcal{G} \mid \Gamma \Rightarrow \diamond A, \Delta \quad \mathcal{G} \mid \Gamma \Rightarrow \diamond B, \Delta}{\mathcal{G} \mid \Gamma \Rightarrow \diamond(A \wedge B), \Delta}\ (\Rightarrow\diamond\wedge)$$

$$\frac{\mathcal{G} \mid \Gamma, \Box B \Rightarrow \Delta \quad \mathcal{G} \mid \Gamma \Rightarrow \diamond A, \Delta}{\mathcal{G} \mid \Gamma, \Box(A \to B) \Rightarrow \Delta}\ (\Box\to\Rightarrow) \qquad \frac{\mathcal{G} \mid \Gamma, \Box A \Rightarrow \Delta \quad \mathcal{G} \mid \Gamma \Rightarrow \diamond B, \Delta}{\mathcal{G} \mid \Gamma \Rightarrow \diamond(A \to B), \Delta}\ (\Rightarrow\diamond\to)$$

$$\frac{\mathcal{G} \mid \Gamma, \Box A \Rightarrow \Delta}{\mathcal{G} \mid \Gamma, \Box\Box A \Rightarrow \Delta}\ (\Box\Box\Rightarrow) \qquad \frac{\mathcal{G} \mid \Gamma \Rightarrow \Box A, \Delta}{\mathcal{G} \mid \Gamma \Rightarrow \diamond\Box A, \Delta}\ (\Rightarrow\diamond\Box)$$

$$\frac{\mathcal{G} \mid \Gamma, \diamond A \Rightarrow \Delta}{\mathcal{G} \mid \Gamma, \Box\diamond A \Rightarrow \Delta}\ (\Box\diamond\Rightarrow) \qquad \frac{\mathcal{G} \mid \Gamma \Rightarrow \diamond A, \Delta}{\mathcal{G} \mid \Gamma \Rightarrow \diamond\diamond A, \Delta}\ (\Rightarrow\diamond\diamond)$$

Fig. 6.10 The hypersequent calculus GCRL

$$v(A \wedge B) = \min(v(A), v(B)) \qquad v(\mathrm{e}) = \tfrac{1}{2}$$

$$v(\Box A) = \begin{cases} 1 & \text{if } v(A) = 1 \\ 0 & \text{if otherwise} \end{cases} \qquad v(\diamond A) = \begin{cases} 0 & \text{if } v(A) = 0 \\ 1 & \text{if otherwise} \end{cases}$$

Our hypersequent calculus GCRL, displayed in Fig. 6.10, has many non-standard aspects. The $(\to\Rightarrow)$ rule fails the subformula property since $\Box$ occurs in its right premise but not its conclusion. Also, the last eight logical rules treat combinations of connectives rather than one principal connective.

Example 6.60. We illustrate GCRL with the following derivation:

$$
\dfrac{\dfrac{\dfrac{}{A \Rightarrow A}\,(\text{ID}) \quad \dfrac{}{B \Rightarrow B}\,(\text{ID})}{A, B \Rightarrow A, B}\,(\text{MIX}) \qquad \dfrac{\dfrac{}{A, \Box B \Rightarrow B}\,(\Box\Rightarrow) \quad \dfrac{}{A \Rightarrow \diamond A, B}\,(\Rightarrow\diamond)}{A, \Box(A \to B) \Rightarrow B}\,(\Box{\to}{\Rightarrow})}{\dfrac{\dfrac{A, A \to B \Rightarrow B}{A \Rightarrow (A \to B) \to B}\,(\Rightarrow\to)}{\Rightarrow A \to ((A \to B) \to B)}\,(\Rightarrow\to)}\,(\to\Rightarrow)_c
$$

Let us just sketch the soundness and completeness proof for GCRL, the details being very similar to corresponding proofs for other logics. First observe that the logical rules of GCRL are invertible and hence can be applied (upwards) exhaustively to valid hypersequents to obtain valid "basic" hypersequents: those containing only occurrences of $\diamond$e, p, and $\Box p$ on the left of sequents, and only occurrences of $\Box$e, p, and $\diamond p$ on the right. Valid basic hypersequents are then derived using similar techniques to those employed for Product Logic. Namely, we can remove the occurrences of $\Box$e and $\diamond p$ while maintaining validity, then prove the derivability of strictly atomic valid hypersequents.

Theorem 6.61. $\models_{\mathsf{CRL}} \mathrm{I}_{\mathsf{P}}(\mathcal{G})$ *iff* $\vdash_{\mathsf{GCRL}} \mathcal{G}$.

6.5 Historical Remarks

Investigations into Łukasiewicz and Gödel logics date back to the early dawn of non-classical logics. The finite-valued logics $\mathsf{G_n}$ $(n = 2,3,\ldots)$ were defined by Gödel in the early 1930s [98] to show that Intuitionistic Logic has no finite characteristic matrix. Much later, in 1959, Dummett axiomatized the infinite-valued version G by adding the axiom schema $(A \to B) \vee (B \to A)$ to Intuitionistic Logic [73], and for that reason this logic is often better known as Gödel-Dummett Logic or simply as Dummett Logic LC. Around the same time as Gödel was introducing his finite-valued logics, the infinite-valued logic Ł was defined by Łukasiewicz [135] as a generalization of his own finite-valued logics introduced in the 1920s [133]. Łukasiewicz also conjectured the axiomatization for Ł presented in Section 6.2 with a fifth axiom schema proved redundant (independently) by Chang [46] and Meredith [141]. A completeness proof was obtained by Wajsberg in the 1930s, but the first published proof by Rose and Rosser appeared in the 1950s [187]. An algebraic completeness proof was given by Chang [45] around the same time that introduced and made crucial use of MV-algebras. The book [58] provides a detailed study of this class of algebras, and also includes details of McNaughton's 1951 representation theorem [139]. We also mention alternative completeness proofs of Łukasiewicz's axiomatization by Scott [197] and Panti [178].

The third fundamental fuzzy logic, Product Logic P, was defined explicitly by Hájek et al. in [112], although product implication had been considered by Goguen much earlier in [100]. The "related logics" considered here, Cancellative Hoop Logic CHL and Cross Ratio Logic CRL, were defined by Esteva, Godo, Hájek, and Montagna [78], and Gabbay and Metcalfe [87] respectively. Good overviews of the historical development of Ł, G, and other many-valued logics may be found in [137, 212], the key reference for the fuzzy logic perspective being Hájek's [105].

The first proof systems for fuzzy logics dealt with Gödel Logic. Avron's elegant hypersequent calculus (GG modulo some inessential changes) was introduced in 1991 [11], and investigated further by Baaz and co-workers in [15, 29]. An earlier (rather complicated) sequent calculus was defined by Sonobe [200] in 1975, and improved – terminating and contraction-free – versions were subsequently developed by Avellone, Ferrari, and Miglioli [8], and Dyckhoff [75]. The sequent calculus presented in this chapter is adapted from calculi introduced by Avron and Konikowska [12], while the sequent of relations approach, which applies to a wider class of "projective" logics, was developed by Baaz and Fermüller in [19].

Developing proof theory for Łukasiewicz and Product logics proved to be more problematic. The first calculi presented for Ł made essential use of either the cut rule [183] or extra syntax, see e.g. the labelled Tableaux of [104, 171] and the Resolution systems of [162, 215]. The sequent and hypersequent calculi for Łukasiewicz Logic presented above were introduced by Metcalfe, Olivetti, and Gabbay in 2005 [148], making use of the translation of Ł into Meyer and Slaney's Abelian Logic [149]. A cut elimination proof for GŁ (and calculi for related logics), similar to that provided for GA in the previous chapter was obtained by Ciabattoni and Metcalfe [54]. The sequent and hypersequent calculi for Product Logic described here were defined by Metcalfe, Olivetti, and Gabbay in 2004 [146]. The relationship of GŁ to Giles game, introduced by Giles in [95, 96] following e.g. [132], was discussed in [50]. We mention also the related work of Mundici on Ulam's game for finite-valued Łukasiewicz logics [160, 161], and work by Fermüller on parallel dialogue games [81].

Chapter 7
Uniformity and Efficiency

The proof systems encountered so far in this book provide an elegant presentation of fuzzy logics and are useful for establishing key properties for this family of logics, not least standard completeness. However, such systems are not always so convenient from an automated reasoning perspective. In particular, for the fundamental logics our calculi lack certain desirable properties:

- *Uniformity*. Suppose that we want to check the validity of a formula in G, $\mathsf{Ł}$, and P *at the same time*. All we can do is attempt separate proofs in different systems. Since each has its own set of rules, there is no hope of obtaining a derivation common to all three.
- *Focus*. There are no restrictions on proof search in our systems – either on the rule to apply next or the formula to apply it to – even though there may be good choices that avoid backtracking, looping, or treating irrelevant formulas.
- *Efficiency*. The proof systems are themselves inefficient, duplicating formulas unnecessarily and repeating steps such as decomposing the same formula several times.

In this chapter we try to remedy these defects. We give a uniform framework for $\mathsf{Ł}$, G, and P by stepping up in complexity from hypersequents to "relational hypersequents" with two types of sequents. We then describe a logic programming style methodology based on the idea that proof search should be directed by what is to be proved: the "goal" of the computation. Finally, we use proof systems to establish lower and upper bounds for the complexity of the set of valid formulas for a selection of fuzzy logics.

7.1 Uniform Systems

The calculi for the fundamental logics defined in the previous chapter are nice in many respects, but they do seem rather arbitrary. For each logic we have simply picked rules that work. In contrast, the systems of Chapter 4 have the same logical rules and differ only at the structural level. Here we define a similar framework for Ł, G, and P. That is, we define uniform logical rules and distinguish the logics using only structural rules. The cost is more structure: hypersequents are generalized to "relational hypersequents" with two kinds of sequent arrow instead of one.

7.1.1 Uniform Logical Rules

We will use the language $\mathcal{L}_F = \{\to, \odot, \wedge, \vee, \bot, \top\}$ (replacing e and f with the more usual $\top$ and $\bot$), recalling its adequacy for Ł, G, and P, and indeed any logic based on a residuated t-norm.

Definition 7.1. A *relational hypersequent (r-hypersequent)* is a finite multiset of ordered triples, written:

$$\Gamma_1 \lhd_1 \Delta_1 \mid \ldots \mid \Gamma_n \lhd_n \Delta_n$$

where $\lhd_i \in \{<, \leq\}$ and Γ_i and Δ_i are finite multisets of formulas for $i = 1 \ldots n$.

A hypersequent can be treated as an r-hypersequent with just one relation symbol, e.g. letting $\lhd_i$ be $\leq$ for $i = 1 \ldots n$. Similarly, a sequent of relations can be treated as an r-hypersequent where all multisets Γ_i, Δ_i contain exactly one formula. All our familiar notions for hypersequents – rules, proof systems etc. – extend to r-hypersequents, and we will use them here without further comment. Also, we let the complexity of an r-hypersequent be the complexity of the hypersequent obtained by treating each symbol $\leq$ or $<$ as $\Rightarrow$.

Validity for r-hypersequents is defined for each logic individually, understanding $\mid$ as before as a meta-level disjunction, where $<$ and $\leq$ denote inequalities between combinations (different for each logic) of truth values of formulas. The symbols $<$ and $\leq$ therefore have two uses: a syntactic one as part of an r-hypersequent, and a semantic one as an inequality holding between mathematical expressions. Often we will use $\lhd$ to stand uniformly for $\leq$ or $<$ (in either sense).

Definition 7.2. An r-hypersequent $\mathcal{G}$ is L-*valid* for $\mathrm{L} \in \{\text{Ł}, \mathrm{G}, \mathrm{P}\}$, written $\models_{\mathrm{L}} \mathcal{G}$, iff for all L-valuations v:

$$\star^v_{\mathrm{L}} \Gamma \lhd \star^v_{\mathrm{L}} \Delta \quad \text{for some} \quad (\Gamma \lhd \Delta) \in \mathcal{G}$$

where $\star^v_{\mathrm{L}}([]) = 1$ for $\mathrm{L} \in \{\text{Ł}, \mathrm{G}, \mathrm{P}\}$, and:

$$\begin{aligned} \star^v_{\text{Ł}}(\Gamma) &= 1 + \textstyle\sum[v(A) - 1 : A \in \Gamma] \\ \star^v_{\mathrm{G}}(\Gamma) &= \min[v(A) : A \in \Gamma] \\ \star^v_{\mathrm{P}}(\Gamma) &= \textstyle\prod[v(A) : A \in \Gamma] \end{aligned}$$

An r-hypersequent rule $\mathcal{G}_1, \ldots, \mathcal{G}_n \,/\, \mathcal{G}$ is L-*sound* if whenever $\models_{\mathrm{L}} \mathcal{G}_i$ for $i = 1 \ldots n$, also $\models_{\mathrm{L}} \mathcal{G}$, and L-*invertible* if the reverse implication holds.

$$\frac{\mathcal{G} \mid \Gamma, B \triangleleft A, \Delta \mid A \leq B \quad \mathcal{G} \mid \Gamma \triangleleft \Delta \mid B < A}{\mathcal{G} \mid \Gamma, A \to B \triangleleft \Delta} \ (\to\triangleleft)$$

$$\frac{\mathcal{G} \mid \Gamma \triangleleft \Delta \quad G \mid \Gamma, A \triangleleft B, \Delta \mid A \leq B}{\mathcal{G} \mid \Gamma \triangleleft A \to B, \Delta} \ (\triangleleft\to)$$

$$\frac{\mathcal{G} \mid \Gamma, A, B \triangleleft \Delta \quad \mathcal{G} \mid \Gamma, \bot \triangleleft \Delta}{\mathcal{G} \mid \Gamma, A \odot B \triangleleft \Delta} \ (\odot\triangleleft) \qquad \frac{\mathcal{G} \mid \Gamma \triangleleft \bot, \Delta \mid \Gamma \triangleleft A, B, \Delta}{\mathcal{G} \mid \Gamma \triangleleft A \odot B, \Delta} \ (<\odot)$$

$$\frac{\mathcal{G} \mid \Gamma, A \triangleleft \Delta \mid \Gamma, B \triangleleft \Delta}{\mathcal{G} \mid \Gamma, A \wedge B \triangleleft \Delta} \ (\wedge\triangleleft) \qquad \frac{\mathcal{G} \mid \Gamma \triangleleft A, \Delta \quad \mathcal{G} \mid \Gamma \triangleleft B, \Delta}{\mathcal{G} \mid \Gamma \triangleleft A \wedge B, \Delta} \ (\triangleleft\wedge)$$

$$\frac{\mathcal{G} \mid \Gamma, A \triangleleft \Delta \quad \mathcal{G} \mid \Gamma, B \triangleleft \Delta}{\mathcal{G} \mid \Gamma, A \vee B \triangleleft \Delta} \ (\vee\triangleleft) \qquad \frac{\mathcal{G} \mid \Gamma < A, \Delta \mid \Gamma \triangleleft B, \Delta}{\mathcal{G} \mid \Gamma \triangleleft A \vee B, \Delta} \ (\triangleleft\vee)$$

Fig. 7.1 Uniform rules

Notice immediately that for any formula A and $\mathrm{L} \in \{\text{Ł}, \mathsf{G}, \mathsf{P}\}$:

$$\models_{\mathrm{L}} \ \leq A \quad \text{iff} \quad \models_{\mathrm{L}} A$$

Hence we can express that a single formula is L-valid, as well as other relationships, such as $\models_{\mathrm{L}} A < B$, that cannot be expressed using just the L-validity of a formula.

Example 7.3. Consider the r-hypersequent:

$$r \leq r, q \mid p, q < p$$

This is Ł-valid since for any Ł-valuation v:

If $v(q) = 1$, then $\star^v_{\text{Ł}}[r] = v(r) = 1 + (v(r) - 1) + (v(q) - 1) = \star^v_{\text{Ł}}[r, q]$.

If $v(q) < 1$, then $\star^v_{\text{Ł}}[p, q] = 1 + (v(p) - 1) + (v(q) - 1) < v(p) = \star^v_{\text{Ł}}[p]$.

However, for $\mathrm{L} \in \{\mathsf{G}, \mathsf{P}\}$, if $v(p) = v(q) = 0$ and $v(r) > 0$, then:

$$\star^v_{\mathrm{L}}[r] = v(r) > 0 = \star^v_{\mathrm{L}}[r, q] \quad \text{and} \quad \star^v_{\mathrm{L}}[p, q] = 0 = v(p) = \star^v_{\mathrm{L}}[p]$$

so the r-hypersequent is not valid in these logics.

Our reward for this greater flexibility – both in the structures and their interpretations – is the set of uniform rules displayed in Fig. 7.1 (recalling that $\triangleleft$ is uniformly either $\leq$ or $<$ in each instance of a rule). Notice that the rules for $\to$, $\wedge$, and $\vee$ have the subformula property, but not the rules for $\odot$ ($\bot$ appears in the premises and possibly not the conclusion). This is one of the costs of uniformity. In the cases of G and P, we could remove the right premise of $(\odot\triangleleft)$, and $(\Gamma < \bot, \Delta)$ in the premise of $(\triangleleft\odot)$, while for Ł we could make do with just rules for $\to$.

Example 7.4. Crucially, if we apply these rules upwards to an r-hypersequent, then we always end up with atomic r-hypersequents, e.g.

$$\dfrac{\dfrac{p \leq q \mid p,q \leq p,q \quad p \leq q \mid q < p}{p,p \to q \leq q}\,(\to\leq) \qquad \bot \leq q}{p \odot (p \to q) \leq q}\,(\odot\leq)$$

Notice that the top r-hypersequents here are valid in all the fundamental logics.

Lemma 7.5. *The uniform rules are sound and invertible for* Ł, G, *and* P.

Proof. We consider only the rules for $\to$, disregarding side r-hypersequents since if $S_1,\ldots,S_n\ /\ S$ is sound and invertible for one of these logics, then so easily is $(\mathcal{G} \mid S_1),\ldots,(\mathcal{G} \mid S_n)\ /\ (\mathcal{G} \mid S)$. Let v be a valuation for Ł, G, or P. If $v(A) \leq v(B)$, then $v(A \to B) = 1$, and for both $(\to\lhd)$ and $(\lhd\to)$, the premises hold iff the conclusion holds. Now suppose that $v(A) > v(B)$. We consider each rule in turn:

- $(\to\lhd)$. The right premise clearly holds. For Ł and P, by simple arithmetic, the conclusion holds iff the left premise holds. For G, $v(A \to B) = v(B)$. Moreover, $\min(\star^v_{\mathrm{G}}(\Gamma), v(B)) \lhd \min(v(A), \star^v_{\mathrm{G}}(\Delta))$ iff $\min(\star^v_{\mathrm{G}}(\Gamma), v(B)) \lhd v(A)$ and $\min(\star^v_{\mathrm{G}}(\Gamma), v(A \to B)) \lhd \star^v_{\mathrm{G}}(\Delta)$. However, $\min(\star^v_{\mathrm{G}}(\Gamma), v(B)) \lhd v(A)$ since $v(A) > v(B)$, so we have that the left premise holds iff the conclusion holds.
- $(\lhd\to)$. If the conclusion holds, then easily both premises hold. For Ł and P, again by simple arithmetic, the conclusion holds if the right premise holds. For G, suppose that $\min(\star^v_{\mathrm{G}}(\Gamma), v(A)) \lhd \min(v(B), \star^v_{\mathrm{G}}(\Delta))$. Then $\min(\star^v_{\mathrm{G}}(\Gamma), v(A)) \lhd v(B)$, and, since $v(A) > v(B)$, $\min(\star^v_{\mathrm{G}}(\Gamma), v(A)) = \star^v_{\mathrm{G}}(\Gamma)$. Hence the conclusion holds if the right premise holds. ☐

As for many of the hypersequent calculi developed in the previous chapter, it is easy to see that the complexity of the premises is strictly less than the complexity of the conclusion. Moreover, there is always a rule available to apply (upwards) to non-atomic formulas occurring in an r-hypersequent. Hence a sound and complete calculus for L where L $\in$ {Ł, G, P} consists of the uniform rules extended with (as initial r-hypersequents) all L-valid atomic r-hypersequents.

7.1.2 Revised Logical Rules

Although our rules are uniform and invertible, they are still far from ideal for proof search. For one thing, they sometimes (reading upwards) multiply occurrences of formulas. Consider, for example, the rule $(\to\lhd)$:

$$\frac{\mathcal{G} \mid \Gamma, B \lhd A, \Delta \mid A \leq B \quad \mathcal{G} \mid \Gamma \lhd \Delta \mid B < A}{\mathcal{G} \mid \Gamma, A \to B \lhd \Delta}\,(\to\lhd)$$

The formulas A and B both occur twice in the left premise, but only once in the conclusion. We solve this problem with a trick already used in the density rule: we introduce new variables. Consider, as an intermediate step:

$$\frac{\mathcal{G} \mid \Gamma, q \lhd \Delta \mid B < q, A}{\mathcal{G} \mid \Gamma, A \to B \lhd \Delta}\ (\to\lhd)^r \qquad \frac{\mathcal{G} \mid \Gamma \lhd \Delta \quad \mathcal{G} \mid \Gamma, p \lhd q, \Delta \mid p \leq q \mid A < p \mid q < B}{\mathcal{G} \mid \Gamma \lhd A \to B, \Delta}\ (\lhd\to)^r$$

$$\frac{\mathcal{G} \mid \Gamma, A, B \lhd \Delta \quad \mathcal{G} \mid \Gamma, \bot \lhd \Delta}{\mathcal{G} \mid \Gamma, A \odot B \lhd \Delta}\ (\odot\lhd) \qquad \frac{\mathcal{G} \mid \Gamma \lhd q, \Delta \mid q < A, B \mid q < \bot}{\mathcal{G} \mid \Gamma \lhd A \odot B, \Delta}\ (\lhd\odot)^r$$

$$\frac{\mathcal{G} \mid \Gamma, p \lhd \Delta \mid A < p \mid B < p}{\mathcal{G} \mid \Gamma, A \wedge B \lhd \Delta}\ (\wedge\lhd)^r \qquad \frac{\mathcal{G} \mid \Gamma \lhd A, \Delta \quad \mathcal{G} \mid \Gamma \lhd B, \Delta}{\mathcal{G} \mid \Gamma \lhd A \wedge B, \Delta}\ (\lhd\wedge)$$

$$\frac{\mathcal{G} \mid \Gamma, A \lhd \Delta \quad \mathcal{G} \mid \Gamma, B \lhd \Delta}{\mathcal{G} \mid \Gamma, A \vee B \lhd \Delta}\ (\vee\lhd) \qquad \frac{\mathcal{G} \mid \Gamma \lhd p, \Delta \mid p < A \mid p < B}{\mathcal{G} \mid \Gamma \lhd A \vee B, \Delta}\ (\lhd\vee)^r$$

where p and q are variables not occurring in $\mathcal{G}$, Γ, Δ, A, or B.

Fig. 7.2 Revised uniform rules

$$\frac{\mathcal{G} \mid \Gamma, q \lhd \Delta \mid A \to B < q}{\mathcal{G} \mid \Gamma, A \to B \lhd \Delta}$$

where q does not occur in the conclusion. This rule is sound for each logic $\mathrm{L} \in \{\text{Ł}, \mathrm{G}, \mathrm{P}\}$, since any valuation v for the variables in the conclusion can be extended with $v(q) = v(A \to B)$. Also it is invertible since if $\star_{\mathrm{L}}^{v}(\Gamma, A \to B) \lhd \star_{\mathrm{L}}^{v}(\Delta)$ and $v(q) \leq v(A \to B)$, then immediately $\star_{\mathrm{L}}^{v}(\Gamma, q) \lhd \star_{\mathrm{L}}^{v}(\Delta)$.

Now we try again applying $(\to\lhd)$:

$$\frac{\mathcal{G} \mid \Gamma, q \lhd \Delta \mid B < q, A \mid A \leq B \quad \mathcal{G} \mid \Gamma, q \lhd \Delta \mid < q \mid B < A}{\mathcal{G} \mid \Gamma, q \lhd \Delta \mid A \to B < q}$$

There seems to be no improvement here, just one more variable to deal with. But look again at the right premise: this is valid iff the left premise is valid. Also $A \leq B$ can be dropped from the left premise since if $v(A) \leq v(B)$, then $v(A \to B) < v(q)$ cannot hold in the conclusion. So we arrive at a single premise $(\mathcal{G} \mid \Gamma, q \lhd \Delta \mid B < q, A)$ and conclusion $(\mathcal{G} \mid \Gamma, A \to B \Rightarrow \Delta)$. Proceeding similarly for the other uniform rules, we arrive at revised versions, displayed in Fig. 7.2, with the same soundness and invertibility properties as the original rules.

Lemma 7.6. *The revised uniform rules are sound and invertible for* Ł, G, *and* P.

Example 7.7. Consider these applications of the revised rules:

$$\cfrac{\cfrac{q \leq D \mid C < q, p \mid p < A, B \mid p < \bot}{q \leq D \mid C < q, A \odot B}\ (<\odot)^r}{(A \odot B) \to C \leq D}\ (\to\leq)^r$$

If we use the original rules to decompose this r-hypersequent, we get (even without expanding the top occurrence of $\odot$):

Initial R-Hypersequents

$$\frac{}{\mathcal{G} \mid A \leq A}\ (\textsc{id}) \qquad \frac{}{\mathcal{G} \mid \Gamma, \bot \leq A}\ (\bot\leq) \qquad \frac{}{\mathcal{G} \mid \Gamma \leq \top}\ (\leq\top)$$

$$\frac{}{\mathcal{G} \mid \leq}\ (\textsc{emp}) \qquad \frac{}{\mathcal{G} \mid \Gamma, \bot <}\ (\bot<) \qquad \frac{}{\mathcal{G} \mid \Gamma, \bot < \top}\ (<\top)$$

Logical Rules: as in Fig. 7.1

Structural Rules:

$$\frac{\mathcal{G}}{\mathcal{G} \mid \mathcal{H}}\ (\textsc{ew}) \qquad \frac{\mathcal{G} \mid \mathcal{H} \mid \mathcal{H}}{\mathcal{G} \mid \mathcal{H}}\ (\textsc{ec}) \qquad \frac{\mathcal{G} \mid \Gamma \lhd \Delta}{\mathcal{G} \mid \Gamma, \Pi \lhd \Delta}\ (\textsc{w}) \qquad \frac{\mathcal{G} \mid \Gamma_1 \lhd \Delta_1 \quad \mathcal{G} \mid \Gamma_2 \lhd \Delta_2}{\mathcal{G} \mid \Gamma_1, \Gamma_2 \lhd \Delta_1, \Delta_2}\ (\textsc{mix})_{\lhd}$$

$$\frac{\mathcal{G} \mid \Gamma_1, \Gamma_2 \leq \Delta_1, \Delta_2}{\mathcal{G} \mid \Gamma_1 \leq \Delta_1 \mid \Gamma_2 \leq \Delta_2}\ (\textsc{split})_{\leq} \qquad \frac{\mathcal{G} \mid \Gamma_1 \leq \Delta_2 \quad \mathcal{G} \mid \Gamma_2 \lhd \Delta_1}{\mathcal{G} \mid \Gamma_1 \lhd \Delta_1 \mid \Gamma_2 < \Delta_2}\ (\textsc{split})_{<}$$

Fig. 7.3 The basic r-hypersequent calculus GRB

$$\dfrac{\dfrac{C \leq \bot, D \mid C \leq A, B, D \mid A \odot B \leq D}{C \leq A \odot B, D \mid A \odot B \leq D}\ (<\odot) \qquad \dfrac{\leq D \mid C < \bot \mid C < A, B}{\leq D \mid C < A \odot B}\ (<\odot)}{(A \odot B) \rightarrow C \leq D}\ (\rightarrow\leq)$$

Notice that in the first derivation, not only are two branches combined with $(\rightarrow\lhd)^r$, but also we avoid duplication of the formulas A, B, C, and D (which could be large).

Each upwards application of a revised rule gives only a constant increase in the size (number of symbol occurrences) of an r-hypersequent. Hence applying these rules upwards to an r-hypersequent $\mathcal{G}$ terminates with atomic r-hypersequents of size linear in the size of $\mathcal{G}$.

7.1.3 Structural Rules

The sets of valid atomic r-hypersequents are polynomial time for Ł, G, and P (see Section 7.3), and hence could be acceptable computationally as initial r-hypersequents of our systems. Here however, we will show that validity in the different logics can be distinguished using very simple initial r-hypersequents together with structural rules. Consider first the basic system GRB presented in Fig. 7.3. It is easy to show that the rules of GRB are sound for Ł, G, and P.

Example 7.8. The following useful initial r-hypersequent is GRB-derivable:

$$\frac{}{\mathcal{G} \mid \Gamma, \Delta \leq \Delta}\ (\textsc{id})^g$$

Just use (EMP), (MIX), and (ID) to derive $(\Delta \leq \Delta)$, then apply (W) and (EW).

Calculi for Ł, G, and P can then be defined by extending GRB with rules reflecting the characteristic properties of each logic. For G, we just add a contraction rule.

Theorem 7.9. $\vdash_{\mathsf{GRG}} \mathcal{G}$ *iff* $\models_{\mathsf{G}} \mathcal{G}$ *where* GRG *is* GRB *extended with:*

$$\frac{\mathcal{G} \mid \Gamma,\Pi,\Pi \lhd \Sigma,\Sigma,\Delta}{\mathcal{G} \mid \Gamma,\Pi \lhd \Sigma,\Delta} \text{ (C)}$$

Proof. For soundness, we just note that the extra rule (C) is clearly G-sound. For completeness, it is enough to show that all G-valid atomic r-hypersequents are GRG-derivable. We note first that the following rules are both GRG-derivable and G-invertible:

$$\frac{\mathcal{G} \mid \Gamma_1 \lhd \Delta \mid \Gamma_2 \lhd \Delta}{\mathcal{G} \mid \Gamma_1,\Gamma_2 \lhd \Delta} \text{ (ML)}_{\lhd} \qquad \frac{\mathcal{G} \mid \Gamma \lhd \Delta_1 \quad \mathcal{G} \mid \Gamma \lhd \Delta_2}{\mathcal{G} \mid \Gamma < \Delta_1,\Delta_2} \text{ (MR)}_{\lhd}$$

Hence we can apply these derived rules upwards to a valid atomic r-hypersequent to obtain a valid r-hypersequent $\mathcal{H}$ in which all multisets on the left or right of some $\lhd$ contain at most one atomic formula (essentially a sequent of relations). Let $\mathcal{H}'$ be $\mathcal{H}$ with $\top$ replacing empty spaces on either side of $\lhd$. Clearly $\models_{\mathsf{G}} \mathcal{H}$ iff $\models_{\mathsf{G}} \mathcal{H}'$ and, by an easy induction, $\vdash_{\mathsf{GRG}} \mathcal{H}$ iff $\vdash_{\mathsf{GRG}} \mathcal{H}'$. Moreover, we can regard $\mathcal{H}'$ simply as a set of inequalities. So by Lemma 6.9, there exists $(a_i \lhd_i a_{i+1}) \in \mathcal{H}'$ for $i = 1 \dots n$ such that one of the following holds:

(1) $a_1 = a_{n+1}$ or $a_1 = \bot$ or $a_{n+1} = \top$, where $\lhd_i$ is $\leq$ for some $i \in \{1,\dots,n\}$.
(2) $a_1 = \bot$ and $a_{n+1} = \top$.

In either case, we can apply (SPLIT)$_{\leq}$ and (MIX) or (SPLIT)$_{<}$ repeatedly to $(a_i \lhd_i a_{i+1})$ for $i = 1 \dots n$. For (1), we obtain one of $(a_1 \leq a_1)$, $(\bot \leq a_{n+1})$, or $(a_1 \leq \top)$, and for (2), $(\bot \lhd \top)$, all of which are GRB-derivable. So $\vdash_{\mathsf{GRG}} \mathcal{H}'$ and hence also $\vdash_{\mathsf{GRG}} \mathcal{H}$. □

For Ł, we need a stronger split rule and a rule for weakening left occurrences of $\bot$.

Theorem 7.10. $\vdash_{\mathsf{GRŁ}} \mathcal{G}$ *iff* $\models_{\mathsf{Ł}} \mathcal{G}$ *where* GRŁ *is* GRB *extended with:*

$$\frac{\mathcal{G} \mid \Gamma_1,\Gamma_2 \lhd \Delta_1,\Delta_2}{\mathcal{G} \mid \Gamma_1 \lhd \Delta_1 \mid \Gamma_2 < \Delta_2} \text{ (SPLIT)}_{\text{Ł}} \qquad \frac{\mathcal{G} \mid \Gamma \leq \Delta}{\mathcal{G} \mid \Gamma,\bot < \Delta} \text{ (W)}_{\bot}$$

Proof. For soundness it is easy to see that the extra rules (SPLIT)$_{\text{Ł}}$ and (W)$_{\bot}$ are Ł-sound. For completeness, we again need consider only the case where $\mathcal{G}$ is an Ł-valid atomic r-hypersequent. We then proceed as in Theorem 6.24 (the completeness of GŁ) by induction on the number of distinct variables occurring on the left of sequents in $\mathcal{G}$. We just show as before that there is an Ł-valid r-hypersequent $\mathcal{G}'$ with one fewer distinct variables on the left of sequents, such that $\mathcal{G}$ is GRŁ-derivable from $\mathcal{G}'$. The details (mostly repeated from the proof of Theorem 6.24) are left as an exercise for the interested reader. □

Finding a calculus for P is a bit more tricky. Here we require restricted splitting and contraction rules to cope with the special case of 0.

Theorem 7.11. $\vdash_{\mathsf{GRP}} \mathcal{G}$ *iff* $\models_{\mathsf{P}} \mathcal{G}$ *where* GRP *is* GRB *extended with:*

$$\frac{\mathcal{G} \mid \Gamma_1,\Gamma_2 \leq \Delta_1,\Delta_2 \quad \mathcal{G} \mid \bot < \Delta_2}{\mathcal{G} \mid \Gamma_1 \leq \Delta_1 \mid \Gamma_2 < \Delta_2}\ (\text{SPLIT})_{\mathsf{P}} \qquad \frac{\mathcal{G} \mid \Gamma,\bot,\bot \lhd \Delta}{\mathcal{G} \mid \Gamma,\bot \lhd \Delta}\ (\text{C})_{\bot}$$

Proof. Clearly $(\text{C})_{\bot}$ is sound for P. For $(\text{SPLIT})_{\mathsf{P}}$, let v be a P-valuation. Suppose contrapositively – disregarding the side r-hypersequent $\mathcal{G}$ – that (1) $\star^v_{\mathsf{P}}(\Gamma_1) > \star^v_{\mathsf{P}}(\Delta_1)$ and (2) $\star^v_{\mathsf{P}}(\Gamma_2) \geq \star^v_{\mathsf{P}}(\Delta_2)$. If $\star^v_{\mathsf{P}}(\Delta_2) = 0$, then we are done. Otherwise we can multiply (1) on both sides by $\star^v_{\mathsf{P}}(\Delta_2)$ to get $\star^v_{\mathsf{P}}(\Gamma_1) \cdot \star^v_{\mathsf{P}}(\Delta_2) > \star^v_{\mathsf{P}}(\Delta_1) \cdot \star^v_{\mathsf{P}}(\Delta_2)$. But then using (2), $\star^v_{\mathsf{P}}(\Gamma_1) \cdot \star^v_{\mathsf{P}}(\Gamma_2) > \star^v_{\mathsf{P}}(\Delta_1) \cdot \star^v_{\mathsf{P}}(\Delta_2)$ as required.

For completeness, we again need consider only a P-valid atomic r-hypersequent:

$$\mathcal{G} = (\Gamma_1 < \Delta_1 \mid \ldots \mid \Gamma_n < \Delta_n \mid \Pi_1 \leq \Sigma_1 \mid \ldots \mid \Pi_m \leq \Sigma_m)$$

The case where $m = 0$ is fairly straightforward, so we will leave it as an exercise, and assume $m > 0$. Let $A_i = \mathrm{I}_{\mathsf{P}}(\Gamma_i \Rightarrow \Delta_i)$ for $i = 1 \ldots n$ and $\mathcal{H} = (\Pi_1 \Rightarrow \Sigma_1 \mid \ldots \mid \Pi_m \Rightarrow \Sigma_m)$. Then it follows from the definition of P-validity, that:

$$A_1,\ldots,A_n \models_{\mathsf{P}} \mathrm{I}_{\mathsf{P}}(\mathcal{H})$$

It now follows from the local deduction theorem for P (Theorem 3.43) and some additional derivabilities in this logic, that $\models_{\mathsf{P}} \mathcal{G}^H$ for some hypersequent $\mathcal{G}^H$ obtained from $\mathcal{H}$ by placing some number of copies of $A_1,\ldots,A_n$ on the left of sequents. Moreover, by the completeness of GP (Theorem 6.53), we have $\vdash_{\mathsf{GP}} \mathcal{G}^H$. We can then prove that $\mathcal{G}$ is GRP-derivable, proceeding by induction on the height of a GP-derivation of any such $\mathcal{G}^H$. The key cases are $(\to\Rightarrow)$ and $(\neg\Rightarrow)$; for the former, we make use of $(\text{SPLIT})_{\mathsf{P}}$ and for the latter, $(\text{SPLIT})_{<}$. □

Example 7.12. The following derivations illustrate how the same r-hypersequent $(q \leq p,p \mid p,q < q)$ is proved differently in each system. In GRG, we make crucial use of the (C) rule to multiply copies of p on the left:

$$\dfrac{\dfrac{}{q \leq q}\,(\text{ID}) \qquad \dfrac{\dfrac{\dfrac{}{p,p \leq p,p}\,(\text{ID})^g}{p \leq p,p}\,(\text{C})}{p,q \leq p,p}\,(\text{W})}{q \leq p,p \mid p,q < q}\,(\text{SPLIT})_{<}$$

In GRŁ, we use the $(\text{SPLIT})_{Ł}$ rule to combine sequents:

$$\dfrac{\dfrac{\dfrac{\dfrac{}{p,q,q,p,q \leq p,p,q,q}\,(\text{ID})^g}{q,p,q \leq p,p,q \mid p,q < q}\,(\text{SPLIT})_{Ł}}{q \leq p,p \mid p,q < q \mid p,q < q}\,(\text{SPLIT})_{Ł}}{q \leq p,p \mid p,q < q}\,(\text{EC})$$

Finally for GRP, we use $(\text{SPLIT})_{\mathsf{P}}$ to combine sequents, but must also take care of the extra premise:

$$
\dfrac{\dfrac{\dfrac{\dfrac{}{p,q<q \mid p,p,q,q,q \leq p,p,q,q}\;(\text{ID})^g \qquad \mathcal{H}}{q \leq p,p \mid p,q<q \mid p,q,q \leq p,p,q}\;(\text{SPLIT})_{\mathsf{P}} \qquad \mathcal{H}}{q \leq p,p \mid q \leq p,p \mid p,q<q \mid p,q<q}\;(\text{SPLIT})_{\mathsf{P}}}{q \leq p,p \mid p,q<q}\;(\text{EC})
$$

where $\mathcal{H} = (q \leq p,p \mid p,q<q \mid \bot < q)$ is derivable as follows, noting that $(\mathcal{G} \mid \Gamma, \bot \leq \Delta)$ is a derived initial r-hypersequent of GRP:

$$
\dfrac{\dfrac{\dfrac{}{q \leq q}\;(\text{ID}) \qquad \bot \leq p,p}{q \leq p,p \mid \bot < q}\;(\text{SPLIT})_{<}}{q \leq p,p \mid p,q<q \mid \bot < q}\;(\text{EW})
$$

7.2 Goal-Directed Methods

Although completeness results guarantee that derivations of valid formulas exist for our systems, and can in many cases be found by applying rules exhaustively, there is no indication of how to apply the rules. In particular, if the structure is large, there is no strategy for focussing on the parts that might be most relevant. In this section, we tackle such problems by introducing a style of proof search that is "goal-directed" in the sense that the next step is guided by the problem to be solved.

7.2.1 The Goal-Directed Methodology

Suppose that Γ is a (possibly large) collection (a set, multiset, or something more structured) of formulas, called rather grandly a *database*, and A is a formula, called the *goal*. Fixing a logic, we write:

$$\Gamma \Rightarrow^{?} A$$

to stand for a query "does A follow from Γ?". Derivations for queries are called *goal-directed* if the next step is determined by the form of the current goal. For example, a complex goal might be decomposed until its atomic constituents are reached. Atomic goals q might then be matched (if possible) with the "head" of a formula $B \to q$ in the database, and its "body" B asked in turn.

Goal-directed proof search can also be refined, e.g. by:

1. putting constraints on databases, restricting formulas available to match goals;
2. adding control mechanisms to ensure termination such as loop-checking or "diminishing resources" – removing formulas used to match a goal;

3. storing goals previously occurring in the derivation in a history and re-asking them using "restart rules".

Goal-directed procedures have been defined for a wide range of logics (see the historical remarks at the end of this chapter). Let us just illustrate the approach here with an algorithm for the implicational fragment of Intuitionistic Logic, adopting the convention throughout this section of writing:

$$[A_1,\dots,A_n] \to q \quad \text{for} \quad A_1 \to (A_2 \to \dots (A_n \to q)\dots)$$

We also make use of the useful definition:

$$\text{Head}([A_1,\dots,A_n] \to q) = q \quad \text{and} \quad \text{Head}(\Gamma) = \{\text{Head}(A) : A \in \Gamma\}$$

Queries for $\mathsf{AIL}^{\to}$ are ordered triples of the form $(\Gamma \Rightarrow^? A; H)$ where Γ is a multiset (a set would also work fine here) of implicational formulas called the database, A is a formula called the goal, and H is a sequence of atomic goals called the history (writing $H * (q)$ for q appended to H). $\mathsf{AIL}^{\to}$ then consists of the following rules, written in goal-directed format with the conclusion first:

(SUCCESS) $\Gamma \Rightarrow^? q; H$ succeeds if $q \in \Gamma$.

(IMPLICATION) From $\Gamma \Rightarrow^? \Pi \to q; H$ step to $\Gamma, \Pi \Rightarrow^? q; H$.

(REDUCTION) From $\Gamma, \Pi \to q \Rightarrow^? q; H$ step to $\Gamma \Rightarrow^? A; H * (q)$ for all $A \in \Pi$.

(BOUNDED-RESTART) From $\Gamma \Rightarrow^? q; H$ step to $\Gamma \Rightarrow^? p; H * (q)$ if p follows q in H.

Derivations of queries are given by the general definitions in Chapter 3, but to reflect the algorithmic nature of the systems, we write the initial query first, stepping to further queries as directed by the rules.

Example 7.13. Consider the following derivation, noting that (BOUNDED-RESTART) is needed at (2) to compensate for the removal of $p \to q$ at (1):

	$\Rightarrow^? [(p \to q) \to p, p \to q] \to q; ()$	(IMPLICATION)
(1)	$(p \to q) \to p, p \to q \Rightarrow^? q; ()$	(REDUCTION)
	$(p \to q) \to p \Rightarrow^? p;\ (q)$	(REDUCTION)
	$\Rightarrow^? p \to q;\ (q, p)$	(IMPLICATION)
(2)	$p \Rightarrow^? q;\ (q, p)$	(BOUNDED-RESTART)
	$p \Rightarrow^? p;\ (q, p, q)$	(SUCCESS)

A goal-directed calculus for the same fragment of Classical Logic is obtained by liberalizing (BOUNDED-RESTART) to allow restarts from *any* previous atomic goal. That is, $\mathsf{ACL}^{\to}$ has the same rules as $\mathsf{AIL}^{\to}$ with (BOUNDED-RESTART) replaced by:

(RESTART) From $\Gamma \Rightarrow^? q; H$ step to $\Gamma \Rightarrow^? p; H * (q)$ if p occurs in H.

Example 7.14. Peirce's axioms (which are not intuitionistically valid) are derivable in $\mathsf{ACL}^{\rightarrow}$ as follows, using (RESTART) to continue the deduction at (1):

	$\Rightarrow^{?} ((p \to q) \to p) \to p; ()$	(IMPLICATION)
	$(p \to q) \to p \Rightarrow^{?} p; ()$	(REDUCTION)
	$\Rightarrow^{?} p \to q; (p)$	(IMPLICATION)
(1)	$p \Rightarrow^{?} q; (p)$	(RESTART)
	$p \Rightarrow^{?} p; (p, q)$	(SUCCESS)

7.2.2 Uniform Rules

Goal-directed queries for fuzzy logics generalize the above structures for Classical Logic and Intuitionistic Logic. They consist of a database together with a multiset of goals (rather than just one), and a history of previous states of the database with goals (rather than just goals). Restarts are also allowed, but limited this time to at most one for each multiset of goals.

Definition 7.15. A *goal-directed query* (query for short) $\mathcal{Q}$ is a structure of the form:

$$\Gamma_1 \Rightarrow^{?} \Delta_1; R_1; \{(\Gamma_2 \Rightarrow^{?} \Delta_2; R_2), \ldots, (\Gamma_n \Rightarrow^{?} \Delta_n; R_n)\}$$

where: $\Gamma_1, \ldots, \Gamma_n$ are multisets of formulas called *databases*;

$\Delta_1, \ldots, \Delta_n$ are multisets of formulas called *goals*;

$R_1, \ldots, R_n$ are multisets of at most one atomic formula called *restarts*.

$\mathcal{Q}$ is L-*valid* for $\mathrm{L} \in \{\mathsf{Ł}, \mathsf{G}, \mathsf{P}\}$, written $\models_{\mathrm{L}} \mathcal{Q}$, if for every L-valuation v:

$$\text{either } \star^{v}_{\mathrm{L}}(\Gamma_i) \leq \star^{v}_{\mathrm{L}}(\Delta_i) \text{ or } \star^{v}_{\mathrm{L}}(\Gamma_i \uplus R_i) < \star^{v}_{\mathrm{L}}(\Delta_i) \text{ for some } i \in \{1, \ldots, n\}.$$

Intuitively, the meaning of a query is that for each valuation for the logic, there should be a state of the database where the associated goals "follow from" that database (possibly using a restart). This will remind the reader of hypersequents – unsurprisingly, since our algorithms are refinements of the Gentzen systems presented earlier. Note, however, that in goal-directed queries, one "sequent" is given priority, and also there are restarts, an essentially algorithmic notion.

We could also define validity for queries by first translating into r-hypersequents. E.g. the query:

$$p \to q \Rightarrow^{?} q \to r, s; [p]; \{(q \to p \Rightarrow^{?} r; [])\}$$

is L-valid iff the following r-hypersequent is L-valid:

$$p \to q \leq q \to r, s \mid p \to q, p < q \to r, s \mid q \to p \leq r \mid q \to p < r$$

Restart formulas give a choice for each state: either the restart is absent from the database and the relation is "less than or equal to", or the restart is present and the

(IMPLICATION) From $\Gamma \Rightarrow^? \Pi \to q, \Delta; R; H$ step to $\Gamma \Rightarrow^? \Delta; R; H$ and $\Gamma, \Pi \Rightarrow^? q, \Delta; R; H$.

(L-REDUCTION) From $\Gamma, \Pi \to q \Rightarrow^? q, \Delta; R; H$ step to
$\Gamma, q \Rightarrow^? \Pi, q, \Delta; R; H \cup \{(\Gamma \Rightarrow^? q, \Delta; R)\}$ and $\Rightarrow^? \Pi; [q]; H \cup \{(\Gamma \Rightarrow^? q, \Delta; R)\}$.

(R-REDUCTION) From $\Gamma \Rightarrow^? q, \Delta; R_1; H \cup \{(\Gamma', \Pi \to q \Rightarrow^? \Delta'; R_2)\}$ step to
$\Gamma', q \Rightarrow^? \Pi, \Delta'; R_2; H \cup \{(\Gamma' \Rightarrow^? \Delta'; R_2), (\Gamma \Rightarrow^? q, \Delta; R_1)\}$ and
$\Rightarrow^? \Pi; [q]; H \cup \{(\Gamma' \Rightarrow^? \Delta'; R_2), (\Gamma \Rightarrow^? q, \Delta; R_1)\}$.

(SWITCH) From $\Gamma_1 \Rightarrow^? \Delta_1; R_1; H \cup \{(\Gamma_2 \Rightarrow^? \Delta_2; R_2)\}$ step to
$\Gamma_2 \Rightarrow^? \Delta_2; R_2; H \cup \{(\Gamma_1 \Rightarrow^? \Delta_1; R_1)\}$.

Fig. 7.4 Goal-directed implication rules

relation is "strictly less than". More simply, however, a formula A is L-valid for $\mathrm{L} \in \{\text{Ł}, \mathsf{G}, \mathsf{P}\}$ iff the query $\Rightarrow^? A; []; \emptyset$ is L-valid.

Uniform goal-directed rules for handling implication in Ł, G, and P are presented in Fig. 7.4. (IMPLICATION) treats a query with an implicational goal $\Pi \to q$, and steps to two further queries: one where this goal is removed, and one where Π is added to the database and q replaces $\Pi \to q$ as a goal. However, if $\Pi \to q$ is the only goal, then a (derived) rule with one premise is sufficient, i.e.

(IMPLICATION)$_1$ From $\Gamma \Rightarrow^? \Pi \to q; R; H$ step to $\Gamma, \Pi \Rightarrow^? q; R; H$.

The presence of an implicational formula in a state of the database is treated by two rules. Local reduction (L-REDUCTION) and remote reduction (R-REDUCTION) treat the cases where a goal matches the head of a formula in the current database, and in a database of a history state, respectively. Finally, unlike previous calculi, we make use of a rule (SWITCH) that switches the current state to one in the history.

Example 7.16. We illustrate these rules with a simple example:

$$\Rightarrow^? \{p, p \to q\} \to q; []; \emptyset \quad \text{(IMPLICATION)}_1$$

$$p, p \to q \Rightarrow^? q; []; \emptyset \quad \text{(L-REDUCTION)}$$

/ \

$$p, q \Rightarrow^? p, q; []; \{(p \Rightarrow^? q; \emptyset)\} \qquad \Rightarrow^? p; [q]; \{(p \Rightarrow^? q; \emptyset)\}$$

Lemma 7.17. *The implication rules are* L*-sound and* L*-invertible for* $\mathrm{L} \in \{\text{Ł}, \mathsf{G}, \mathsf{P}\}$.

Proof. We just check (L-REDUCTION), assuming (harmlessly) that the common part H of the histories of the premises and conclusion is empty. Let v be an L-valuation for $\mathrm{L} \in \{\text{Ł}, \mathsf{G}, \mathsf{P}\}$. Suppose first that $\star_{\mathrm{L}}^{v}(\Pi) \leq v(q)$, and hence that

$v(\Pi \to q) = 1$. Clearly, both premises hold if the conclusion holds, since $(\Gamma \Rightarrow^? q, \Delta; R)$ is present in both histories. Moreover, if the premises hold, then, using the first premise for $\star^v_{\mathrm{L}}(\Pi) = v(q) = 1$, and the second otherwise, the conclusion holds. Now assume that $\star^v_{\mathrm{L}}(\Pi) > v(q)$. Clearly, the second premise holds. For G, $v(\Pi \to q) = v(q)$, and for Ł and P, we can use the fact that for any Γ', Δ', and $\lhd \in \{\leq, <\}$:

$$\star^v_{\mathrm{L}}(\Gamma' \uplus [\Pi \to q]) \lhd \star^v_{\mathrm{L}}(\Delta') \quad \text{iff} \quad \star^v_{\mathrm{L}}(\Gamma' \uplus [q]) \lhd \star^v_{\mathrm{L}}(\Delta' \uplus \Pi) \text{ or } \star^v_{\mathrm{L}}(\Gamma') \lhd \star^v_{\mathrm{L}}(\Delta')$$

So in all cases the conclusion holds iff the first premise holds. □

The implication rules decompose formulas occurring in the query into subformulas. Hence, assuming a loop-checking mechanism to stop (SWITCH) repeating ad infinitum, these rules terminate with queries where all goals are atomic and fail to match the head of any non-atomic database formula. Recall that for a multiset of formulas Γ, $\mathrm{Set}(\Gamma)$ is just the *set* of members of Γ.

Definition 7.18. A query $\Gamma_1 \Rightarrow^? \Delta_1; R_1; \{(\Gamma_2 \Rightarrow^? \Delta_2; R_2), \ldots, (\Gamma_n \Rightarrow^? \Delta_n; R_n)\}$ is *irreducible* if for each $i = 1 \ldots n$:

1. Δ_i is atomic.
2. $\Gamma_i = \Pi_i \uplus \Sigma_i$ where Σ_i is atomic and $\mathrm{Head}(\Pi_1 \uplus \ldots \uplus \Pi_n) \cap \mathrm{Set}(\Delta_1 \uplus \ldots \uplus \Delta_n) = \emptyset$.

Proposition 7.19. *Applying the implication rules with* (SWITCH) *to queries (using loop-checking for applications of* (SWITCH)*) terminates with irreducible queries.*

Moreover, if an irreducible query is valid for some logic, then by removing non-atomic database formulas from the query, we obtain an atomic query that is also valid. That is, we can disregard "irrelevant" parts of the databases.

Lemma 7.20. *Let* $\mathrm{L} \in \{\text{Ł}, \mathrm{G}, \mathrm{P}\}$ *and suppose that the following conditions hold:*

1. $\models_{\mathrm{L}} \Gamma_1, \Pi_1 \Rightarrow^? \Delta_1; R_1; \{(\Gamma_2, \Pi_2 \Rightarrow^? \Delta_2; R_2), \ldots, (\Gamma_n, \Pi_n \Rightarrow^? \Delta_n; R_n)\}$.
2. Γ_i *and* Δ_i *are atomic for* $i = 1 \ldots n$.
3. $\mathrm{Head}(\Pi_1 \uplus \ldots \uplus \Pi_n) \cap \mathrm{Set}(\Delta_1 \uplus \ldots \uplus \Delta_n) = \emptyset$.

Then $\models_{\mathrm{L}} \Gamma_1 \Rightarrow^? \Delta_1; R_1; \{(\Gamma_2 \Rightarrow^? \Delta_2; R_2), \ldots, (\Gamma_n \Rightarrow^? \Delta_n; R_n)\}$.

Proof. Suppose that $\not\models_{\mathrm{L}} \Gamma_1 \Rightarrow^? \Delta_1; R_1; \{(\Gamma_2 \Rightarrow^? \Delta_2; R_2), \ldots, (\Gamma_n \Rightarrow^? \Delta_n; R_n)\}$ for a contradiction. This means that there is an L-valuation v such that $\star^v_{\mathrm{L}}(\Gamma_i) > \star^v_{\mathrm{L}}(\Delta_i)$ and $\star^v_{\mathrm{L}}(\Gamma_i \uplus R_i) \geq \star^v_{\mathrm{L}}(\Delta_i)$ for $i = 1 \ldots n$. We define a new L-valuation w as follows:

$$w(q) = \begin{cases} 1 & \text{if } q \in \mathrm{Head}(\Pi_1 \uplus \ldots \uplus \Pi_n) \\ v(q) & \text{otherwise} \end{cases}$$

Using the fact that $\mathrm{Head}(\Pi_1 \uplus \ldots \uplus \Pi_n) \cap \mathrm{Set}(\Delta_1 \uplus \ldots \uplus \Delta_n) = \emptyset$, we get $\star^w_{\mathrm{L}}(\Delta_i) = \star^v_{\mathrm{L}}(\Delta_i)$ for $i = 1 \ldots n$. It also follows that $\star^w_{\mathrm{L}}(\Pi_i) = 1$, $\star^w_{\mathrm{L}}(\Gamma_i) \geq \star^v_{\mathrm{L}}(\Gamma_i)$, and $\star^w_{\mathrm{L}}(R_i) \geq \star^v_{\mathrm{L}}(R_i)$ for $i = 1 \ldots n$. Hence $\star^w_{\mathrm{L}}(\Pi_i \uplus \Gamma_i) > \star^w_{\mathrm{L}}(\Delta_i)$ and $\star^w_{\mathrm{L}}(\Pi_i \uplus \Gamma_i \uplus R_i) \geq \star^w_{\mathrm{L}}(\Delta_i)$ for $i = 1 \ldots n$. So $\not\models_{\mathrm{L}} \Gamma_1, \Pi_1 \Rightarrow^? \Delta_1; R_1; \{(\Gamma_2, \Pi_2 \Rightarrow^? \Delta_2; R_2), \ldots, (\Gamma_n, \Pi_n \Rightarrow^? \Delta_n; R_n)\}$, a contradiction. □

Since the implication rules are both sound and invertible for Ł, G, and P, and terminating (modulo loop-checking for (SWITCH)), we are able to reduce checking the validity of a query in these logics to checking the validity of irreducible queries, which reduces in turn (by Lemma 7.20) to checking the validity of atomic queries.

An exponential growth in the size of queries is avoided, as for r-hypersequents, by defining rules that introduce new variables during the reduction process:

(L-REDUCTION)$_r$ From $\Gamma, \Pi \to q \Rightarrow^? q, \Delta; R; H$ step to
$\Gamma, q \Rightarrow^? q, \Delta; R; H$ and $\Rightarrow^? p, \Pi; [q]; H \cup \{(\Gamma, p \Rightarrow^? q, \Delta; R)\}$
where p is a new variable.

(R-REDUCTION)$_r$ From $\Gamma \Rightarrow^? q, \Delta; R_1; H \cup \{(\Gamma', \Pi \to q \Rightarrow^? \Delta'; R_2)\}$ step to
$\Gamma \Rightarrow^? q, \Delta; R_1; H \cup \{(\Gamma', q \Rightarrow^? \Delta'; R_2)\}$ and
$\Rightarrow^? \Pi, p; [q]; H \cup \{(\Gamma \Rightarrow^? q, \Delta; R_1), (\Gamma', p \Rightarrow^? \Delta'; R_2)\}$
where p is a new variable.

7.2.3 Goal-Directed Systems

The uniform goal-directed rules given above provide algorithms for reducing implicational queries to atomic queries which can then be tested for validity using some method such as linear programming (see Section 7.3). However, to obtain a full algorithmic interpretation for the implicational fragments of Ł, G, and P we require further rules. Let us begin by defining a basic stock of rules that work for all three logics, featuring a new rule that allows the combination of databases and goals.

Definition 7.21. $\mathsf{AB}^{\to}$ consists of the implication rules and:

(COMBINE) From $\Gamma_1 \Rightarrow^? q, \Delta_1; R_1; H \cup \{(\Gamma_2, q \Rightarrow^? \Delta_2; R_2)\}$ step to
$\Gamma_1, \Gamma_2 \Rightarrow^? \Delta_1, \Delta_2; []; H \cup \{(\Gamma_1 \Rightarrow^? q, \Delta_1; R_1), (\Gamma_2, q \Rightarrow^? \Delta_2; R_2)\}$.

Calculi for Ł, G, and P may be defined by extending $\mathsf{AB}^{\to}$ with different restart and success rules. We first consider G.

Theorem 7.22. *$\vdash_{\mathsf{AG}^{\to}} \mathcal{Q}$ iff $\models_{\mathsf{G}} \mathcal{Q}$ where $\mathsf{AG}^{\to}$ is $\mathsf{AB}^{\to}$ extended with:*

(SUCCESS)$_\mathsf{G}$ $\Gamma \Rightarrow^? \Delta; R; H$ if $\mathrm{Set}(\Delta) \subseteq \mathrm{Set}(\Gamma)$.

(RESTART)$_\mathsf{G}$ From $\Gamma_1 \Rightarrow^? q, \Delta_1; R; H \cup \{(\Gamma_2 \Rightarrow^? \Delta_2; [q])\}$ step to
$\Gamma_1, \Gamma_2 \Rightarrow^? \Delta_1, \Delta_2; []; H \cup \{(\Gamma_1 \Rightarrow^? q, \Delta_1; R), (\Gamma_2 \Rightarrow^? \Delta_2; [q])\}$.

Proof. For the left-to-right direction, it is sufficient to check the $\mathsf{AG}^{\to}$-soundness of (COMBINE), (SUCCESS)$_\mathsf{G}$, and (RESTART)$_\mathsf{G}$ (an easy exercise). For the right-to-left direction, we first note the following fact, proved by induction on the height of an $\mathsf{AG}^{\to}$-derivation:

If $\vdash_{\mathsf{AG}^{\to}} \Gamma \Rightarrow^? \Delta_1, R; H$ and $\vdash_{\mathsf{AG}^{\to}} \Gamma \Rightarrow^? \Delta_2, R; H$, then $\vdash_{\mathsf{AG}^{\to}} \Gamma \Rightarrow^? \Delta_1, \Delta_2, R; H$.

This means that not only can we assume, using Proposition 7.19 and Lemma 7.20 that $\mathcal{Q} = (\Gamma_1 \Rightarrow^? \Delta_1; R_1; \{(\Gamma_2 \Rightarrow^? \Delta_2; R_2), \ldots, (\Gamma_n \Rightarrow^? \Delta_n; R_n)\})$ is atomic, but also that $|\Delta_i| = 1$ for $i = 1 \ldots n$.

Suppose now that $\models_{\mathsf{G}} \mathcal{Q}$. Then the following set of inequalities is valid:

$$\bigcup_{i=1}^{n} \{p \leq q : p \in \Gamma_i \text{ and } q \in \Delta_i\} \cup \bigcup_{i=1}^{n} \{p < q : p \in R_i \text{ and } q \in \Delta_i\}$$

Hence using Lemma 6.9, there is a sequence $(p_i \lhd_i p_{i+1})$ with $\lhd_i \in \{\leq, <\}$ for $i = 1 \ldots m$ such that $p_1 = p_{m+1}$ and without loss of generality, $\lhd_1$ is $\leq$. We prove that $\mathcal{Q}$ is derivable by induction on m. If $m = 1$, then we have $\Delta_j = [p_1]$ and $p_1 \in \Gamma_j$ for some $j \in \{1, \ldots, n\}$ and $\mathcal{Q}$ is derivable using (SWITCH) and (SUCCESS)$_{\mathsf{G}}$. If $m > 1$, then we have $(p_1 \leq p_2)$ and also $(p_2 < p_3)$ or $(p_2 \leq p_3)$. If we have the first, then there is $p_2 \in \Delta_i$ and $p_2 \in R_j$ for some $i, j \in \{1, \ldots, m\}$. But then by applying (SWITCH) and (RESTART)$_{\mathsf{G}}$, we obtain $(\Gamma_i, \Gamma_j \Rightarrow^? p_3; R'; H')$. Since $p_1 \in \Gamma_i$, we can shorten the sequence by replacing $(p_1 \leq p_2)$ and $(p_2 < p_3)$ with $(p_1 \leq p_3)$, and apply the induction hypothesis. The other case is very similar, making use of (COMBINE). □

Theorem 7.23. $\vdash_{\mathsf{A\L}^{\to}} \mathcal{Q}$ *iff* $\models_{\text{Ł}} \mathcal{Q}$ *where* $\mathsf{A\L}^{\to}$ *is* $\mathsf{AB}^{\to}$ *extended with:*

(SUCCESS)$_{\text{Ł}}$ $\Gamma, \Delta \Rightarrow^? \Delta; R; H$

(RESTART)$_{\text{Ł}}$ *From* $\Gamma_1 \Rightarrow^? q, \Delta_1; R_1; H \cup \{(\Gamma_2 \Rightarrow^? \Delta_2; R_2)\}$
where $q \in R_1 \uplus R_2$, *let* $R = (R_1 \uplus R_2) \ominus [q]$, *and step to*
$\Gamma_1, \Gamma_2 \Rightarrow^? \Delta_1, \Delta_2; R; H \cup \{(\Gamma_1 \Rightarrow^? q, \Delta_1; R_1), (\Gamma_2 \Rightarrow^? \Delta_2; R_2)\}$.

Proof. Proving the soundness of (COMBINE), (SUCCESS)$_{\text{Ł}}$, and (RESTART)$_{\text{Ł}}$, and hence $\mathsf{A\L}^{\to}$ for Ł, is left as an exercise in elementary arithmetic. For the right-to-left direction, it is sufficient to consider an Ł-valid atomic query $\mathcal{Q} = (\Gamma_1 \Rightarrow^? \Delta_1; R_1; \{(\Gamma_2 \Rightarrow^? \Delta_2; R_2), \ldots, (\Gamma_n \Rightarrow^? \Delta_n; R_n)\})$. Using the fact that the corresponding r-hypersequent is Ł-valid, and the completeness of the calculus GRŁ, we obtain $\lambda_1, \mu_1, \ldots, \lambda_n, \mu_n \in \mathbb{N}$ such that $\lambda_i > 0$ for some $1 \leq i \leq n$ and:

$$\biguplus_{i=1}^{n} \Delta_i^{\lambda_i + \mu_i} \subseteq \biguplus_{i=1}^{n} \Gamma_i^{\lambda_i + \mu_i} \uplus R_i^{\mu_i}$$

We show that $\mathcal{Q}$ succeeds in $\mathsf{A\L}^{\to}$ by induction on $\gamma = \sum_{i=1}^{n} (\lambda_i + \mu_i)$. If $\gamma = 1$ then $\Delta_i \subseteq \Gamma_i$ for some $1 \leq i \leq n$. So $\mathcal{Q}$ succeeds by an application of (SWITCH) if necessary and (SUCCESS)$_{\text{Ł}}$. For $\gamma > 1$ we consider i such that $\lambda_i > 0$. If $\Delta_i \subseteq \Gamma_i$, then again we are done by (SWITCH) if necessary and (SUCCESS)$_{\text{Ł}}$. Otherwise, we have $q \in \Delta_i$ where one of the following cases occurs:

1. $q \in \Gamma_j$ for some $j \neq i$ with $\lambda_j > 0$. Since we can always apply (SWITCH), we assume without loss of generality that $i = 1$ and $j = 2$. By applying (COMBINE) to $\mathcal{Q}$ we obtain a query $\mathcal{Q}'$:

$$\Gamma_1, \Gamma_2 \ominus [q] \Rightarrow^? \Delta_1 \ominus [q], \Delta_2; []; \{(\Gamma_1 \Rightarrow^? \Delta_1; R_1), \ldots, (\Gamma_n \Rightarrow^? \Delta_n; R_n)\}$$

If $\lambda_1 \geq \lambda_2$, then:

$$\Delta_1^{\lambda_1-\lambda_2} \uplus (\Delta_1 \ominus [q] \uplus \Delta_2)^{\lambda_2} \uplus \Delta' \subseteq \Gamma_1^{\lambda_1-\lambda_2} \uplus (\Gamma_1 \uplus \Gamma_2 \ominus [q])^{\lambda_2} \uplus \Gamma'$$

for $\Delta' = \bigcup_{i=3}^n \Delta_i^{\lambda_i} \uplus \bigcup_{i=1}^n \Delta_i^{\mu_i}$ and $\Gamma' = \bigcup_{i=3}^n \Gamma_i^{\lambda_i} \uplus \bigcup_{i=1}^n (\Gamma_i \uplus R_i)^{\mu_i}$. But $(\lambda_1 - \lambda_2) + \lambda_2 + \sum_{i=3}^n \lambda_i + \sum_{i=1}^n \mu_i < \lambda$. So by the induction hypothesis, $\mathcal{Q}'$ succeeds in AŁ$^{\to}$ and we are done. The case where $\lambda_2 \geq \lambda_1$ is very similar.

2. $q \in R_j$ for some j with $\mu_j > 0$. If $i \neq j$, then without loss of generality we assume that $i = 1$ and $j = 2$. Otherwise $i = j$ and observe that either $\Delta_i \subseteq \Gamma_i$ and we can apply (SUCCESS)$_{\text{Ł}}$, or there exists $k \neq i$ such that either $\lambda_k > 0$ or $\mu_k > 0$. If the latter, then assume without loss of generality that $i = 1$ and $k = 2$. Now, in both cases, by applying (RESTART)$_{\text{Ł}}$ we obtain:

$$\mathcal{Q}' = \Gamma_1, \Gamma_2 \Rightarrow^? \Delta_1 \ominus [q], \Delta_2; R_1; \{(\Gamma_1 \Rightarrow^? \Delta_1; R_1), \ldots, (\Gamma_n \Rightarrow^? \Delta_n; R_n)\}$$

We then proceed similarly to above, considering $\lambda_1 \leq \mu_2$ and $\mu_2 \leq \lambda_1$. □

A similar but more complicated system can also be provided for P. We leave finding suitable rules as an exercise for the interested reader.

Example 7.24. Consider the following atomic query:

$$p \Rightarrow^? q; [q]; \{(q \Rightarrow^? p, p; [])\}$$

For AG$^{\to}$, we apply (COMBINE) to obtain:

$$p, q \Rightarrow^? q, p, p; []; \{(p \Rightarrow^? q; [q]), (q \Rightarrow^? p, p; [])\}$$

which succeeds by (SUCCESS)$_{\text{G}}$ since $\{q, p, p\} \subseteq \{p, q\}$. For AŁ$^{\to}$ on the other hand, we first apply the (RESTART)$_{\text{Ł}}$ rule of AŁ$^{\to}$ to obtain:

$$p, q \Rightarrow^? p, p; []; \{(p \Rightarrow^? q; [q]), (q \Rightarrow^? p, p; [])\}$$

We then apply (COMBINE) to get:

$$p, q, p \Rightarrow^? p, p, q; []; \{(p \Rightarrow^? q; [q]), (q \Rightarrow^? p, p; []), (p, q \Rightarrow^? p, p; [])\}$$

which succeeds by (SUCCESS)$_{\text{Ł}}$.

7.3 Complexity

A good proof system can often be useful in establishing the complexity class for the *validity problem* of a logic L, namely, the problem of determining whether a formula in the appropriate language is or is not L-valid. Here we show that this is indeed the case for several important fuzzy logics, referring to [92] for basic definitions on complexity.

7.3.1 Co-NP-Hardness

We show first that all the fuzzy logics considered here are co-NP-hard. This is accomplished as usual by giving a linear translation of a problem already known to be co-NP-hard (in this case validity in Classical Logic) into a validity checking problem for the logic in question.

Theorem 7.25. *Let* HL *be any* HUL^-*-extension such that* $\models_{\mathsf{CL}} A$ *whenever* $\models_{\mathrm{L}} A$. *Then the validity problem for* L *is* co-NP-hard.

Proof. Consider a formula of the form:

$$A = \bigvee_{i \in I} \bigwedge_{j \in J_i} L_{ij} \quad \text{where each } L_{ij} \text{ is of the form } p \text{ or } \neg p.$$

It is well known that the problem of checking the validity of such formulas for Classical Logic is co-NP-hard [63]. Here we define a linear translation from formulas of this form to formulas in the language $\mathcal{L}_{\mathrm{F}} = \{\wedge, \vee, \odot, \to, \mathrm{f}, \mathrm{e}\}$ as follows:

$$A^c = \bigvee_{i \in I} \bigwedge_{j \in J_i} L^c_{ij} \quad \text{where } p^c = p \oplus p \text{ and } (\neg p)^c = \neg(p \odot p).$$

Then to establish the theorem it is sufficient to show:

$$\models_{\mathsf{CL}} A \quad \text{iff} \quad \models_{\mathrm{L}} A^c$$

For the right-to-left direction, observe that if $\models_{\mathrm{L}} A^c$, then by assumption, $\models_{\mathsf{CL}} A^c$. But for all CL-algebras (term equivalent to Boolean algebras), $x \oplus x = x$ and $x \odot x = x$, so $\models_{\mathsf{CL}} A$. For the other direction, we use distributivity properties of both CL and L to obtain formulas:

$$B = \bigwedge_{m \in M} \bigvee_{n \in N_m} L'_{mn} \quad \text{where each } L'_{mn} \text{ is of the form } p \text{ or } \neg p;$$

$$B^c = \bigwedge_{m \in M} \bigvee_{n \in N_m} L'^c_{mn} \quad \text{where } p^c = p \oplus p \text{ and } (\neg p)^c = \neg(p \odot p);$$

such that $\models_{\mathsf{CL}} A$ iff $\models_{\mathsf{CL}} B$ and $\models_{\mathrm{L}} A^c$ iff $\models_{\mathrm{L}} B^c$.

Suppose that $\models_{\mathsf{CL}} A$. Then $\models_{\mathsf{CL}} B$. So for every $m \in M$, there exists $n, n' \in N_m$ such that L'_{mn} is p and $L'_{mn'}$ is $\neg p$. But $\models_{\mathsf{UL}^-} (p \oplus p) \vee \neg(p \odot p)$, so also $\models_{\mathrm{L}} B^c$. Hence finally $\models_{\mathrm{L}} A^c$ as required. □

A couple of the logics that we have considered, A and CHL, do not quite fit the pattern of the above theorem. However, recall the embedding of Ł into A defined in Theorem 6.35: $\models_{\text{Ł}} A$ iff $\models_{\mathsf{A}} A^*$. Since we have just shown that checking Ł-validity is co-NP-hard, and the translation $*$ is $O(\text{cp}(A))$, it follows that checking A-validity must also be co-NP hard. Moreover, there exists a very similar embedding of Ł into CHL. Hence we can extend our results to obtain:

Theorem 7.26. *The validity problems for* A *and* CHL *are* co-NP-hard.

We will show below that several important fuzzy logics (including the three fundamental logics, A, and CHL) are in fact co-NP-complete, although this is not expected to be the case in general for this family of logics.

7.3.2 Bi-Coloured Graphs

We turn our attention first to Gödel Logic, the logic of order. In this case, as seen in Chapter 6, checking the validity of formulas can be reduced to checking the validity of sets of inequalities. A nice "visual" way of showing that this latter problem is polynomial time is to view each set of inequalities W as a *bi-coloured graph* $G_W = \langle V, E \rangle$ where:

- the *vertices* V of G_W are $\top$, $\bot$, and the variables occurring in W;
- the *edges* E are ordered triples $\langle a, b, \lhd \rangle$ for each $a \lhd b \in W$, representing a directed arrow between vertices a and b with "colour" $\lhd \in \{\leq, <\}$.

A sequence of vertices $a_1, \ldots, a_n$ such that $\langle a_i, a_{i+1}, \lhd \rangle \in E$ for $i = 1 \ldots n-1$ is called a *path*, and a *cycle* if additionally $a_1 = a_n$. If there is a path from a to b, then b is called *reachable* from a.

Recall that a set of inequalities W is valid iff it satisfies one of the conditions in Lemma 6.9, or in terms of bi-coloured graphs, iff G_W has one of the following:

(1) A cycle passing through a $\leq$-edge.
(2) A path from $\bot$ passing through a $\leq$-edge.
(3) A path leading to $\top$ passing through a $\leq$-edge.
(4) A path from $\bot$ to $\top$.

Theorem 7.27. *Checking the validity of a set of inequalities is polynomial time.*

Proof. Let n be the number of symbols occurring in W – called the *size* of W. We describe an $O(n)$-time algorithm (i.e. requiring a number of steps linear in the size of W) that returns TRUE if G_W satisfies one of the above conditions (1)–(4) and FALSE otherwise. As a preliminary step we first make the following changes to W to obtain a new set of inequalities W'.

1. We add $\top \leq \bot$ and $p < \bot$ and $\top < p$ for every variable p occurring in W.
2. We remove all inequalities of the form $p < p$.

It is easy to see that this can be accomplished in polynomial time and that the size of W' is linear in the size of W. Moreover, conditions (1)–(4) above hold for G_W iff condition (1) holds for $G_{W'}$.

To check the latter, we use a standard linear-time algorithm to compute the strongly-connected components (SCCs) of the graph: maximal subgraphs where every vertex is reachable from every other vertex. Briefly, the algorithm makes a depth-first traversal of $G_{W'} = (V,E)$ and then a depth-first traversal of the transposed graph $G^T_{W'}$ of $G_{W'}$ (that is, the graph containing the edges $(b,a,\triangleleft)$ such that $(a,b,\triangleleft) \in E$) considering the vertices in a special order determined by the first traversal. The complexity of the SCC algorithm is $O(n)$. For our purposes, we assume that the algorithm returns a function $scc : V \to V$ where for each vertex $a \in V$, $scc(a)$ is one vertex in the same SCC of a and $scc(a) = scc(b)$ iff a and b are in the same SCC.

We then observe that there is a cycle in $G_{W'}$ passing through a $\leq$-edge iff there is a $\leq$-edge connecting two vertices in the same SCC. Thus for each edge $(a,b,\leq)$, we just check whether $scc(a) = scc(b)$, returning TRUE if we find one edge satisfying this condition and FALSE otherwise. Clearly this step is $O(n)$, and so the complexity of the whole algorithm is $O(n)$. □

Hence checking G-validity for atomic sequent of relations, and by Lemma 6.8, implicational sequents, is polynomial time.

Theorem 7.28. *The validity problem for* G *is* co-NP-complete.

Proof. We have already established co-NP-hardness. To show co-NP-inclusion, we make use of the sequent of relations calculus $\mathrm{GG_s}$ in the language with connectives $\wedge$, $\to$, and $\bot$. First observe that the following rules are derivable in this calculus:

$$\frac{\mathcal{G} \mid A_1 \leq q \mid \ldots \mid A_n \leq q \mid C \leq q}{\mathcal{G} \mid C \leq A_1 \to A_2 \to \ldots A_n \to q} \qquad \frac{\mathcal{G} \mid A_1 \leq q \mid \ldots \mid A_n \leq q \mid C < q \quad \mathcal{G} \mid C < \top}{\mathcal{G} \mid C < A_1 \to A_2 \to \ldots A_n \to q}$$

We check the non-validity of a formula A by applying the rules of $\mathrm{GG_s}$ upwards, using the above derived rules instead of $(<\to)$ and $(\leq\to)$ and restricting $(\wedge\triangleleft)$ to cases where the right hand side is atomic. Choosing any branch of the resulting derivation tree non-deterministically, the leaf is an atomic sequent of relations S. Moreover, since each application of a rule strictly reduces the number of connectives in the sequent of relations, we get that both the length of the branch and $|\mathrm{S}|$ are $O(\mathrm{cp}(A))$. But the previous theorem tells us that checking the validity of S is polynomial time. So since the soundness and invertibility of the rules preserves validity in both directions, checking the non-validity of A is in NP. □

7.3.3 Linear Programming

Just as checking validity for Gödel Logic – the logic of order – can be translated into a graph problem, so checking validity in Łukasiewicz Logic – the logic of magnitude – can be translated into a linear programming problem. More precisely, the Ł-validity of a strictly atomic hypersequent corresponds to the existence of a solution for a linear programming problem involving linear inequalities over the reals. As we will see, similar correspondences hold also for a number of related logics, including Product Logic and Abelian Logic.

Definition 7.29. Let $\bar{x} = (x_1,\ldots,x_n)$ be a vector of variables, $\bar{b} = (b_1,\ldots,b_m)$ a vector of integers, and A an $m \times n$ integer matrix. The *strict inequalities feasibility problem (SIF-problem)* is to check whether $\mathsf{A}\bar{x} > \bar{b}$ has a solution over $\mathbb{R}$.[1]

The key to our complexity proofs is the following result from the linear programming literature, based on the proof of Khachiyan [128] that the standard linear programming problem is polynomial (see also [193] for details).

Theorem 7.30 ([128]). *The SIF-problem is polynomial time.*

So to show that checking the L-validity of a strictly atomic hypersequent $\mathcal{G}$ for some logic L is polynomial, it is sufficient to find an integer matrix A and integer vector $\bar{b}$ linear in the size of $\mathcal{G}$ such that $\not\models_{\mathrm{L}} \mathcal{G}$ iff $\mathsf{A}\bar{x} > \bar{b}$ for some $\bar{x} \in \mathbb{R}$. Consider then:

$$\mathcal{G} = (\Gamma_1 \Rightarrow \Delta_1 \mid \ldots \mid \Gamma_m \Rightarrow \Delta_m) \quad \text{containing variables } p_1,\ldots,p_n.$$

For CHL, Ł, or P, the strictly atomic hypersequent $\mathcal{G}$ is valid iff the formula $\mathrm{I}(\mathcal{G})$ is valid in the CHL-algebra (since the hypersequent calculi for these logics share the same structural rules):

$$\mathbf{A}_{\mathsf{CHL}} = \langle(-\infty,0],\min,\max,+,\rightarrow_{\mathsf{CHL}},0,0\rangle \quad \text{where } x \rightarrow_{\mathsf{CHL}} y = \min(0, y-x).$$

The idea now is to convert each sequent of $\mathcal{G}$ into an inequality over the variables $p_1,\ldots,p_n$. Since the coefficients in these inequalities come from the number of occurrences of each variable, we revisit the original definition of multisets and suppose that $\Gamma_i = \langle \mathrm{Fm}_{\mathcal{L}_{\mathsf{F}}}, f_i\rangle$ and $\Delta_i = \langle \mathrm{Fm}_{\mathcal{L}_{\mathsf{F}}}, g_i\rangle$. We will also need to represent – using additional inequalities – the fact that the values of the variables should be in $(-\infty,0]$ rather than the whole of $\mathbb{R}$.

Let $\bar{b}$ be the zero $m+n$-vector, and define $\mathsf{A} = (a_{ij})$ as the $(m+n)\times n$ matrix:

$$a_{ij} = \begin{cases} f_i(p_j) - g_i(p_j) & \text{for } i = 1\ldots m \\ -1 & \text{for } i = m+j \\ 0 & \text{otherwise} \end{cases}$$

Then $\mathsf{A}\bar{x} > \bar{b}$ for some $\bar{x} \in \mathbb{R}$ iff for some $\mathbf{A}_{\mathsf{CHL}}$-valuation v:

$$\sum[v(q) : q \in \Gamma_j] > \sum[v(q) : q \in \Delta_j] \quad \text{for } j = 1\ldots m.$$

But easily this latter condition holds iff $\mathrm{I}(\mathcal{G})$ is not $\mathbf{A}_{\mathsf{CHL}}$-valid.

A similar translation works also for A and CRL using the fact that in these cases, a strictly atomic hypersequent $\mathcal{G}$ is valid iff it is valid in $\mathbf{R}$. Hence:

---

[1] The *standard linear programming* problem refers to maximizing or minimizing a linear function with respect to (non-strict) linear inequalities over the reals. However, the problem that we consider here can be treated using the same techniques; we refer to [193] for details.

Proposition 7.31. *The problem of checking* $\models_{\mathrm{L}} \mathcal{G}$ *for a strictly atomic hypersequent* $\mathcal{G}$ *is polynomial time for* $\mathrm{L} \in \{\text{Ł}, \mathsf{CHL}, \mathsf{P}, \mathsf{A}, \mathsf{CRL}\}$.

Co-NP-hardness for these logics was established in Theorems 7.25 and 7.26. To show co-NP-inclusion we make use of the above proposition and a relevant hypersequent calculus. First, to ensure that backward application of logical rules does not increase the size of hypersequents exponentially, we consider the following revised rules for implication left:

$$\frac{\mathcal{G} \mid \Gamma, p \Rightarrow \Delta \mid B \Rightarrow p, A}{\mathcal{G} \mid \Gamma, A \to B \Rightarrow \Delta} \qquad \frac{\mathcal{G} \mid \Gamma, \neg A \Rightarrow \Delta \quad \mathcal{G} \mid \Gamma, p \Rightarrow \Delta \mid B \Rightarrow p, A}{\mathcal{G} \mid \Gamma, A \to B \Rightarrow \Delta}$$

where p does not occur in the conclusion in either rule.

The left rule is sound and invertible for CHL and Ł, and the right rule is sound and invertible for P. Hence to check the non-validity of a formula A for one of these logics, we apply the appropriate rule together with the remaining sound and invertible logical rules of GCHL, GŁ, or GP^+ backwards exhaustively, choosing a branch non-deterministically. This branch will have length $O(\mathrm{cp}(A))$ and end with a hypersequent $\mathcal{G}$ of size $O(\mathrm{cp}(A))$ that is strictly atomic for CHL, atomic for Ł, and atomic except for occurrences of the form $\neg p$ on the left of sequents for P. By the previous proposition, checking the validity of $\mathcal{G}$ is polynomial time for CHL. For Ł, we obtain the same result by adding extra sequents $(p \Rightarrow \bot)$ to $\mathcal{G}$ for each variable p occurring in $\mathcal{G}$ and treating $\bot$ as a variable. For P, we can follow the completeness proof of Theorem 6.51: essentially we just have to check whether p also occurs on the left in some sequent in which case the hypersequent is valid, or not, in which case $\neg p$ can be removed without affecting validity. The cases of A and CRL are very similar, so we have:

Theorem 7.32. *The validity problem for* $\mathrm{L} \in \{\text{Ł}, \mathsf{A}, \mathsf{P}, \mathsf{CHL}, \mathsf{CRL}\}$ *is* co-NP-complete.

We have used certain proof systems for our logics to establish these complexity results. We can think of the systems as providing algorithms of "optimal complexity" (in some sense) for the logics. Finally, we remark that it is not hard using the above methods to show that checking the validity of atomic r-hypersequents for Ł and P is polynomial. This is also the case for G, but requires the more complicated techniques of [124]. This is important because it means that applying the uniform rules of Section 7.1 has some value: it reduces co-NP-complete problems to polynomial time problems.

7.4 Historical Remarks

The uniform relational hypersequent approach for the fundamental fuzzy logics was introduced by Ciabattoni, Fermüller, and Metcalfe in their 2005 paper [50]. Also presented in this work was a dialogue game interpretation of the rules based

Table 7.1 Decidability and complexity results for fuzzy logics

Logic	Validity problem	References
Ł	co-NP-complete	[104, 159]
G	co-NP-complete	[105]
P	co-NP-complete	[105]
A	co-NP-complete	[219]
CHL	co-NP-complete	[78]
CRL	co-NP-complete	[87]
BL	co-NP-complete	[23]
SBL	co-NP-complete	[23]
IUML	co-NP-complete	[138]
UML	co-NP-complete	[138]
MTL	Decidable	[172]
IMTL	Decidable	[172]
SMTL	Decidable	[172]
MTL_n	Decidable	[122]
IMTL_n	Decidable	[122]
PMTL	Decidable	[120]
UL	Unknown	N/A
IUL	Unknown	N/A

on the "Giles game" reading of GŁ treated in Chapter 6. Generalizations of the r-hypersequent approach for Basic Logic BL and related logics have been proposed by Bova and Montagna [37] and Vetterlein [213]; we will discuss these briefly in Chapter 9.

The main ideas underlying the goal-directed methodology originated in a 1984 paper by Gabbay and Reyle [89] that introduced extensions of logic programming capable of hypothetical reasoning. These ideas were further developed and extended to a wide range of non-classical logics in the 2000 monograph of Gabbay and Olivetti [88], and to fuzzy logics (as presented in this chapter) by the current authors in [145, 147]. The closely related Uniform Proof paradigm, developed by Miller and others in the early 1990s (see e.g. [115, 150]), has been used as the basis for logic programming in various non-classical (but not fuzzy) logics.

The theory of NP-completeness and complexity has its roots in the work on computability of Turing, Church, Gödel, and others in the 1930s. NP-completeness itself was defined in the fundamental paper of Cook in 1971 [63], and developed further by a number of researchers, including Karp and Levin (see [92] for details). For fuzzy logics, co-NP-completeness for Łukasiewicz Logic was established by Mundici in 1987 [159], essentially by reducing the validity of a formula in Ł to validity in a finite-valued Łukasiewicz logic. A further proof using mixed integer programming, more closely related to the one given in this chapter, appears in the 1993 monograph of Hähnle [104]. Co-NP-completeness for Gödel Logic was folklore, but proofs of this and the same result for Product Logic may be found in Hájek's [105]. The proof given for Gödel Logic in this chapter, and the correspondence with bi-coloured graphs, bears some resemblance to the approach of Larchey-Wendling in [131]. In [23] it was shown by Baaz et al. that Hájek's Basic Logic BL is also co-NP-complete. Similar techniques were subsequently used by Hanikova to show that the same result holds for any continuous t-norm based logic [114].

Results for other fuzzy logics have a variety of sources. MTL, IMTL, and SMTL, were proved to be decidable (with no known complexity bound) by Ono in [172] using the Finite Embeddability Property technique developed by Blok and Van Alten [35]. Regarding logics without weakening, co-NP-completeness was established for abelian ℓ-groups and hence also for A by Weispfenning [219], for CHL by Esteva et al. [78], for CRL by Gabbay and Metcalfe in [87], and for IUML and UML by Marchioni and Montagna [138]. Finally, we remark that the decidability or otherwise of the validity problems for UL and IUL is unknown. Table 7.1 summarizes the known situation.

Chapter 8
First-Order Logics

So far – mostly for clarity – we have considered logics and proof systems only at the propositional level. However, as we will see below, many of the methods and results described in previous chapters transfer unscathed to first-order logics. That is, for many fuzzy logics, we can extend Hilbert and Gentzen systems with axioms and rules for the universal and existential quantifiers $\forall$ "for all" and $\exists$ "there exists", and obtain completeness with respect to classes of residuated lattices and even uninorm-based standard algebras.[1] Moreover, although (as we show) all these logics are undecidable, analogues of Herbrand's theorem and Skolemization – the twin pillars of theorem proving in Classical Logic – can be established for their prenex fragments.

Not all of our efforts for propositional logics transfer so easily, however. Gödel Logic excepted, the first-order versions of the logics of Chapter 6 – Łukasiewicz Logic, Product Logic, etc. – are not recursively axiomatizable. There is no hope in these cases of defining complete Hilbert or Gentzen systems by adding finitely many rule schema. Nevertheless, for first-order Łukasiewicz Logic at least, some interesting results can be obtained. Although the usual Herbrand theorem fails in this case, we show that an "approximate" version holds instead. Also, we use this fact first to prove that Skolemization holds, and then to establish completeness for a cut-free hypersequent calculus with an infinitary rule.

8.1 Syntax and Semantics

The core elements of our first-order languages will be taken directly from Classical Logic, including here also the connectives of an underlying propositional language. To simplify our presentation, we will assume as in the propositional case that our first-order languages are all countable. Recall that the arity n of a symbol is just a

[1] The fascinating topic of generalized quantifiers is not considered here, although operations formalizing linguistic notions such as *many* and *almost all* are certainly very relevant for Fuzzy Logic.

natural number (possibly 0) representing the number of arguments accepted by the symbol, which is then called n-ary.

Definition 8.1. A (countable) first-order language $\mathbb{L}$ comprises:

- A (countable) non-empty set $\mathcal{P}$ of *predicate symbols* with associated arities.
- A (countable) set $\mathcal{F}$ of *function symbols* with associated arities.
- A (countable) set $\mathcal{X}$ of *object variables*.
- A (countable) set $\mathcal{C}$ of *connectives* with associated arities.

The set of $\mathbb{L}$-*terms* is the smallest set such that:

(1) Each $x \in \mathcal{X}$ is an $\mathbb{L}$-term.
(2) If $t_1, \ldots, t_n$ are $\mathbb{L}$-terms, then $f(t_1, \ldots, t_n)$ is an $\mathbb{L}$-term for each n-ary $f \in \mathcal{F}$.

The set of $\mathbb{L}$-*formulas* $\mathrm{Fm}_{\mathbb{L}}$ is the smallest set such that:

(1) $p(t_1, \ldots, t_n) \in \mathrm{Fm}_{\mathbb{L}}$ for each n-ary $p \in \mathcal{P}$ and $\mathbb{L}$-terms $t_1, \ldots, t_n$.
(2) $\star(A_1, \ldots, A_n) \in \mathrm{Fm}_{\mathbb{L}}$ for each n-ary $\star \in \mathcal{C}$ and $A_1, \ldots, A_n \in \mathrm{Fm}_{\mathbb{L}}$.
(3) $(\forall x)A \in \mathrm{Fm}_{\mathbb{L}}$ and $(\exists x)A \in \mathrm{Fm}_{\mathbb{L}}$ for each $A \in \mathrm{Fm}_{\mathbb{L}}$ and $x \in \mathcal{X}$.

The *subformulas* of a formula are defined inductively by:

(1) A is a subformula of A for all $A \in \mathrm{Fm}_{\mathbb{L}}$.
(2) Each subformula of A_i for $i = 1 \ldots n$ is a subformula of $\star(A_1, \ldots, A_n)$ for each n-ary $\star \in \mathcal{L}$ and $A_1, \ldots, A_n \in \mathrm{Fm}_{\mathbb{L}}$.
(3) Each subformula of B is a subformula of $(\forall x)B$ and $(\exists x)B$.

The *scope* of quantified subformulas $(\forall x)B$ and $(\exists x)B$ in a formula A is B, where all occurrences of x in B are *bound* by the occurrence of $(\forall x)$ or $(\exists x)$, respectively. Any variable occurrence in A that is not bound is *free*. An $\mathbb{L}$-formula A is an $\mathbb{L}$-*sentence* if there are no free occurrences of variables in A. A set of $\mathbb{L}$-sentences T is called an $\mathbb{L}$-*theory*.

The *complexity* of an $\mathbb{L}$-formula is defined inductively by:

(1) $\mathrm{cp}(p(t_1, \ldots, t_n)) = 1$ for all n-ary $p \in \mathcal{P}$ and $\mathbb{L}$-terms $t_1, \ldots, t_n$.
(2) $\mathrm{cp}(\star(A_1, \ldots, A_n)) = 1 + \sum_{i=1}^{n} \mathrm{cp}(A_i)$ for all n-ary $\star \in \mathcal{C}$ and $A_1, \ldots, A_n \in \mathrm{Fm}_{\mathbb{L}}$.
(3) $\mathrm{cp}((\forall x)A) = \mathrm{cp}((\exists x)A) = \mathrm{cp}(A) + 1$ for all $A \in \mathrm{Fm}_{\mathbb{L}}$ and $x \in \mathcal{X}$.

We call formulas with complexity 1, *atoms* or *atomic* formulas.

For convenience, we call a predicate symbol p with arity 0, a *propositional variable*, and a function symbol c with arity 0, an *(object) constant*. We also write $\mathbb{L}' \geq \mathbb{L}$ to mean that the language $\mathbb{L}'$ is an extension of the language $\mathbb{L}$ with at most countably many new propositional variables and constants.

We write $\bar{x}$ for a sequence of variables $x_1, \ldots, x_n$ and $(\mathrm{Q}\bar{x})$ to mean a sequence of quantifiers $(\mathrm{Q}_1 x_1) \ldots (\mathrm{Q}_n x_n)$ where $\mathrm{Q}_i \in \{\forall, \exists\}$ for $i = 1 \ldots n$. We write $A(x_1, \ldots, x_n)$ or $A(\bar{x})$ to mean that the free variables of A are among $x_1, \ldots, x_n$. We also denote the result of replacing all free occurrences of x_i by t_i for $i = 1 \ldots n$ in $A(\bar{x}, \bar{y})$ by $A(\bar{t}, \bar{y})$. Finally, an *existential formula* is of the form $(\exists \bar{x})P(\bar{x})$ and a *prenex formula* is of the form $(\mathrm{Q}\bar{x})P(\bar{x})$ where P is quantifier-free.

Formulas are interpreted using algebras – here, (bounded) pointed commutative residuated lattices – as in the propositional case.

Definition 8.2. For a first-order language $\mathbb{L}$, an algebra $\mathbf{A}$ for $\mathbb{L}$ is an algebra with operations of the same arity as the connectives in $\mathcal{C}$. An $\mathbf{A}$-*structure* for $\mathbb{L}$ is a triple:

$$\mathbf{M} = (M, (p_{\mathbf{M}})_{p \in \mathcal{P}}, (f_{\mathbf{M}})_{f \in \mathcal{F}})$$

where M is a non-empty set called the *domain*, and:

1. $f_{\mathbf{M}}$ is a function $M^n \to M$ for each n-ary $f \in \mathcal{F}$.
2. $p_{\mathbf{M}}$ is a function $M^n \to L_{\mathbf{A}}$ for each n-ary $p \in \mathcal{P}$.

Note that this definition includes the cases where f is a constant (nullary function symbol) and $f_{\mathbf{M}}$ is a member of M, and where p is a propositional variable (nullary predicate symbol) and $p_{\mathbf{M}}$ is a member of $L_{\mathbf{A}}$.

We take care of free variables by defining additional assignments from variables into the domain.

Definition 8.3. Let $\mathbf{A}$ be an algebra for $\mathbb{L}$ and let $\mathbf{M}$ be an $\mathbf{A}$-structure for $\mathbb{L}$:

- An $\mathbf{M}$-*assignment* is a function $m : \mathcal{X} \to M$.
- For an $\mathbf{M}$-assignment m, $x \in \mathcal{X}$, and $d \in M$, let:

$$m[x \to d](y) = \begin{cases} d & \text{if } x = y \\ m(y) & \text{otherwise} \end{cases}$$

Let us assume from now on that $\mathbb{L}$ is a first-order language with connectives from one of the propositional languages $\mathcal{L}_{\mathrm{B}} = \{\wedge, \vee, \odot, \to, \mathrm{f}, \mathrm{e}, \bot, \top\}$ or $\mathcal{L}_{\mathrm{F}} = \{\wedge, \vee, \odot, \to, \mathrm{f}, \mathrm{e}\}$, so that algebras for $\mathbb{L}$ have a natural order defined by $x \leq y$ iff $x \wedge y = x$. We define the value of a formula in a structure inductively as in the classical case, interpreting the quantifiers using suprema and infima. The only problem here is that certain suprema and infima of sets in $L_{\mathbf{A}}$ might not exist in $L_{\mathbf{A}}$. One possibility then is to consider only structures where they do exist (e.g. based on complete algebras). Here instead we just leave such values undefined.

Definition 8.4. For an algebra $\mathbf{A}$ for $\mathbb{L}$, an $\mathbf{A}$-structure $\mathbf{M}$, and $\mathbf{M}$-assignment m:

$$\begin{aligned} \|x\|^{\mathbf{A}}_{\mathbf{M},m} &= m(x) \\ \|f(t_1, \ldots, t_n)\|^{\mathbf{A}}_{\mathbf{M},m} &= f_{\mathbf{M}}(\|t_1\|^{\mathbf{A}}_{\mathbf{M},m}, \ldots, \|t_n\|^{\mathbf{A}}_{\mathbf{M},m}) \text{ for all } n\text{-ary } f \in \mathcal{F} \\ \|p(t_1, \ldots, t_n)\|^{\mathbf{A}}_{\mathbf{M},m} &= p_{\mathbf{M}}(\|t_1\|^{\mathbf{A}}_{\mathbf{M},m}, \ldots, \|t_n\|^{\mathbf{A}}_{\mathbf{M},m}) \text{ for all } n\text{-ary } p \in \mathcal{P} \end{aligned}$$

For each n-ary $\star \in \mathcal{C}$, if $\|A_1\|^{\mathbf{A}}_{\mathbf{M},m}, \ldots, \|A_n\|^{\mathbf{A}}_{\mathbf{M},m}$ are all defined, then:

$$\| \star (A_1, \ldots, A_n)\|^{\mathbf{A}}_{\mathbf{M},m} = \star(\|A_1\|^{\mathbf{A}}_{\mathbf{M},m}, \ldots, \|A_n\|^{\mathbf{A}}_{\mathbf{M},m})$$

otherwise $\| \star (A_1, \ldots, A_n)\|^{\mathbf{A}}_{\mathbf{M},m}$ is undefined.

If $\|A\|^{\mathbf{A}}_{\mathbf{M},m[x\to d]}$ is defined for all $d \in M$, then:

$$\|(\forall x)A\|^{\mathbf{A}}_{\mathbf{M},m} = \begin{cases} \inf_{d\in M} \|A\|^{\mathbf{A}}_{\mathbf{M},m[x\to d]} & \text{if } \inf_{d\in M} \|A\|^{\mathbf{A}}_{\mathbf{M},m[x\to d]} \text{ exists in } M \\ \text{undefined} & \text{otherwise} \end{cases}$$

$$\|(\exists x)A\|^{\mathbf{A}}_{\mathbf{M},m} = \begin{cases} \sup_{d\in M} \|A\|^{\mathbf{A}}_{\mathbf{M},m[x\to d]} & \text{if } \sup_{d\in M} \|A\|^{\mathbf{A}}_{\mathbf{M},m[x\to d]} \text{ exists in } M \\ \text{undefined} & \text{otherwise} \end{cases}$$

otherwise $\|(\forall x)A\|^{\mathbf{A}}_{\mathbf{M},m}$ and $\|(\exists x)A\|^{\mathbf{A}}_{\mathbf{M},m}$ are both undefined.

$\mathbf{M}$ is called *safe* if $\|A\|^{\mathbf{A}}_{\mathbf{M},m}$ is defined for each $\mathbb{L}$-formula A and $\mathbf{M}$-assignment m.

Notice that for a sentence A and safe $\mathbf{A}$-structure $\mathbf{M}$ for an algebra $\mathbf{A}$, $\|A\|^{\mathbf{A}}_{\mathbf{M},m}$ is the same element of $L_{\mathbf{A}}$ for any $\mathbf{M}$-assignment m. In general, we define a fixed value for a formula in a safe $\mathbf{A}$-structure as follows.

Definition 8.5. For an algebra $\mathbf{A}$ for $\mathbb{L}$, safe $\mathbf{A}$-structure $\mathbf{M}$, and $\mathbb{L}$-formula A:

$$\|A\|^{\mathbf{A}}_{\mathbf{M}} =_{\text{def}} \inf\{\|A\|^{\mathbf{A}}_{\mathbf{M},m} : m \text{ is an } \mathbf{M}\text{-assignment}\}$$

A is $\mathbf{A}$-*valid* if $\|A\|^{\mathbf{A}}_{\mathbf{M}} \geq \mathrm{e}$ for every safe $\mathbf{A}$-structure $\mathbf{M}$.

- A safe $\mathbf{A}$-structure $\mathbf{M}$ is an $\mathbf{A}$-*model* of a set of $\mathbb{L}$-formulas T if $\|A\|^{\mathbf{A}}_{\mathbf{M}} \geq \mathrm{e}$ for every $A \in T$, and an $\mathbf{A}$-model of an $\mathbb{L}$-formula A if it is an $\mathbf{A}$-model of $\{A\}$.
- We write $T \models_{\mathbf{A}} A$ if every $\mathbf{A}$-model of T is an $\mathbf{A}$-model of $\{A\}$.

For a class of algebras $\mathcal{K}$ for $\mathbb{L}$, we write $T \models_{\mathcal{K}} A$ if $T \models_{\mathbf{A}} A$ for all $\mathbf{A} \in \mathcal{K}$.

When the algebra is clear from the context, we write $\|A\|_{\mathbf{M}}$ rather than $\|A\|^{\mathbf{A}}_{\mathbf{M}}$.

8.2 Hilbert Systems

First-order versions of the Hilbert systems in Chapter 3 can be defined uniformly by adding extra axiom schema and rules to deal with $\forall$ and $\exists$. Let us assume for the remainder of this section that HL is an HUL^--extension and that $\mathbb{L}$ is a fixed first-order language with the same connectives as HL. We will write "HL extended to $\mathbb{L}$" to mean the Hilbert system with axioms/rules consisting of the substitution instances of axiom/rule schema of HL with $\mathbb{L}$-formulas.

Definition 8.6. HL$\forall$ consists of HL extended to $\mathbb{L}$ plus the axioms:

$(\forall 1)$	$(\forall x)A(x) \to A(t)$	(t substitutable for x in A)
$(\forall 2)$	$(\forall x)(A \to B) \to (A \to (\forall x)B)$	(x not free in A)
$(\forall 3)$	$(\forall x)(A \lor B) \to (A \lor (\forall x)B)$	(x not free in A)
$(\exists 1)$	$A(t) \to (\exists x)A(x)$	(t substitutable for x in A)
$(\exists 2)$	$(\forall x)(A \to B) \to ((\exists x)A \to B)$	(x not free in B)

and the *generalization rule*:

$$\frac{A}{(\forall x)A} \ (\textsc{gen})$$

where a term t is *substitutable* for an object variable x in a formula A iff no free occurrence of x in A is in the scope of $(\forall y)$ or $(\exists y)$ for some variable y in t.

Adding the axiom schema $(\forall 1)$, $(\forall 2)$, $(\exists 1)$, and $(\exists 2)$ to a propositional Hilbert system is enough for axiomatizing many first-order logics, including first-order Intuitionistic Logic and Classical Logic, and first-order versions of substructural logics such as Linear Logic or Monoidal Logic. However, in the fuzzy case, we want something more: completeness with respect to linearly ordered algebras. To achieve this, as we will see, we require the "infinite distributivity" or "shifting law of quantifiers" axiom schema $(\forall 3)$.

Example 8.7. We can derive the "opposite direction" of $(\forall 2)$ in HL$\forall$ for any HUL^--extension HL as follows, assuming again that x is not free in A:

1. $(\forall x)B \to B$ — $(\forall 1)$
2. $(A \to (\forall x)B) \to (((\forall x)B \to B) \to (A \to B))$ — (B)
3. $((\forall x)B \to B) \to ((A \to (\forall x)B) \to (A \to B))$ — Lemma 3.23
4. $(A \to (\forall x)B) \to (A \to B)$ — (MP) 1, 3
5. $(\forall x)((A \to (\forall x)B) \to (A \to B))$ — (GEN)
6. $(\forall x)((A \to (\forall x)B) \to (A \to B)) \to ((A \to (\forall x)B) \to (\forall x)(A \to B))$ — $(\forall 2)$
7. $(A \to (\forall x)B) \to (\forall x)(A \to B)$ — (MP) 5, 6

Let us collect together some other useful theorems of these systems.

Lemma 8.8. *The following are theorems in* HL$\forall$*, assuming that x is not free in A:*

(i) $(\forall x)(A \to B) \leftrightarrow (A \to (\forall x)B)$
(ii) $(\forall x)(B \to A) \leftrightarrow ((\exists x)B \to A)$
(iii) $(\exists x)(A \to B) \to (A \to (\exists x)B)$
(iv) $(\exists x)(B \to A) \to ((\forall x)B \to A)$
(v) $(\forall x)(B \to C) \to ((\forall x)B \to (\forall x)C)$
(vi) $(\forall x)(B \to C) \to ((\exists x)B \to (\exists x)C)$
(vii) $((\exists x)B \odot (\exists x)C) \to (\exists x)(B \odot C)$
(viii) $(\forall x)B(x) \leftrightarrow (\forall y)B(y)$ *(y substitutable for x in B)*
(ix) $(\exists x)B(x) \leftrightarrow (\exists y)B(y)$ *(y substitutable for x in B)*
(x) $(\exists x)(B \odot A) \leftrightarrow ((\exists x)B \odot A)$
(xi) $(\exists x)(B \odot B) \leftrightarrow ((\exists x)B \odot (\exists x)B)$
(xii) $(\exists x)B \to \neg(\forall x)\neg B$
(xiii) $\neg(\exists x)B \leftrightarrow (\forall x)\neg B$
(xiv) $(\exists x)(A \wedge B) \leftrightarrow (A \wedge (\exists x)B)$
(xv) $(\exists x)(A \vee B) \leftrightarrow (A \vee (\exists x)B)$
(xvi) $(\forall x)(A \wedge B) \leftrightarrow (A \wedge (\forall x)B)$
(xvii) $(\exists x)(B \vee C) \leftrightarrow ((\exists x)B \vee (\exists x)C)$
(xviii) $(\forall x)(B \wedge C) \leftrightarrow ((\forall x)B \wedge (\forall x)C)$

Note also that just as in Classical Logic where the existential quantifier can be defined using the universal quantifier, the same holds here for fuzzy logics possessing an involutive negation. Just observe that for any HIUL$^-$-extension HL:

$$\vdash_{\text{HL}\forall} (\exists x)A \leftrightarrow \neg(\forall x)\neg A$$

That is, $\vdash_{\text{HL}\forall} \neg\neg(\exists x)A \leftrightarrow \neg(\forall x)\neg A$ by Lemma 8.8 (xiii) and Lemma 3.24 (xiv), and the result then follows using the involution axiom schema $\neg\neg A \to A$.

Many properties of HUL$^-$-extensions transfer smoothly from the propositional to the first-order level, including the local deduction theorem and the proof-by-cases property. Recall Definition 3.41 of a confusion of a set of formulas.

Theorem 8.9. *For any $\mathbb{L}$-theory T and $\mathbb{L}$-sentence A:*

(a) $T \cup \{A\} \vdash_{\text{HL}\forall} B$ *iff* $T \vdash_{\text{HL}\forall} C \to B$ *for some confusion* C *of* $\{A\}$.
(b) $T \vdash_{\text{HL}\forall} A$ *iff* $\vdash_{\text{HL}\forall} C \to A$ *for some confusion* C *of* T.
(c) HL∀ *has the proof-by-cases property, i.e. for any $\mathbb{L}$-theory T and $\mathbb{L}$-sentences A, B: if $T \cup \{A\} \vdash_{\text{HL}\forall} C$ and $T \cup \{B\} \vdash_{\text{HL}\forall} C$, then $T \cup \{A \vee B\} \vdash_{\text{HL}\forall} C$.*

Proof. For (a) and (b), we show that $T_1 \cup T_2 \vdash_{\text{HL}\forall} B$ iff $T_1 \vdash_{\text{HL}\forall} C \to B$ for some confusion C of T_2. The right-to-left direction follows exactly as in Theorem 3.43. For the left-to-right direction, we also proceed as in the proof of Theorem 3.43 by induction on the height of a derivation for $T_1 \cup T_2 \vdash_{\text{HL}} B$. The only new case occurs when the last step is $T_1 \cup T_2 \vdash_{\text{HL}\forall} B'(x)$ and B is $(\forall x)B'(x)$. Then by the induction hypothesis, $T_1 \vdash_{\text{HL}\forall} C \to B'(x)$ for some confusion C of T_2. So by (GEN), $T_1 \vdash_{\text{HL}\forall} (\forall x)(C \to B'(x))$. But a confusion of a set of sentences must also be a sentence; i.e C does not contain any free occurrences of x. Hence $T \vdash_{\text{HL}\forall} C \to (\forall x)B'(x)$ using ($\forall$2) as required. The proof of (c) then proceeds exactly as in Lemma 3.54. □

A useful substitution property also holds for HL∀, proved by an easy induction on the height of a derivation.

Lemma 8.10. *If $d;T \vdash_{\text{HL}\forall} A(x)$, then for any term t substitutable for x in A, $d';T \vdash_{\text{HL}\forall} A(t)$ for some derivation d' with* $\text{ht}(d') = \text{ht}(d)$.

We now turn our attention to the question of completeness for HL∀, beginning with L-chains, then moving on as before to dense L-chains and standard L-algebras. The method is very similar to the propositional case. Namely, we show that if $T \nvdash_{\text{HL}\forall} A$ for some theory T and sentence A, then we can extend T to a linear theory $\hat{T}$ where $\hat{T} \nvdash_{\text{HL}\forall} A$. To this end, we recall (in a first-order setting) the notions of a linear and dense theory, and also introduce the idea of a Henkin theory.

Definition 8.11. An $\mathbb{L}$-theory T is:

- $\mathbb{L}$-*linear* if for all $\mathbb{L}$-sentences A and B, either $T \vdash_{\text{HL}\forall} A \to B$ or $T \vdash_{\text{HL}\forall} B \to A$.
- $\mathbb{L}$-*dense* if for all $\mathbb{L}$-sentences A and B, whenever $T \nvdash_{\text{HL}} A \to B$, then $T \nvdash_{\text{HL}} A \to C$ and $T \nvdash_{\text{HL}} C \to B$ for some $\mathbb{L}$-sentence C.

- $\mathbb{L}$-*Henkin* if for each $\mathbb{L}$-formula $A(x)$ with one free variable x, whenever $T \nvdash_{\mathrm{HL}\forall} (\forall x)A(x)$, then $T \nvdash_{\mathrm{HL}\forall} A(c)$ for some $\mathbb{L}$-constant c.

As in the propositional case, the crucial step is to show that if a sentence is not derivable from a theory, then it is not derivable from a theory with special properties, in this case, being both linear and Henkin.

Lemma 8.12. *Let T be an $\mathbb{L}$-theory and let C be an $\mathbb{L}$-sentence. If $T \nvdash_{\mathrm{HL}\forall} C$, then $\hat{T} \nvdash_{\mathrm{HL}\forall} C$ for some $\hat{\mathbb{L}}$-linear $\hat{\mathbb{L}}$-Henkin $\hat{\mathbb{L}}$-theory $\hat{T} \supseteq T$ such that $\mathbb{L} \leq \hat{\mathbb{L}}$.*

Proof. Let $\hat{\mathbb{L}}$ be $\mathbb{L}$ extended with countably infinitely many new constants. We enumerate all pairs of $\hat{\mathbb{L}}$-sentences $\langle A_i, B_i\rangle$ and $\hat{\mathbb{L}}$-formulas with one free-variable F_i for $i \in \mathbb{N}$. We then define sequences T_n of $\hat{\mathbb{L}}$-theories and C_n of $\hat{\mathbb{L}}$-sentences inductively starting with $T_0 = T$ and $C_0 = C$. For each $n \in \mathbb{N}$:

(1) If $n = 2i$ for some $i \in \mathbb{N}$, then define:

$$C_{n+1} = C_n \quad \text{and} \quad T_{n+1} = \begin{cases} T_n \cup \{A_i \to B_i\} & \text{if } T_n \cup \{A_i \to B_i\} \nvdash_{\mathrm{HL}\forall} C_n \\ T_n \cup \{B_i \to A_i\} & \text{otherwise} \end{cases}$$

(2) If $n = 2i+1$ for some $i \in \mathbb{N}$, then let c_i be a $\hat{\mathbb{L}}$-constant not occurring in T_n, C_n, or F_i, and define:

$$C_{n+1} = \begin{cases} C_n \vee F_i(c_i) & \text{if } T_n \nvdash_{\mathrm{HL}\forall} C_n \vee F_i(c_i) \\ C_n & \text{otherwise} \end{cases}$$

$$T_{n+1} = \begin{cases} T_n & \text{if } T_n \nvdash_{\mathrm{HL}\forall} C_n \vee F_i(c_i) \\ T_n \cup \{C_n \to (\forall x)F_i(x)\} & \text{otherwise} \end{cases}$$

Claim. $T_n \nvdash_{\mathrm{HL}\forall} C_n$ for all $n \in \mathbb{N}$.

Proof of claim. We proceed by induction on n. The base case is immediate. Suppose that the claim holds for $n = 2i$ for some $i \in \mathbb{N}$. If $T_n \cup \{A_i \to B_i\} \nvdash_{\mathrm{HL}\forall} C_n$ or $T_n \cup \{B_i \to A_i\} \nvdash_{\mathrm{HL}\forall} C_n$, then the claim holds for $n+1$. Otherwise, using the proof-by-cases property and prelinearity as in the propositional case, $T_n \vdash_{\mathrm{HL}\forall} C_n$, a contradiction.

Now assume that the claim holds for $n = 2i+1$ for some $i \in \mathbb{N}$. The case where $T_n \nvdash_{\mathrm{HL}\forall} C_n \vee F_i(c_i)$ is immediate. Suppose then that $T_n \vdash_{\mathrm{HL}\forall} C_n \vee F_i(c_i)$ and hence, replacing the new constant c_i in the derivation by a new variable y, $T_n \vdash_{\mathrm{HL}\forall} C_n \vee F_i(y)$. Using (GEN) and a change of variables, $T_n \vdash_{\mathrm{HL}\forall} (\forall x)(C_n \vee F_i(x))$. So using ($\forall$3), $T_n \vdash_{\mathrm{HL}\forall} C_n \vee (\forall x)F_i(x)$. It follows easily that $T_n \cup \{(\forall x)F_i(x) \to C_n\} \vdash_{\mathrm{HL}\forall} C_n$. Hence $T_n \cup \{C_n \to (\forall x)F_i(x)\} \nvdash_{\mathrm{HL}\forall} C_n$, since otherwise using the proof-by-cases property and prelinearity, $T_n \vdash_{\mathrm{HL}\forall} C_n$, a contradiction.

Now let $\hat{T} = \bigcup_{n\in\mathbb{N}} T_n$. Notice that from the previous claim, $\hat{T} \nvdash_{\mathrm{HL}\forall} C_n$ for all $n \in \mathbb{N}$. Moreover, $\hat{T}$ is clearly $\hat{\mathbb{L}}$-linear by construction. To see that $\hat{T}$ is $\hat{\mathbb{L}}$-Henkin, suppose that $\hat{T} \nvdash_{\mathrm{HL}\forall} (\forall x)F_i(x)$ for some $i \in \mathbb{N}$. If $T_n \vdash_{\mathrm{HL}\forall} C_n \vee F_i(c_i)$

for $n = 2i+1$, then $T_{n+1} = T_n \cup \{C_n \to (\forall x)F_i(x)\}$ and $C_{n+1} = C_n$. So as reasoned in the proof of the claim above, $T_n \vdash_{\mathrm{HL}\forall} C_n \vee (\forall x)F_i(x)$. But then easily, $T_n \cup \{C_n \to (\forall x)F_i(x)\} \vdash_{\mathrm{HL}\forall} (\forall x)F_i(x)$. So $\hat{T} \vdash_{\mathrm{HL}\forall} (\forall x)F_i(x)$, a contradiction. Hence $T_n \nvdash_{\mathrm{HL}\forall} C_n \vee F_i(c_i)$, $T_{n+1} = T_n$, and $C_{n+1} = C_n \vee F_i(c_i)$. So $\hat{T} \nvdash_{\mathrm{HL}\forall} F_i(c_i)$, since otherwise $\hat{T} \vdash_{\mathrm{HL}\forall} C_{n+1}$, contradicting the above claim. □

The Lindenbaum algebra for HL∀ with respect to an $\mathbb{L}$-theory T is defined as for the propositional case. That is, let $[A]_T^{\mathrm{L}\forall} =_{\mathrm{def}} \{B \in \mathrm{Fm}_{\mathbb{L}} : T \vdash_{\mathrm{HL}\forall} A \leftrightarrow B\}$. Then:

$$\mathsf{LIND}_T^{\mathrm{L}\forall} =_{\mathrm{def}} \langle L_T^{\mathrm{L}\forall}, \{\star_T^{\mathrm{L}\forall} : \star \in \mathcal{C}\}\rangle$$

where $L_T^{\mathrm{L}\forall} =_{\mathrm{def}} \{[A]_T^{\mathrm{L}\forall} : A \in \mathrm{Fm}_{\mathbb{L}}\}$ and $\star_T^{\mathrm{L}\forall}([A_1]_T^{\mathrm{L}\forall}, \ldots, [A_n]_T^{\mathrm{L}\forall}) =_{\mathrm{def}} [\star(A_1, \ldots, A_n)]_T^{\mathrm{L}\forall}$.

We then define a special structure for this algebra for each $\mathbb{L}$-linear $\mathbb{L}$-Henkin $\mathbb{L}$-theory, using variable-free terms as their own interpretation and mapping predicates to functions from terms to equivalence classes of formulas.

Definition 8.13. Let T be an $\mathbb{L}$-linear $\mathbb{L}$-Henkin $\mathbb{L}$-theory:

$$\mathbf{CM}_T =_{\mathrm{def}} (M, (p_{\mathbf{CM}_T})_{p \in \mathcal{P}}, (f_{\mathbf{CM}_T})_{f \in \mathcal{F}})$$

where M is the set of variable-free $\mathbb{L}$-terms, $f_{\mathbf{CM}_T}(t_1, \ldots, t_n) = f(t_1, \ldots, t_n)$ for all n-ary $f \in \mathcal{F}$, and $p_{\mathbf{CM}_T}(t_1, \ldots, t_n) = [p(t_1, \ldots, t_n)]_T$ for all n-ary $p \in \mathcal{P}$.

Lemma 8.14. *Let T be an $\mathbb{L}$-Henkin $\mathbb{L}$-theory and $A(x)$ an $\mathbb{L}$-formula with one free-variable x:*

(a) $[(\forall x)A(x)]_T = \inf\{[A(c)]_T : c \text{ is an } \mathbb{L}\text{-constant}\}$.

(b) $[(\exists x)A(x)]_T = \sup\{[A(c)]_T : c \text{ is an } \mathbb{L}\text{-constant}\}$.

Proof. For (a), easily $T \vdash_{\mathrm{HL}\forall} (\forall x)A(x) \to A(c)$ for all $\mathbb{L}$-constants c. Now suppose that $[C]_T \leq [A(c)]_T$ for all $\mathbb{L}$-constants c but $[C]_T \not\leq [(\forall x)A(x)]_T$. We get $T \nvdash_{\mathrm{HL}\forall} C \to (\forall x)A(x)$, so also $T \nvdash_{\mathrm{HL}\forall} (\forall x)(C \to A(x))$. But then by the $\mathbb{L}$-Henkin property $T \nvdash_{\mathrm{HL}\forall} C \to A(c')$ for some $\mathbb{L}$-constant c', a contradiction.

For (b), also $T \vdash_{\mathrm{HL}\forall} A(c) \to (\exists x)A(x)$ for all $\mathbb{L}$-constants c. Suppose then that $[A(c)]_T \leq [C]_T$ for all $\mathbb{L}$-constants c but $[(\exists x)A(x)]_T \not\leq [C]_T$. It follows that $T \nvdash_{\mathrm{HL}\forall} (\exists x)A(x) \to C$. So also $T \nvdash_{\mathrm{HL}\forall} (\forall x)(A(x) \to C)$. But then by the Henkin property $T \nvdash_{\mathrm{HL}\forall} A(c') \to C$ for some constant c', a contradiction. □

Lemma 8.15. *Let T be an $\mathbb{L}$-linear $\mathbb{L}$-Henkin $\mathbb{L}$-theory, and A an $\mathbb{L}$-sentence. Then:*

(a) $\|A\|_{\mathbf{CM}_T} = [A]_T$.

(b) $T \vdash_{\mathrm{HL}\forall} A$ *iff A is $\mathsf{LIND}_T^{\mathrm{L}\forall}$-valid.*

Proof. We prove (a) by induction on $\mathrm{cp}(A)$. Clearly $\|t\|_{\mathbf{CM}_T} = t$ for any term t, so also $\|p(t_1, \ldots, t_n)\|_{\mathbf{CM}_T} = [p(t_1, \ldots, t_n)]_T$ for any n-ary $p \in \mathcal{P}$. The cases of the propositional connectives follow as in the propositional case. Suppose now that A is $(\forall x)A'(x)$. Then, using Lemma 8.14 and the fact that the theory is $\mathbb{L}$-linear and $\mathbb{L}$-Henkin:

$$\|A\|_{\mathbf{CM}_T} = \inf_{d \in M} \|A'(x)\|_{\mathbf{CM}_T, m[x \to d]} = \inf_{d \in M} [A'(d)]_T = [(\forall x)A'(x)]_T$$

The case where A is $(\exists x)A'(x)$ is very similar. Finally, (b) follows from (a) since $T \vdash_{\mathrm{HL}\forall} A$ iff $T \vdash_{\mathrm{HL}\forall} \mathrm{e} \to A$ iff $[\mathrm{e}]_T \leq [A]_T$ iff $\mathrm{e} \leq \|A\|_{\mathbf{CM}_T}$ in $\mathbf{CM}_T$. □

Suppose now that T is an $\mathbb{L}$-theory and A is an $\mathbb{L}$-sentence. By Lemma 8.12, if $T \nvdash_{\mathrm{HL}\forall} A$, then $\hat{T} \nvdash_{\mathrm{HL}\forall} A$ for some $\hat{\mathbb{L}}$-linear $\hat{\mathbb{L}}$-Henkin $\hat{\mathbb{L}}$-theory $\hat{T} \supseteq T$ such that $\mathbb{L} \leq \hat{\mathbb{L}}$. But by (b) of the previous lemma, $\hat{T} \nvdash_{\mathrm{HL}\forall} A$ iff A is $\mathsf{LIND}_{\hat{T}}^{\mathrm{L}\forall}$-valid. So, since $\mathsf{LIND}_{\hat{T}}^{\mathrm{L}\forall}$ is an L-chain, we obtain the following:

Theorem 8.16. *For any $\mathbb{L}$-theory T and $\mathbb{L}$-sentence A: $T \vdash_{\mathrm{HL}\forall} A$ iff $T \models_{\mathsf{LIN(L)}} A$.*

Moreover, just as in the propositional case, adding the density rule to HL∀ for any HUL^--extension HL, gives a calculus $\mathrm{HL}\forall^{\mathrm{D}}$ that is complete with respect to dense L-chains. We need the following results, proved by combining the proofs for first-order logics above with the previous proofs for propositional logics with density.

Lemma 8.17. *For any $\mathbb{L}$-theory T:*

(a) *$T \cup \{A\} \vdash_{\mathrm{HL}\forall^{\mathrm{D}}} B$ iff $T \vdash_{\mathrm{HL}\forall^{\mathrm{D}}} C \to B$ for some confusion C of $\{A\}$.*
(b) *$T \vdash_{\mathrm{HL}\forall^{\mathrm{D}}} A$ iff $\vdash_{\mathrm{HL}\forall^{\mathrm{D}}} C \to A$ for some confusion C of T.*
(c) *$\mathrm{HL}\forall^{\mathrm{D}}$ has the proof-by-cases property.*
(d) *If $T \nvdash_{\mathrm{HL}\forall^{\mathrm{D}}} A$ for some $\mathbb{L}$-sentence A, then there exists a countable $\hat{\mathbb{L}}$-linear $\hat{\mathbb{L}}$-Henkin $\hat{\mathbb{L}}$-dense $\hat{\mathbb{L}}$-theory $\hat{T} \supseteq T$ such that $\mathbb{L} \leq \hat{\mathbb{L}}$ and $\hat{T} \nvdash_{\mathrm{HL}\forall^{\mathrm{D}}} A$.*

The rest of the completeness proof then proceeds as for chains.

Theorem 8.18. *For any $\mathbb{L}$-theory T and $\mathbb{L}$-sentence A: $T \vdash_{\mathrm{HL}\forall^{\mathrm{D}}} A$ iff $T \models_{\mathsf{DEN(L)}} A$.*

We can take a step further for some logics and obtain completeness with respect to standard L-algebras. We just follow the proof of Theorem 3.65, noting that we need also that the Dedekind-MacNeille embedding Φ into a standard algebra given by Theorem 2.58 is complete (i.e. $\Phi(\inf \alpha) = \inf \Phi(\alpha)$ and $\Phi(\sup \alpha) = \sup \Phi(\alpha)$).

Theorem 8.19. *Let HL be any extension of HUL^- with axiom schema taken from the set $\{(\bot), (\top), (\mathrm{e}), (\mathrm{f}), (\mathrm{INV}), (\mathrm{W}), (\mathrm{M}), (\mathrm{EM}), (\mathrm{NC})\} \cup \{(\mathrm{C_n}) : n \geq 3\}$. Then for any $\mathbb{L}$-theory T and $\mathbb{L}$-sentence A: $T \vdash_{\mathrm{HL}\forall^{\mathrm{D}}} A$ iff $T \models_{\mathsf{STAN(L)}} A$.*

Finally, as an interesting aside to these completeness results, consider the formula:

$$C = (\forall x)(p(x) \odot q) \to ((\forall x)p(x) \odot q)$$

C is $\mathbf{A}$-valid for a BL-chain $\mathbf{A}$ iff:

$$\inf_{i \in I}(a_i \odot b) \leq (\inf_{i \in I} a_i) \odot b \quad \text{for all } \{a_i\}_{i \in I} \cup \{b\} \subseteq L_{\mathbf{A}} \tag{8.1}$$

We will define a (non-dense) BL-chain where this fails. Let:

$$\mathbf{A} = \langle \{\bot, d\} \cup (0,1], \min, \max, \odot, \to, \bot, 1 \rangle$$

where $\bot < d < x$ for all $x \in (0,1]$ and:

$$x \odot y = \begin{cases} x \cdot y & \text{if } x, y \in (0,1] \\ \bot & \text{if } x = y = d \\ \min(x,y) & \text{otherwise} \end{cases}$$

It is easy to check that **A** is a BL-chain. Moreover, C is not **A**-valid since for $\{a_i\}_{i \in I} = (0,1]$ and $b = d$ in Condition 8.1:

$$\inf_{r \in (0,1]} (r \odot d) = d > \bot = d \odot d = (\inf_{r \in (0,1]} r) \odot d$$

However C *is* valid in all dense HBL-chains. Suppose otherwise. Then for some dense BL-chain **B**, we have $\{a_i\}_{i \in I} \cup \{b\} \subseteq L_{\mathbf{B}}$ such that Condition 8.1 fails. So using density and linearity, there exists $c \in L_{\mathbf{B}}$ such that:

$$\inf_{i \in I} (a_i \odot b) > c > (\inf_{i \in I} a_i) \odot b$$

So for all $i \in I$, $a_i \odot b > c$. By residuation, $a_i > b \to c$. Hence $\inf_{i \in I} \geq b \to c$. Since also $b > c$, by divisibility, $(b \to c) \odot b = c$. But then we can deduce the following contradiction:

$$\inf_{i \in I} (a_i \odot b) > c > (\inf_{i \in I} a_i) \odot b \geq (b \to c) \odot b = c$$

So by Theorem 8.18, C is derivable in HBL$\forall^{\mathrm{D}}$ but not HBL$\forall$. The density rule is not admissible for HBL$\forall$: we could not eliminate it from HBL$\forall^{\mathrm{D}}$ however hard we tried. It therefore remains an intriguing question as to whether HBL$\forall^{\mathrm{D}}$ can be obtained as an *axiomatic* extension of HBL$\forall$. That is, can we add axiom schema – perhaps the substitution instances of C – to obtain HBL$\forall^{\mathrm{D}}$?

8.3 Gentzen Systems

First-order versions of the Gentzen systems introduced in Chapter 4 are easy to define. We just add hypersequent versions of the usual quantifier rules for first-order Classical Logic and Intuitionistic Logic. To be more precise, suppose that we have a matching calculus GL for some HUL$^-$-extension HL and $\mathbb{L}$ is a fixed first-order language with the same connectives as HL. Then to simplify matters, we make a syntactic distinction between bound variables, denoted by x, y, z and free variables, denoted by a, b, and assume for the rest of this section that only bound variables are quantified and only free variables occur freely. As for Hilbert systems, we will write "GL extended to $\mathbb{L}$" to mean the Gentzen system consisting of substitution instances of GL rule schema with $\mathbb{L}$-formulas.

Definition 8.20. GL$\forall$ consists of GL extended to $\mathbb{L}$ plus:

$$\frac{\mathcal{G} \mid \Gamma, A(t) \Rightarrow \Delta}{\mathcal{G} \mid \Gamma, (\forall x)A(x) \Rightarrow \Delta}\ (\forall\Rightarrow) \qquad \frac{\mathcal{G} \mid \Gamma \Rightarrow A(a), \Delta}{\mathcal{G} \mid \Gamma \Rightarrow (\forall x)A(x), \Delta}\ (\Rightarrow\forall)$$

$$\frac{\mathcal{G} \mid \Gamma, A(a) \Rightarrow \Delta}{\mathcal{G} \mid \Gamma, (\exists x)A(x) \Rightarrow \Delta}\ (\exists\Rightarrow) \qquad \frac{\mathcal{G} \mid \Gamma \Rightarrow A(t), \Delta}{\mathcal{G} \mid \Gamma \Rightarrow (\exists x)A(x), \Delta}\ (\Rightarrow\exists)$$

where a does not occur in the lower hypersequent in $(\Rightarrow\forall)$ or $(\exists\Rightarrow)$

Note that unlike in the extra quantifier axioms for HL$\forall$, here we do not require that t is substitutable for x in $(\forall\Rightarrow)$ and $(\Rightarrow\exists)$: no free variable in t can be bound in $A(t)$ since free variables and bound variables are distinguished.

Example 8.21. We can use these quantifier rules and the standard rules for $\rightarrow$ to give sequent derivations of the $(\exists 2)$ axioms as follows:

$$\dfrac{\dfrac{\dfrac{\dfrac{\dfrac{\dfrac{}{A(a) \Rightarrow A(a)}\ (\text{ID}) \qquad \dfrac{}{B \Rightarrow B}\ (\text{ID})}{A(a) \rightarrow B, A(a) \Rightarrow B}\ (\rightarrow\Rightarrow)}{(\forall x)(A \rightarrow B), A(a) \Rightarrow B}\ (\forall\Rightarrow)}{(\forall x)(A \rightarrow B), (\exists x)A \Rightarrow B}\ (\exists\Rightarrow)}{(\forall x)(A \rightarrow B) \Rightarrow (\exists x)A \rightarrow B}\ (\Rightarrow\rightarrow)}{\Rightarrow (\forall x)(A \rightarrow B) \rightarrow ((\exists x)A \rightarrow B)}\ (\Rightarrow\rightarrow)$$

The axioms $(\forall 1)$–$(\forall 2)$ and $(\exists 1)$–$(\exists 2)$ are all derivable in GMAILL$\forall$ using just sequents. However, to derive $(\forall 3)$, hypersequents and the communication rule are essential. Recall that $(\Rightarrow\vee)$ and $(\wedge\Rightarrow)$ are derived rules of GUL, and from Example 4.35, that $\vdash_{\text{GUL}} A \vee B \Rightarrow A \mid A \vee B \Rightarrow B$. Then we have a derivation:

$$\dfrac{\dfrac{\dfrac{\dfrac{\dfrac{A \vee B(a) \Rightarrow A \mid A \vee B(a) \Rightarrow B(a)}{A \vee B(a) \Rightarrow A \mid (\forall x)(A \vee B) \Rightarrow B(a)}\ (\forall\Rightarrow)}{(\forall x)(A \vee B) \Rightarrow A \mid (\forall x)(A \vee B) \Rightarrow B(a)}\ (\forall\Rightarrow)}{(\forall x)(A \vee B) \Rightarrow A \mid (\forall x)(A \vee B) \Rightarrow (\forall x)B}\ (\Rightarrow\forall)}{(\forall x)(A \vee B) \Rightarrow A \vee (\forall x)B}\ (\Rightarrow\vee)}{\Rightarrow (\forall x)(A \vee B) \rightarrow (A \vee (\forall x)B)}\ (\Rightarrow\rightarrow)$$

The $(\forall 3)$ axioms are derivable in sequent calculi for Classical Logic making use of contraction and weakening right rules, but not in calculi for Intuitionistic Logic or many other substructural logics. Intriguingly, while it may be interesting to investigate first-order fuzzy logics lacking $(\forall 3)$, the hypersequent framework presented here does not allow it. Communication ensures that these axioms are derivable along with prelinearity and distributivity.

Soundness and completeness are established for first-order Gentzen systems with respect to Hilbert systems in similar fashion to the propositional case. Note that since all formulas, sequents, and hypersequents are assumed to have free and bound variables distinguished, we must be careful to take this into account in our proofs.

Theorem 8.22. *Let* HL *and* GL *be matching. Then* $\vdash_{\mathrm{GL}\forall} \mathcal{G}$ *iff* $\vdash_{\mathrm{HL}\forall} \mathrm{I}(\mathcal{G})$.

Proof. For the right-to-left direction, suppose that $d \vdash_{\mathrm{HL}\forall} A$. We prove that $\vdash_{\mathrm{GL}\forall} \Rightarrow A$ by induction on $\mathrm{ht}(d)$. For the base cases, observe that the extra axioms $(\forall 1)$–$(\forall 3)$ and $(\exists 1)$–$(\exists 2)$ are all derivable in GL∀. For the induction step, (MP) and (ADJ) are admissible for GL∀ as in the propositional case. So suppose now that A is of the form $(\forall x)B(x)$ and follows from $d' \vdash_{\mathrm{HL}\forall} B(x)$. Note that $B(x)$ is not a formula in the refined sense of this section. However, by Lemma 8.10, we have that $d'' \vdash_{\mathrm{HL}\forall} B(a)$ for some new free variable a not occurring in B or d' where $\mathrm{ht}(d'') = \mathrm{ht}(d')$. Hence by the induction hypothesis, $\vdash_{\mathrm{GL}\forall} \Rightarrow B(a)$. But then by $(\Rightarrow\forall)$, $\vdash_{\mathrm{GL}\forall} \Rightarrow (\forall x)B(x)$ as required. Finally, if $\vdash_{\mathrm{HL}\forall} \mathrm{I}(\mathcal{G})$, then $\vdash_{\mathrm{GL}\forall} \Rightarrow \mathrm{I}(\mathcal{G})$. So by Proposition 4.61, $\vdash_{\mathrm{GL}\forall} \mathcal{G}$.

For the other direction, suppose that $d \vdash_{\mathrm{GL}\forall} \mathcal{G}$. We prove as in the propositional case, that $\vdash_{\mathrm{HL}\forall} \mathrm{I}(\mathcal{G})$ by induction on $\mathrm{ht}(d)$. The only new cases are where the last application is of a quantifier rule. For $(\forall\Rightarrow)$, suppose that $\vdash_{\mathrm{GL}\forall} \mathcal{G} \mid \Gamma, A(t) \Rightarrow \Delta$. Then by the induction hypothesis, $\vdash_{\mathrm{HL}\forall} \mathrm{I}(\mathcal{G}) \vee \mathrm{I}(\Gamma, A(t) \Rightarrow \Delta)$. But by $(\forall 1)$, $\vdash_{\mathrm{HL}\forall} (\forall x)A(x) \to A(t)$. Hence using (B), $\vdash_{\mathrm{HL}\forall} (\mathrm{I}(\mathcal{G}) \vee \mathrm{I}(\Gamma, A(t) \Rightarrow \Delta)) \to (\mathrm{I}(\mathcal{G}) \vee \mathrm{I}(\Gamma, (\forall x)A(x) \Rightarrow \Delta))$ and by an application of modus ponens we are done. For $(\Rightarrow\forall)$, suppose that $\vdash_{\mathrm{GL}\forall} \mathcal{G} \mid \Gamma \Rightarrow A(a), \Delta$. Then by the induction hypothesis, $\vdash_{\mathrm{HL}\forall} \mathrm{I}(\mathcal{G}) \vee \mathrm{I}(\Gamma \Rightarrow A(a), \Delta)$. By Lemma 8.10, $\vdash_{\mathrm{HL}\forall} \mathrm{I}(\mathcal{G}) \vee \mathrm{I}(\Gamma \Rightarrow A(x), \Delta)$, so by (GEN), $\vdash_{\mathrm{HL}\forall} (\forall x)(\mathrm{I}(\mathcal{G}) \vee \mathrm{I}(\Gamma \Rightarrow A(x), \Delta))$. Now using $(\forall 2)$ and $(\forall 3)$ and the fact that a does not occur in $\mathcal{G}$, Γ, Δ, or $A(x)$, we get $\vdash_{\mathrm{HL}\forall} \mathrm{I}(\mathcal{G}) \vee \mathrm{I}(\Gamma \Rightarrow (\forall x)A(x), \Delta)$. The rules for $\exists$ are very similar. □

Cut elimination follows the same pattern as the propositional case. Let us assume that GL is a regular hypersequent calculus. The quantifier rules are nearly but not quite substitutive and reductive. For substitutivity, the problem is that substituting may spoil the restriction on a in $(\Rightarrow\forall)$ or $(\exists\Rightarrow)$. For reductivity, we cannot cut the premises of $(\forall\Rightarrow)$ and $(\Rightarrow\forall)$ (or $(\exists\Rightarrow)$ and $(\Rightarrow\exists)$), $(\mathcal{G} \mid \Gamma, A(t) \Rightarrow \Delta)$ and $(\mathcal{G} \mid \Pi \Rightarrow A(a), \Sigma)$, respectively, to get $(\mathcal{G} \mid \Gamma, \Pi \Rightarrow \Sigma, \Delta)$. We take care of these problems, however, with a substitution lemma, proved by a simple induction on the height of a derivation.

Lemma 8.23. *If* $d; \mathcal{G}_1(a), \ldots, \mathcal{G}_n(a) \vdash_{\mathrm{GL}\forall} \mathcal{G}(a)$ *and* t *is a term with variables not occurring in* d*, then* $d'; \mathcal{G}_1(t), \ldots, \mathcal{G}_n(t) \vdash_{\mathrm{GL}\forall} \mathcal{G}(t)$*, for some derivation* d' *with* $\mathrm{ht}(d') = \mathrm{ht}(d)$.

Theorem 8.24. *Cut elimination holds for* GL∀.

Proof. As in Theorem 5.9, it is sufficient to prove that for any hypersequent $\mathcal{G}$ and hypersequent $\mathcal{H}$ with marked formula A:

Claim. If $d_{\mathcal{G}} \vdash_{\mathrm{GL}\forall^\circ} \mathcal{G}$ and $d_{\mathcal{H}} \vdash_{\mathrm{GL}\forall^\circ} \mathcal{H}$, then $\vdash_{\mathrm{GL}\forall^\circ} \mathrm{CUT}(\mathcal{G}, \mathcal{H})$.

We prove the claim as before by induction on the lexicographically ordered triple $\langle \mathrm{cp}(A), \mathrm{e}(d_{\mathcal{H}}), \mathrm{ht}(d_{\mathcal{G}}) \rangle$, recalling that $\mathrm{e}(d) = 0$ if d ends with a rule application whose principal formula is marked, and $\mathrm{e}(d) = 1$ otherwise. Note first that using

Lemma 8.23, we can assume that any new free variables introduced (upwards) by $(\Rightarrow\forall)$ or $(\exists\Rightarrow)$ in $d_{\mathcal{G}}$ $(d_{\mathcal{H}})$ do not occur in $d_{\mathcal{H}}$ $(d_{\mathcal{G}})$.

If $d_{\mathcal{G}}$ ends with a rule application where the principal formula is not an occurrence of A on the opposite side to $d_{\mathcal{H}}$, then (as in the propositional case), we can make use of the substitutivity of the rule and apply the induction hypothesis. The assumption that new free variables are distinct in $d_{\mathcal{G}}$ from those in $d_{\mathcal{H}}$ and vice versa ensures that this also works for the quantifier rules. Otherwise, let us assume – since propositional connectives are treated in the proof of Theorem 5.9 – that A is of the form $(\forall x)A'(x)$ and $d_{\mathcal{G}}$ ends with:

$$\frac{\mathcal{G}' \mid \Gamma, [(\forall x)A'(x)]^{\lambda-1}, A'(t) \Rightarrow \Delta}{\mathcal{G}' \mid \Gamma, [(\forall x)A'(x)]^{\lambda} \Rightarrow \Delta} \quad \text{or} \quad \frac{\mathcal{G}' \mid \Gamma \Rightarrow A'(a), [(\forall x)A'(x)]^{\lambda-1}, \Delta}{\mathcal{G}' \mid \Gamma \Rightarrow [(\forall x)A'(x)]^{\lambda}, \Delta}$$

where $A \notin \Gamma$ or $A \notin \Delta$ as appropriate and $\mathcal{H}$ is of the form

$$\mathcal{H}' \mid \Pi \Rightarrow \underline{(\forall x)A'(x)}, \Sigma \qquad \text{or} \qquad \mathcal{H}' \mid \Pi, \underline{(\forall x)A'(x)} \Rightarrow \Sigma$$

Let $\mathcal{G}^{\mathcal{H}} \in \text{CUT}(\mathcal{G},\mathcal{H})$. The only tricky case is when $\mathcal{G}^{\mathcal{H}}$ is of the form $(\mathcal{H}' \mid \mathcal{G}'' \mid \Gamma, \Pi^{\lambda} \Rightarrow \Sigma^{\lambda}, \Delta)$ where $\mathcal{G}'' \in \text{CUT}(\mathcal{G}',\mathcal{H})$. But then also:

$$\text{either} \quad \frac{\mathcal{H}' \mid \mathcal{G}'' \mid \Gamma, \Pi^{\lambda-1}, A'(t) \Rightarrow \Sigma^{\lambda-1}, \Delta}{\mathcal{H}' \mid \mathcal{G}'' \mid \Gamma, \Pi^{\lambda-1}, (\forall x)A'(x) \Rightarrow \Sigma^{\lambda-1}, \Delta}$$

$$\text{or} \quad \frac{\mathcal{H}' \mid \mathcal{G}'' \mid \Gamma, \Pi^{\lambda-1} \Rightarrow A'(a), \Sigma^{\lambda-1}, \Delta}{\mathcal{H}' \mid \mathcal{G}'' \mid \Gamma, \Pi^{\lambda-1} \Rightarrow (\forall x)A'(x), \Sigma^{\lambda-1}, \Delta}$$

is an instance of the appropriate rule. Moreover, by the induction hypothesis, the premise is derivable so we have a derivation d ending with such a rule application.

If $\text{e}(d_{\mathcal{H}}) = 1$: i.e. $d_{\mathcal{H}}$ does not end with the application of a logical rule to the marked occurrence of A, then we proceed as in the propositional case. Suppose therefore that $\text{e}(d_{\mathcal{H}}) = 0$: i.e. $d_{\mathcal{H}}$ ends with an application of $(\forall\Rightarrow)$ or $(\Rightarrow\forall)$ to the marked occurrence of A, and is of the form:

$$\frac{\mathcal{H}' \mid \Pi \Rightarrow A(a), \Sigma}{\mathcal{H}' \mid \Pi \Rightarrow (\forall x)A'(x), \Sigma} \quad \text{or} \quad \frac{\mathcal{H}' \mid \Pi, A(t) \Rightarrow \Sigma}{\mathcal{H}' \mid \Pi, (\forall x)A'(x) \Rightarrow \Sigma}$$

By Lemma 8.23, there is a derivation of the same height of $(\mathcal{H}' \mid \Pi \Rightarrow A(t), \Sigma)$ or $(\mathcal{H}' \mid \mathcal{G}'' \mid \Gamma, \Pi^{\lambda-1} \Rightarrow A'(t), \Sigma^{\lambda-1}, \Delta)$. But then by the induction hypothesis, since $\text{cp}(A'(t)) < \text{cp}((\forall x)A'(x))$, using a further application of (EC), $\vdash_{\text{GL}^\circ} \mathcal{G}^{\mathcal{H}}$. □

It follows from cut elimination that GL∀ and therefore also HL∀ are conservative extensions of GL and HL, respectively. Just observe that any GL∀-derivable propositional hypersequent has a cut-free derivation in GL∀ which does not involve any quantifiers. Cut elimination will not help us with decidability – indeed, all these logics are undecidable. However, as we will see shortly, we can use this result instead to prove versions of Herbrand's theorem and Skolemization for the prenex fragments of these logics.

It is also straightforward to extend density elimination to first-order Gentzen systems: the quantifier rules are treated just like the rules for other connectives.

Theorem 8.25. *Let* GL *be a regular and local single-conclusion hypersequent calculus with weakening, or one of* GUL, GIMTL, GUML, GIUML. *Then density elimination holds for* $\mathrm{GL}\forall^{\mathrm{D}}$.

Moreover, as before we can use density elimination to establish equivalence for Hilbert systems with and without density.

Proposition 8.26. *Let* HL *be an* HUL*-extension and let* GL *be a matching hypersequent calculus where* $\mathrm{GL}\forall$ *admits density elimination. Then* $T \vdash_{\mathrm{HL}\forall} A$ *iff* $T \vdash_{\mathrm{HL}\forall^{\mathrm{D}}} A$.

Proof. We have the following chain of reasoning:

$T \vdash_{\mathrm{HL}\forall} A$	iff	$\vdash_{\mathrm{HL}\forall} C \to A$ for some confusion C of T	Theorem 8.22
	iff	$\vdash_{\mathrm{GL}\forall} C \Rightarrow A$ for some confusion C of T	Theorem 8.9.(b)
	iff	$\vdash_{\mathrm{GL}\forall^{\mathrm{D}}} C \Rightarrow A$ for some confusion C of T	Density Elimination
	iff	$\vdash_{\mathrm{HL}\forall^{\mathrm{D}}} C \to A$ for some confusion C of T	Theorem 8.22
	iff	$T \vdash_{\mathrm{HL}\forall^{\mathrm{D}}} A$	Theorem 8.17 □

Hence using Theorem 3.65, we obtain the following standard completeness results.

Theorem 8.27. *For* $\mathrm{L} \in \{\mathsf{UL}, \mathsf{MTL}, \mathsf{SMTL}, \mathsf{IMTL}, \mathsf{G}, \mathsf{UML}, \mathsf{IUML}, \mathsf{MTL_n}\ (n \geq 2)\}$*:* $T \vdash_{\mathrm{HL}\forall} A$ *iff* $T \models_{\mathsf{STAN(L)}} A$.

8.4 Herbrand's Theorem and Skolemization

We will be using our systems in a nice way to establish some fundamental properties of first-order fuzzy logics. However, first we prove a negative result: these logics are all *undecidable*. In fact, we can show something stronger, the undecidability of the existential fragment (with function symbols). Note that this result is not entirely predictable. Contraction-free first-order logics such as first-order Monoidal Logic and Multiplicative Additive Linear Logic *are* decidable.[2]

Let is assume in what follows that P stands always for a quantifier-free formula, and recall that existential formulas are of the form $(\exists\bar{x})P(\bar{x})$.

Theorem 8.28. *For any* HUL^-*-extension* HL *such that* $\vdash_{\mathrm{HL}} A$ *implies* $\vdash_{\mathsf{HCL}} A$*, the* L*-validity problem for existential formulas is undecidable.*

---

[2] In the function-free case, we just restrict the term t in $(\forall\Rightarrow)$ and $(\Rightarrow\exists)$ to variables occurring in lower sequents or the first new variable, and notice that every rule of the corresponding sequent calculus, as in the propositional case, reduces the complexity of the sequent.

Proof. We just adapt slightly the embedding used in Theorem 7.25 to establish co-NP-hardness for propositional logics. Let P be any formula of the form $\bigvee_{i\in I}\bigwedge_{j\in J_i} L_{ij}$ for index sets I and J_i for $i \in I$ where each L_{ij} is of the form $p(\bar{t})$ or $\neg p(\bar{t})$ with variables among $\bar{x}$. Then we define the translation:

$$P^c = \bigvee_{i\in I}\bigwedge_{j\in J_i} L^c_{ij} \quad \text{where } p(\bar{t})^c = p(\bar{t}) \oplus p(\bar{t}) \text{ and } (\neg p(\bar{t}))^c = \neg(p(\bar{t}) \odot p(\bar{t})).$$

Claim. $\vdash_{\mathsf{HCL}\forall} (\exists\bar{x})P(\bar{x})$ iff $\vdash_{\mathsf{HL}\forall} (\exists\bar{x})P^c(\bar{x})$.

Proof of claim. The right-to-left direction follows as in Theorem 7.25 from the fact that $\vdash_{\mathsf{HCL}} p(\bar{t}) \leftrightarrow (p(\bar{t}) \odot p(\bar{t}))$, $\vdash_{\mathsf{HCL}} p(\bar{t}) \leftrightarrow (p(\bar{t}) \oplus p(\bar{t}))$, and every HL-derivable formula is HCL-derivable. For the other direction, we make us of the classical Herbrand theorem (see below). If $\vdash_{\mathsf{HCL}\forall} (\exists\bar{x})P(\bar{x})$, then $\vdash_{\mathsf{HCL}\forall} \bigvee_{i\in I} P(\bar{t}_i)$ for some finite index set I and terms $\bar{t}_i$. But then following the proof of Theorem 7.25 (using distributivity in CL and L), $\vdash_{\mathsf{HL}\forall} \bigvee_{i\in I} P^c(\bar{t}_i)$. So also $\vdash_{\mathsf{HL}\forall} (\exists\bar{x})P^c(\bar{x})$ as required.

The result then follows from the classical result of Church [47] that the existential fragment (with function symbols) of Classical Logic is undecidable. □

Although undecidable, the prenex fragments of fuzzy logics often have nice properties. In particular, a version of *Herbrand's theorem* holds for these logics. The validity of a prenex formula is equivalent to the validity of a set of propositional formulas. The key technical tool for establishing this result is a "mid-hypersequent theorem", an analogue of Gentzen's mid-sequent theorem. This tells us that any GL∀-derivable prenex formula has a GL∀-derivation where the propositional connective inferences precede all quantifier inferences. It then follows that there exist hypersequents in the derivation with no propositional connective inference below or quantifier inference above. We establish the result by performing syntactic manipulations on a (cut-free) GL∀°-derivation, pushing propositional connective inferences upwards and quantifier inferences downwards.

Full generality is tricky here, so let us assume that GL is GUL or GIUL extended with rules from $\{(\text{W}), (\text{C}), (\text{MIX}), (\text{MINGLE}), (\text{SPLIT}), (\text{EMP})\} \cup \{(\text{C}_n) : n \geq 2\}$.

Theorem 8.29. *Let $\mathcal{G}$ be a hypersequent containing only prenex formulas. If $\vdash_{\mathrm{GL}\forall} \mathcal{G}$, then $d \vdash_{\mathrm{GL}\forall} \mathcal{G}$ for some derivation d where no propositional inference is below a quantifier inference.*

Proof. It will save us some effort to perform manipulations on a slightly different calculus. Let $\mathrm{GL}\forall^{\vee}$ be $\mathrm{GL}\forall^{\circ}$ with (ID) restricted to strictly atomic instances, and $(\vee\Rightarrow)$ and $(\wedge\Rightarrow)$ replaced with:

$$\frac{\mathcal{G} \mid \Gamma_1, A \Rightarrow \Delta_1 \quad \mathcal{G} \mid \Gamma_2, B \Rightarrow \Delta_2}{\mathcal{G} \mid \Gamma_1, A \vee B \Rightarrow \Delta_1 \mid \Gamma_2, A \vee B \Rightarrow \Delta_2} \ (\vee\Rightarrow)^{\vee}$$

$$\frac{\mathcal{G} \mid \Gamma_1 \Rightarrow A, \Delta_1 \quad \mathcal{G} \mid \Gamma_2 \Rightarrow B, \Delta_2}{\mathcal{G} \mid \Gamma_1 \Rightarrow A \wedge B, \Delta_1 \mid \Gamma_2 \Rightarrow A \wedge B, \Delta_2} \ (\Rightarrow\wedge)^{\vee}$$

These rules are easily derived in GL∀ using (CUT), and we leave it as an (easy) exercise to show that $\vdash_{\mathrm{GL}\forall} \mathcal{G}$ iff $\vdash_{\mathrm{GL}\forall^{\vee}} \mathcal{G}$.

Now let the order $o(d)$ of a derivation d be the multiset containing the lengths (cardinalities) of any sub-branches of d that start (working upwards) with a propositional connective inference and end with a quantifier inference. Then it is sufficient to establish the following:

Claim. If $d \vdash_{\mathrm{GL}\forall^{\vee}} \mathcal{G}$, then $d' \vdash_{\mathrm{GL}\forall^{\vee}} \mathcal{G}$ for some derivation d' where $o(d') = []$.

Proof of claim. We proceed by induction on $o(d)$ using the multiset ordering $\leq_{\mathsf{m}}$. The base case where $o(d) = []$ is immediate. For the inductive step we have a number of possibilities.

Suppose that a quantifier inference occurs directly above a propositional connective inference. Then we can rearrange the derivation so that the quantifier rule application is below the propositional connective rule application, and use the induction hypothesis. The only tricky cases are where $(\Rightarrow\exists)$ appears above $(\vee\Rightarrow)^{\vee}$, or $(\forall\Rightarrow)$ above $(\Rightarrow\wedge)^{\vee}$. Let us consider the former, i.e. d ends with:

$$\dfrac{\dfrac{\dfrac{\vdots\, d_1}{\mathcal{H} \mid \Gamma_1, A \Rightarrow C(t), \Delta_1}}{\mathcal{H} \mid \Gamma_1, A \Rightarrow (\exists x)C(x), \Delta_1}\,(\Rightarrow\exists) \qquad \dfrac{\vdots\, d_2}{\mathcal{H} \mid \Gamma_2, B \Rightarrow \Delta_2}}{\mathcal{H} \mid \Gamma_1, A \vee B \Rightarrow (\exists x)C(x), \Delta_1 \mid \Gamma_2, A \vee B \Rightarrow \Delta_2}\,(\vee\Rightarrow)^{\vee}$$

We can replace this with the following derivation d':

$$\dfrac{\dfrac{\dfrac{\vdots\, d_1}{\mathcal{H} \mid \Gamma_1, A \Rightarrow C(t), \Delta_1} \qquad \dfrac{\vdots\, d_2}{\mathcal{H} \mid \Gamma_2, B \Rightarrow \Delta_2}}{\mathcal{H} \mid \Gamma_1, A \vee B \Rightarrow C(t), \Delta_1 \mid \Gamma_2, A \vee B \Rightarrow \Delta_2}\,(\vee\Rightarrow)^{\vee}}{\mathcal{H} \mid \Gamma_1, A \vee B \Rightarrow (\exists x)C(x), \Delta_1 \mid \Gamma_2, A \vee B \Rightarrow \Delta_2}\,(\Rightarrow\exists)$$

Since $o(d') <_{\mathsf{m}} o(d)$ the induction hypothesis can be applied.

If the previous case does not occur, then there must be a sub-branch with structural inferences occurring between the propositional connective inference and quantifier inferences. We note without proof that applications of the "contraction rules" (EC), (C), (MINGLE), or (C_n) can be pushed downwards over the other structural rules in derivations, and (EW), (W), (SPLIT), and (MIX) can be pushed upwards. If a quantifier inference is directly above (EW), (COM), (SPLIT), (W), or (MIX), then we can move the application of the structural rule above the quantifier inference. In the (most complicated) case of (COM) and $(\forall\Rightarrow)$, we have the following situation:

$$\dfrac{\dfrac{\dfrac{\vdots\, d_1}{\mathcal{H} \mid \Gamma_1, B(t), \Pi_1 \Rightarrow \Sigma_1, \Delta_1}}{\mathcal{H} \mid \Gamma_1, (\forall x)B(x), \Pi_1 \Rightarrow \Sigma_1, \Delta_1}\,(\forall\Rightarrow) \qquad \dfrac{\vdots\, d_2}{\mathcal{H} \mid \Gamma_2, \Pi_2 \Rightarrow \Sigma_2, \Delta_2}}{\mathcal{H} \mid \Gamma_1, (\forall x)B(x), \Gamma_2 \Rightarrow \Delta_1, \Delta_2 \mid \Pi_1, \Pi_2 \Rightarrow \Sigma_1, \Sigma_2}\,(\mathrm{COM})$$

We can replace this with the following derivation d':

$$\dfrac{\dfrac{\dfrac{\vdots\, d_1}{\mathcal{H} \mid \Gamma_1, B(t), \Pi_1 \Rightarrow \Sigma_1, \Delta_1} \quad \dfrac{\vdots\, d_2}{\mathcal{H} \mid \Gamma_2, \Pi_2 \Rightarrow \Sigma_2, \Delta_2}}{\mathcal{H} \mid \Gamma_1, B(t), \Gamma_2 \Rightarrow \Delta_1, \Delta_2 \mid \Pi_1, \Pi_2 \Rightarrow \Sigma_1, \Sigma_2}\;(\textsc{com})}{\mathcal{H} \mid \Gamma_1, (\forall x)B(x), \Gamma_2 \Rightarrow \Delta_1, \Delta_2 \mid \Pi_1, \Pi_2 \Rightarrow \Sigma_1, \Sigma_2}\;(\forall\Rightarrow)$$

Since $o(d') <_{\mathrm{m}} o(d)$ the induction hypothesis can be applied.

The final possibility is that an application of (EC), (C), (MINGLE), or ($\mathrm{C_n}$) occurs directly above a logical connective inference. Again we are always able to push the relevant rule application upwards. Suppose for example that d ends with:

$$\dfrac{\dfrac{\dfrac{\vdots\, d_1}{\mathcal{H} \mid \Gamma_1, \Gamma_2^2 \Rightarrow [A]^2, \Delta_1, \Delta_2^2}}{\mathcal{H} \mid \Gamma_1, \Gamma_2 \Rightarrow A, \Delta_1, \Delta_2}\;(\textsc{c}) \quad \dfrac{\vdots\, d_2}{\mathcal{H} \mid \Gamma_3, B \Rightarrow \Delta_3}}{\mathcal{H} \mid \Gamma_1, \Gamma_2, \Gamma_3, A \to B \Rightarrow \Delta_1, \Delta_2 . \Delta_3}\;(\to\Rightarrow)$$

We can replace this with the following derivation d':

$$\dfrac{\dfrac{\dfrac{\dfrac{\vdots\, d_1}{\mathcal{H} \mid \Gamma_1, \Gamma_2^2 \Rightarrow [A]^2, \Delta_1, \Delta_2^2} \quad \dfrac{\vdots\, d_2}{\mathcal{H} \mid \Gamma_3, B \Rightarrow \Delta_3}}{\mathcal{H} \mid \Gamma_1, \Gamma_2^2, \Gamma_3, A \to B \Rightarrow A, \Delta_1, \Delta_2^2, \Delta_3}\;(\to\Rightarrow) \quad \dfrac{\vdots\, d_2}{\mathcal{H} \mid \Gamma_3, B \Rightarrow \Delta_3}}{\mathcal{H} \mid \Gamma_1, \Gamma_2^2, \Gamma_3^2, [A \to B]^2 \Rightarrow \Delta_1, \Delta_2^2, \Delta_3^2}\;(\to\Rightarrow)}{\mathcal{H} \mid \Gamma_1, \Gamma_2, \Gamma_3, A \to B \Rightarrow \Delta_1, \Delta_2, \Delta_3}\;(\textsc{c})$$

Again, since $o(d') <_{\mathrm{m}} o(d)$ the induction hypothesis can be applied. □

We now recall some basic notions relating to Herbrand's theorem.

Definition 8.30. For a formula A, let $\mathcal{C}_A$, $\mathcal{F}_A$, and $\mathcal{P}_A$ be, respectively, the constants, non-nullary function symbols, and predicate symbols occurring in A, adding a constant to $\mathcal{C}_A$ if it is empty. The *Herbrand universe* of A is $\mathcal{U}(A) = \bigcup_{n=0}^{\infty} \mathcal{U}_n(A)$ where:

$$\begin{aligned} \mathcal{U}_0(A) &= \mathcal{C}_A \\ \mathcal{U}_{n+1}(A) &= \mathcal{U}_n(A) \cup \{f(t_1, \ldots, t_k) : t_1, \ldots, t_k \in \mathcal{U}_n(A) \text{ and } f \in \mathcal{F}_A \text{ with arity } k\} \end{aligned}$$

The *Herbrand base* of A is then defined as follows:

$$\mathcal{B}(A) = \{p(t_1, \ldots, t_n) : t_1, \ldots, t_n \in \mathcal{U}(A) \text{ and } p \in \mathcal{P}_A \text{ with arity } n\}$$

For many fuzzy logics, we obtain the following existential Herbrand theorem.

Theorem 8.31. *Let* HL *be* HUL *extended with some subset of the axiom schema* $\{(\mathrm{INV}), (\mathrm{W}), (\mathrm{e}), (\mathrm{M}), (\mathrm{EM}), (\mathrm{f})\} \cup \{(\mathrm{C_n}) : n \geq 2\}$. *Then:*

$$\models_{\mathsf{GEN(L)}} (\exists\bar{x})P(\bar{x}) \quad \textit{iff} \quad \models_{\mathsf{GEN(L)}} \bigvee_{i=1}^{n} P(\bar{t}_i) \ \textit{for some } \bar{t}_1,\ldots,\bar{t}_n \in \mathcal{U}(P).$$

Proof. The right-to-left direction is immediate. For the other direction, suppose that $\models_{\mathsf{GEN(L)}} (\exists\bar{x})P(\bar{x})$. So $\vdash_{\mathrm{GL}\forall} \Rightarrow (\exists\bar{x})P(\bar{x})$. Hence by Theorem 8.29, we have a derivation of $(\Rightarrow (\exists\bar{x})P(\bar{x}))$ where no propositional inference follows a quantifier inference. Moreover, as in the proof of this theorem, we can push applications of all structural rules except (EC) up over the quantifier rule applications. So we have a derivation of $(\Rightarrow (\exists\bar{x})P(\bar{x}) \mid \ldots \mid \Rightarrow (\exists\bar{x})P(\bar{x}))$ that ends with the only occurring applications of $(\Rightarrow\exists)$. Hence there exists a GL$\forall$-derivable hypersequent $(\Rightarrow P(\bar{t}_1) \mid \ldots \mid \Rightarrow P(\bar{t}_n))$ for some $\bar{t}_1,\ldots,\bar{t}_n \in \mathcal{U}(P)$. But then by the soundness of the calculus $\models_{\mathsf{GEN(L)}} \bigvee_{i=1}^{n} P(\bar{t}_i)$ as required. □

Herbrand's theorem allows us to reduce the validity problem for an existential formula to the validity of formulas that are (essentially) propositional. We can also make use of this theorem (as in Classical Logic) to do the same for prenex formulas. The idea here is to remove universal quantifiers iteratively, replacing the variables that they bind with terms consisting of a new function symbol with variables bound by preceding existential quantifiers. This process is called Skolemization for fuzzy logics, although it is worth noting that for Classical Logic, the usual process involves removing existential quantifiers and preserving satisfiability.

Definition 8.32. Let A be a prenex formula and assume harmlessly that the ith occurrence of $\forall$ is labelled $\forall^i$ and that no function symbol f_i of any arity occurs in A. Then the *Skolem form* A^S of A is defined by induction as follows:

(1) If A is of the form $(\exists\bar{x})P(\bar{x})$ where P is quantifier-free, then A^S is $(\exists\bar{x})P(\bar{x})$.
(2) If A is of the form $(\exists\bar{x})(\forall^i y)B(\bar{x},y)$, then A^S is $((\exists\bar{x})B(\bar{x},f_i(\bar{x})))^S$.

Example 8.33. Consider the prenex formula $A = (\exists x)(\forall y)(\exists z)(\forall u)p(x,y,z,u)$. Skolemizing in two steps, we obtain that $A^S = (\exists x)(\exists z)p(x,f(x),z,g(x,z))$.

Lemma 8.34. *Let $(\exists\bar{x})P^F(\bar{x})$ be the Skolem form of $(\mathrm{Q}\bar{y})P(\bar{y})$. Then $(\Rightarrow (\mathrm{Q}\bar{y})P(\bar{y}))$ is derivable from any hypersequent $\mathcal{G} \subseteq [P^F(\bar{t}) : \bar{t} \in \mathcal{U}((\exists\bar{x})P^F(\bar{x}))]$ using* (EW), (EC), $(\Rightarrow\forall)$, *and* $(\Rightarrow\exists)$, *where in* $(\Rightarrow\forall)$, *any variable-free term not occurring in the conclusion may be used in the premise.*

Proof. Let each occurrence of $\forall$ in $(\Rightarrow (\mathrm{Q}\bar{y})P(\bar{y}))$ be labelled $f(\bar{z})$ where:

(i) f is the constant or n-ary function symbol in $(\exists\bar{x})P^F(\bar{x})$ introduced by Skolemization for this occurrence of $\forall$;
(ii) $z_1,\ldots,z_n$ are the existentially bound variables preceding this occurrence of $\forall$ in $(\mathrm{Q}\bar{y})P(\bar{y})$.

More generally, we will allow this occurrence of $\forall$ to be labelled $f(t_1,\ldots,t_n)$ where $t_1,\ldots,t_n$ are terms. We will also suppose that substituting for a variable in such labelled formulas extends to substituting also in the labels. In particular, given a labelled formula $A(x)$, $A(t)$ is obtained by replacing all free occurrences of x in $A(x)$

by t, including all those in the labels. We then define a sequence of hypersequents as follows. Let $\mathcal{G}_0$ be $(\Rightarrow (\mathrm{Q}\bar{y})P(\bar{y}))$, and given $\mathcal{G}_j$, let $\mathcal{G}_{j+1}$ be the smallest hypersequent satisfying:

(1) $\mathcal{G}_j \subseteq \mathcal{G}_{j+1}$.
(2) If $(\Rightarrow (\forall x)B(x)) \in \mathcal{G}_{j+1}$ and $f(\bar{t})$ labels $\forall$, then $(\Rightarrow B(f(\bar{t}))) \in \mathcal{G}_{j+1}$.
(3) If $(\Rightarrow (\exists x)B(x)) \in \mathcal{G}_{j+1}$, then $(\Rightarrow B(s)) \in \mathcal{G}_{j+1}$ for all $s \in \mathcal{U}_j((\exists \bar{x})P^F(\bar{x}))$.

Notice first that each $\mathcal{G}_j$ can be derived from $\mathcal{G}_{j+1}$ using the given rules. The only difficulty could be that for (2), the term $f(\bar{t})$ occurs already in the conclusion. However, each occurrence of $\forall$ is labelled with a different constant or function symbol f with arguments determined uniquely by the terms chosen for the preceding occurrences of $\exists$. Hence the new formula in the premise for such a case must already occur in the conclusion. The desired result is then a consequence of (EW) and the following:

Claim. If $\bar{t} \in \mathcal{U}_k((\exists \bar{x})P^F(\bar{x}))$ for some $k \in \mathbb{N}$, then $(\Rightarrow P^F(\bar{t})) \in \mathcal{G}_j$ for some $j \in \mathbb{N}$.

Proof of claim. A simple induction on the number of quantifiers in $(\mathrm{Q}\bar{y})P(\bar{y})$. □

Example 8.35. Consider again the formula $A = (\exists x)(\forall y)(\exists z)(\forall u)p(x,y,z,u)$ with Skolem form $A^S = (\exists x)(\exists z)p(x,f(x),z,g(x,z))$. Following the procedure of the previous proof (but only adding the sequents we need), we derive A from the hypersequent $(\Rightarrow p(a,f(a),a,g(a,a)) \mid \Rightarrow p(a,f(a),g(a,a),g(a,g(a,a))))$:

$$\dfrac{\dfrac{\dfrac{\dfrac{\dfrac{\dfrac{\dfrac{\dfrac{\Rightarrow p(a,f(a),a,g(a,a)) \mid \Rightarrow p(a,f(a),g(a,a),g(a,g(a,a)))}{\Rightarrow p(a,f(a),a,g(a,a)) \mid \Rightarrow (\forall_{g(a,g(a,a))}u)p(a,f(a),g(a,a),u)}(\Rightarrow\forall)}{\Rightarrow (\forall_{g(a,a)}u)p(a,f(a),a,u) \mid \Rightarrow (\forall_{g(a,g(a,a))}u)p(a,f(a),g(a,a),u)}(\Rightarrow\forall)}{\Rightarrow (\forall_{g(a,a)}u)p(a,f(a),a,u) \mid \Rightarrow (\exists z)(\forall_{g(a,z)}u)p(a,f(a),z,u)}(\Rightarrow\exists)}{\Rightarrow (\exists z)(\forall_{g(a,z)}u)p(a,f(a),z,u) \mid \Rightarrow (\exists z)(\forall_{g(a,z)}u)p(a,f(a),z,u)}(\Rightarrow\exists)}{\Rightarrow (\exists z)(\forall_{g(a,z)}u)p(a,f(a),z,u)}(\mathrm{EC})}{\Rightarrow (\forall_{f(a)}y)(\exists z)(\forall_{g(a,z)}u)p(a,y,z,u)}(\Rightarrow\forall)}{\Rightarrow (\exists x)(\forall_{f(x)}y)(\exists z)(\forall_{g(x,z)}u)p(x,y,z,u)}(\Rightarrow\exists)$$

Theorem 8.36. *Let* HL *be* HUL *extended with some subset of the axiom schema* $\{(\mathrm{INV}),(\mathrm{W}),(\mathrm{e}),(\mathrm{M}),(\mathrm{EM}),(\mathrm{f})\} \cup \{(\mathrm{C_n}) : n \geq 2\}$ *and let* $(\exists \bar{x})P^F(\bar{x})$ *be the Skolem form of* $(\mathrm{Q}\bar{y})P(\bar{y})$*. Then the following are equivalent:*

(i) $\models_{\mathsf{GEN(L)}} (\mathrm{Q}\bar{y})P(\bar{y})$
(ii) $\models_{\mathsf{GEN(L)}} (\exists \bar{x})P^F(\bar{x})$
(iii) $\models_{\mathsf{GEN(L)}} \bigvee_{i=1}^{n} P^F(\bar{t}_i)$ *for some* $\bar{t}_1,\ldots,\bar{t}_n \in \mathcal{U}(P^F)$

Proof. The equivalence of (ii) and (iii) is just (the existential Herbrand) Theorem 8.31. So it is sufficient to show that (iii) implies (i), and (i) implies (ii). First, suppose that $\models_{\mathsf{GEN(L)}} \bigvee_{i=1}^{n} P(\bar{t}_i)$ for some $\bar{t}_1,\ldots,\bar{t}_n \in \mathcal{U}(P)$. Then by Lemma 8.34, $(\Rightarrow (\mathrm{Q}\bar{y})P(\bar{y}))$ is derivable from $(\Rightarrow P(\bar{t}_1) \mid \ldots \mid P(\bar{t}_n))$ using (EW), (EC), $(\Rightarrow\forall)$, and

$(\Rightarrow \exists)$. Since these rules are sound, it follows that $\models_{\mathsf{GEN(L)}} (\mathrm{Q}\bar{y})P(\bar{y})$ as required. To show that (ii) implies (i), it is sufficient to show that $\models_{\mathsf{GEN(L)}} (\exists\bar{x})(\forall y)B(\bar{x},y)$ implies $\models_{\mathsf{GEN(L)}} (\exists\bar{x})B(\bar{x},f(\bar{x}))$ where f does not occur in $B(\bar{x},y)$. But it is easily checked (e.g. in GL$\forall$) that $\models_{\mathsf{GEN(L)}} (\exists\bar{x})(\forall y)B(\bar{x},y) \to (\exists\bar{x})B(\bar{x},f(\bar{x}))$ so we are done. □

8.5 Łukasiewicz Logic

The first-order situation for Łukasiewicz Logic and its relatives P, A, and CHL, is more complicated than the cases we have just encountered. For these logics, the set of valid formulas is not recursively enumerable (for P, not even arithmetical). Finite sets of rule schema such as the Hilbert and Gentzen systems of the previous section cannot be enough. Here we take a closer look at this problem for the particular case of first-order Łukasiewicz Logic Ł$\forall$.

For simplicity, let us use a more restricted first-order language with connectives $\forall$, $\exists$, $\to$, and $\bot$. Also, since we are interested here only in the bpcrl $\mathbf{A}(*_{Ł},0)$, let us assume without further comment that all structures, models, and so on refer just to this algebra. In particular, $T \models_{Ł} A$ denotes that all models of T are models of A. Let us also write $A \equiv_{Ł} B$ to mean that A and B are Ł$\forall$-*equivalent*, i.e. $\|A\|_{\mathbf{M}} = \|B\|_{\mathbf{M}}$ for all structures $\mathbf{M}$.

Before moving on to the deficiencies of Ł$\forall$, let us consider one of its more attractive features. Like Classical Logic, Ł$\forall$ has the full quota of "quantifier shifts". That is, when x does not occur free in A:

$$A \to (\forall x)B \equiv_{Ł} (\forall x)(A \to B) \qquad (\forall x)B \to A \equiv_{Ł} (\exists x)(B \to A)$$
$$A \to (\exists x)B \equiv_{Ł} (\exists x)(A \to B) \qquad (\exists x)B \to A \equiv_{Ł} (\forall x)(B \to A)$$

This means that for any first-order formula A, we can rewrite all bound variables to new variables not occurring elsewhere, then use the above equivalences (left-to-right) as rewrite rules to push all quantifiers to the outside; i.e.

Theorem 8.37. *Any formula is* Ł$\forall$*-equivalent to a prenex formula.*

8.5.1 An Approximate Herbrand Theorem

We show now that the Herbrand Theorem of the previous section cannot hold for Ł$\forall$. Note first that $\models_{Ł} (\exists x)p(x) \to (\exists y)p(y)$. So using the quantifier-shifting equivalences listed above:

$$\models_{Ł} (\exists y)(\forall x)(p(x) \to p(y))$$

and by the easy direction of Skolemization:

$$\models_{Ł} (\exists y)(p(f(y)) \to p(y))$$

So if Herbrand's theorem holds for Ł∀, then for some constant c and $n \in \mathbb{N}^+$:

$$\models_{\text{Ł}} \bigvee_{i=1}^{n} (p(f^i(c)) \rightarrow p(f^{i-1}(c)))$$

where $f^0(c) = c$ and $f^{i+1}(c) = f(f^i(c))$ for all $i \in \mathbb{N}$. But now let us define a structure $\mathbf{M}$ with $\|p(f^i(c))\|_{\mathbf{M}} = i/n$ for $i = 0 \ldots n$, so that:

$$\|p(f^i(c))\|_{\mathbf{M}} > \|p(f^{i-1}(c))\|_{\mathbf{M}} \text{ for } i = 1 \ldots n.$$

Then $\| \bigvee_{i=1}^{n} (p(f^i(c)) \rightarrow p(f^{i-1}(c)))\|_{\mathbf{M}} < 1$, a contradiction. So the Herbrand theorem must fail. In fact this argument holds for a wide class of infinite-valued logics with quantifier shifts, including A and CHL.

Take another look at the formula $\bigvee_{i=1}^{n} (p(f^i(c)) \rightarrow p(f^{i-1}(c)))$, however. Although this is not a valid formula of Ł∀, it comes within "one nth" of being one. Observe that for any $r_0, r_1, \ldots, r_n \in [0,1]$:

$$\min_{i \in \{1,\ldots,n\}} \{r_{i-1} - r_i\} \leq 1/n \quad \text{and so also} \quad \max_{i \in \{1,\ldots,n\}} \{1 - r_{i-1} + r_i\} \geq 1 - 1/n.$$

Let us write for $\triangleright \in \{>, \geq\}$ and $r \in [0,1]$:

$$\models_{\text{Ł}}^{\triangleright r} A \quad \text{iff} \quad \|A\|_{\mathbf{M}} \triangleright r \text{ for all structures } \mathbf{M}.$$

Then for any $r < 1 - 1/n$:

$$\models_{\text{Ł}}^{>r} \bigvee_{i=1}^{n} (p(f^i(c)) \rightarrow p(f^{i-1}(c)))$$

That is, we have "Herbrand approximations" of $(\exists y)(p(f(y)) \rightarrow p(y))$ that come arbitrarily close to 1. This illustrates a more general phenomenon, captured by the following approximate Herbrand theorem:

Theorem 8.38. $\models_{\text{Ł}} (\exists \bar{x}) P(\bar{x})$ *iff for all* $r < 1$*:*

$$\models_{\text{Ł}}^{>r} \bigvee_{i=1}^{n} P(\bar{t}_i) \quad \textit{for some } \bar{t}_1, \ldots, \bar{t}_n \in \mathcal{U}(P).$$

Proof. We refer to [220] for topological terminology. Suppose that for all $r < 1$: $\models_{\text{Ł}}^{>r} \bigvee_{i=1}^{n} P(\bar{t}_i)$ for some $\bar{t}_1, \ldots, \bar{t}_n \in \mathcal{U}(P)$. Then $\models_{\text{Ł}}^{>r} (\exists \bar{x}) P(\bar{x})$ for all $r < 1$, so clearly $\models_{\text{Ł}} (\exists \bar{x}) P(\bar{x})$. For the other direction, suppose that $\models_{\text{Ł}} (\exists \bar{x}) P(\bar{x})$ and fix $r < 1$. Notice that each $\|\|_{\mathbf{M}}$ can be viewed as a mapping from $\mathcal{B}(P)$ into $[0,1]$: a member either of $[0,1]^k$ for some k if $\mathcal{B}(P)$ is finite, or of the Hilbert cube $[0,1]^{\omega}$ if $\mathcal{B}(P)$ is countably infinite. In either case ($[0,1]^{\omega}$ using the Tychonoff Theorem), $[0,1]^{\mathcal{B}(P)}$ is a compact space with respect to the product topology. Now for each $\bar{t} \in \mathcal{U}(P)$, define:

$$V(\bar{t}) = \{\|\|_{\mathbf{M}} \in [0,1]^{\mathcal{B}(P)} : \|P(\bar{t})\|_{\mathbf{M}} \leq r\}$$

Since P is quantifier-free and the propositional connectives $\rightarrow$ and $\bot$ are interpreted by continuous functions on $[0,1]$, each $V(\bar{t})$ is a closed subset of $[0,1]^{\mathcal{B}(P)}$. Consider:

$$V = \{V(\bar{t}) : \bar{t} \in \mathcal{U}(P)\}$$

By assumption, $\|(\exists \bar{x})P(\bar{x})\|_{\mathbf{M}} > r$ for all $\|\|_{\mathbf{M}} \in [0,1]^{\mathcal{B}(P)}$, so the intersection of the members of V is empty. Hence by the finite intersection property for compact spaces, some $\{V(\bar{t}_1), \ldots, V(\bar{t}_n)\} \subseteq V$ has an empty intersection; i.e. for each $\|\|_{\mathbf{M}} \in [0,1]^{\mathcal{B}(P)}$, $\|P(\bar{t}_i)\|_{\mathbf{M}} > r$ for some $i \in \{1, \ldots, n\}$. So $\models_{Ł}^{>r} \bigvee_{i=1}^{n} P(\bar{t}_i)$ as required. □

This approximate Herbrand theorem has a nice corollary. Let $F = (\forall \bar{x})(\exists \bar{y})P(\bar{x}, \bar{y})$ where P is both quantifier-free and function-free. Then $\models_{Ł} F$ iff $\models_{Ł} (\exists \bar{y})P(\bar{c}, \bar{y})$ for some new constants $\bar{c}$. Let $\mathcal{C}$ be the (finite) set of constants occurring in $(\exists \bar{y})P(\bar{c}, \bar{y})$, adding one if the set is empty. Using the previous theorem:

$$\begin{aligned} \models_{Ł} F \ \text{ iff } \ &\text{for all } r < 1: \ \models_{Ł}^{>r} \textstyle\bigvee_{i=1}^{n} P(\bar{c}, \bar{t}_i) \ \text{ for some } \bar{t}_1, \ldots, \bar{t}_n \in \mathcal{C} \\ \text{iff } \ &\models_{Ł} \textstyle\bigvee_{\bar{d} \in \mathcal{C}} P(\bar{c}, \bar{d}) \end{aligned}$$

But the validity problem for propositional Łukasiewicz Logic is decidable (Theorem 6.25), so we have established the following result:

Proposition 8.39. *The validity problem for function-free formulas* $(\forall \bar{x})(\exists \bar{y})P(\bar{x}, \bar{y})$ *is decidable.*

Moreover, suppose that A is a function-free formula with at most one bound variable. Then we can find a function-free existential formula $(\exists \bar{x})P(\bar{x})$ such that $\models_{Ł} A$ iff $\models_{Ł}$ $(\exists \bar{x})P(\bar{x})$. Just consider the following translations A^+ and A^-, assuming harmlessly that the ith occurrence of a quantifier Q in A is annotated as Q^i and that each a_i is a free variable not occurring in A:

$$\begin{aligned} p(\bar{x})^+ &= p(\bar{x}) & p(\bar{x})^- &= p(\bar{x}) \\ \bot^+ &= \bot & \bot^- &= \bot \\ (B \rightarrow C)^+ &= B^- \rightarrow C^+ & (B \rightarrow C)^- &= B^+ \rightarrow C^- \\ ((\forall^i x)B(x))^+ &= B(a_i)^+ & ((\forall x)^i B(x))^- &= (\forall x)B(x)^- \\ ((\exists^i x)B(x))^+ &= (\exists x)B(x)^+ & ((\exists x)^i B(x))^- &= B(a_i)^- \end{aligned}$$

A is a one-variable formula, so in each subformula $(\forall x)B(x)$ or $(\exists x)B(x)$ of A, any bound variable must be bound by a quantifier in the subformula. Hence using the continuity of the connectives in Ł∀ (pushing sups and infs outwards), it follows by a simple inductive proof that $\models_{Ł} A$ iff $\models_{Ł} A^+$. But using Ł∀-equivalences to push out the remaining quantifiers, A^+ is equivalent to an existential function-free formula. Hence, by the previous proposition:

Corollary 8.40. *The validity problem for function-free one-bound-variable formulas of Ł∀ is decidable.*

Example 8.41. Consider the one-bound-variable formula:

$$A = ((\exists x)p(x) \to (\exists x)q(x)) \to (\forall x)(p(x) \to q(x))$$

We annotate A and calculate:

$$\begin{aligned} A^+ &= ((\exists^1 x)p(x) \to (\exists^2 x)q(x))^- \to ((\forall^3 x)(p(x) \to q(x)))^+ \\ &= (((\exists^1 x)p(x))^+ \to ((\exists^2 x)q(x))^-) \to (p(a_3) \to q(a_3)) \\ &= ((\exists x)p(x) \to q(a_2)) \to (p(a_3) \to q(a_3)) \end{aligned}$$

But then, using quantifier-shifting equivalences:

$$A^+ \equiv_{Ł} (\exists x)((p(x) \to q(a_2)) \to (p(a_3) \to q(a_3)))$$

Now we can use the approximate Herbrand theorem to establish Skolemization for Ł∀. Since by Theorem 8.37, any formula has an equivalent prenex formula, both Skolemization and the approximate Herbrand theorem hold for all formulas of Ł∀.

Theorem 8.42. *Let A be a formula and $(\mathrm{Q}\bar{y})P(\bar{y})$ an equivalent prenex form for A. Let $(\exists\bar{x})P^F(\bar{x})$ be the Skolem form of $(\mathrm{Q}\bar{y})P(\bar{y})$. Then the following are equivalent:*

(i) $\models_{Ł} A$
(ii) $\models_{Ł} (\mathrm{Q}\bar{y})P(\bar{y})$
(iii) $\models_{Ł} (\exists\bar{x})P^F(\bar{x})$
(iv) *For all $r < 1$: $\models_{Ł}^{>r} \bigvee_{i=1}^{n} P^F(\bar{t}_i)$ for some $\bar{t}_1, \ldots, \bar{t}_n \in \mathcal{U}(P^F)$.*

Proof. The equivalence of (i) and (ii) is Theorem 8.37 and the equivalence of (iii) and (iv) is (the approximate Herbrand) Theorem 8.38. Since (ii) implies (iii) as shown in the proof of Theorem 8.36, it is enough to show that (iv) implies (ii). Suppose that for all $r < 1$, there exist $\bar{t}_1, \ldots, \bar{t}_n \in \mathcal{U}(P^F)$ such that $\models_{Ł}^{>r} \bigvee_{i=1}^{n} P^F(\bar{t}_i)$. Then by Lemma 8.34, $(\Rightarrow (\mathrm{Q}\bar{y})P(\bar{y}))$ is derivable from $(\Rightarrow P^F(\bar{t}_1) \mid \ldots \mid \Rightarrow P^F(\bar{t}_n))$ using (EC), (EW), $(\Rightarrow\exists)$, and the revised version of $(\Rightarrow\forall)$. Moreover, it is easily checked that for each instance of these rules with premise $\mathcal{G}$ and conclusion $\mathcal{H}$: if $\models_{Ł}^{\geq r} \mathcal{G}$, then $\models_{Ł}^{\geq r} \mathcal{H}$. So by a simple induction, $\models_{Ł}^{\geq r} (\mathrm{Q}\bar{y})P(\bar{y})$ for all $r < 1$. Hence $\models_{Ł} (\mathrm{Q}\bar{y})P(\bar{y})$ as required. □

Finally, we can use this approximate Herbrand theorem to sketch a proof of a special "approximate completeness" result for the Hilbert system HŁ∀. First notice that for any formula B and $k \in \mathbb{N}^+$:

$$\begin{aligned} 1 - 1/k \leq \|B\|^{\mathbf{M}} \quad &\text{iff} \quad \|\neg B\|^{\mathbf{M}} \leq 1/k \\ &\text{iff} \quad k\|\neg B\|^{\mathbf{M}} \leq 1 \\ &\text{iff} \quad (k-1)\|\neg B\|^{\mathbf{M}} \leq \|B\|^{\mathbf{M}} \\ &\text{iff} \quad 1 \leq \|((k-1).\neg B) \to B\|^{\mathbf{M}} \\ &\text{iff} \quad 1 \leq \|B \oplus B^{k-1}\|^{\mathbf{M}} \end{aligned}$$

Now for any formula A, let $(\exists\bar{x})P^F(\bar{x})$ be the Skolem form of a prenex formula equivalent to A. We note without proof that A is HŁ∀-derivable from $(\exists\bar{x})P^F(\bar{x})$.

Moreover, if $\models_{Ł} A$, then by the approximate Herbrand theorem, for all $k \in \mathbb{N}^+$:

$$\models_{Ł}^{\geq 1-1/k} \bigvee_{i=1}^{n} P^F(\bar{t}_i) \quad \text{for some } \bar{t}_1,\ldots,\bar{t}_n \in \mathcal{U}(P^F).$$

But then by the preceding bit of reasoning and the completeness of HŁ∀ with respect to propositional formulas, for all $k \in \mathbb{N}^+$:

$$\vdash_{\mathrm{HŁ\forall}} B \oplus B^{k-1} \quad \text{where } B = \bigvee_{i=1}^{n} P^F(\bar{t}_i) \quad \text{for some } \bar{t}_1,\ldots,\bar{t}_n \in \mathcal{U}(P^F).$$

However, $(\exists \bar{x})P^F(\bar{x})$ is HŁ∀-derivable from any such B. So $A \oplus A^{k-1}$ is HŁ∀-derivable from $B \oplus B^{k-1}$ for all $k \in \mathbb{N}^+$. Hence we arrive at the following result.

Theorem 8.43. $\models_{Ł} A$ *iff* $\vdash_{\mathrm{HŁ\forall}} A \oplus A^k$ *for all* $k \in \mathbb{N}^+$.

8.5.2 Gentzen Systems

A similar situation occurs when dealing with Gentzen systems for Ł∀. Consider the system GŁ∀, obtained by extending GŁ from Chapter 6 with the rules for the existential and universal quantifiers.

Example 8.44. It is easy to check that the axioms (∀1), (∀2), (∀3), (∃1), and (∃2) are GŁ∀-derivable. For example, for (∀2):

$$\dfrac{\dfrac{\dfrac{\dfrac{\dfrac{\dfrac{\dfrac{}{B(a) \Rightarrow B(a)}\,(\mathrm{ID}) \quad \dfrac{}{A \Rightarrow A}\,(\mathrm{ID})}{B(a), A \Rightarrow A, B(a)}\,(\mathrm{MIX})}{A \to B(a), A \Rightarrow B(a)}\,(\to\Rightarrow)_{\mathrm{A}}}{(\forall x)(A \to B), A \Rightarrow B(a)}\,(\forall\Rightarrow)}{(\forall x)(A \to B), A \Rightarrow (\forall x)B}\,(\Rightarrow\forall)}{(\forall x)(A \to B) \Rightarrow A \to (\forall x)B}\,(\Rightarrow\to)}{\Rightarrow (\forall x)(A \to B) \to (A \to (\forall x)B)}\,(\Rightarrow\to)$$

Moreover, GŁ∀ is sound with respect to the non-standard interpretation of Chapter 6, extended to the first-order level as follows:

Definition 8.45. $\models_{Ł} \mathcal{G}$ iff for all structures $\mathbf{M}$:

$$\sum_{C \in \Gamma} (\|C\|^{\mathbf{M}} - 1) \leq \sum_{D \in \Delta} (\|D\|^{\mathbf{M}} - 1) \qquad \text{for some } (\Gamma \Rightarrow \Delta) \in \mathcal{G}.$$

Theorem 8.46. *If* $d \vdash_{\mathrm{GŁ\forall}} \mathcal{G}$, *then* $\models_{Ł} \mathcal{G}$.

Proof. We proceed by induction on $\mathrm{ht}(d)$. The rules of GŁ were treated in the proof of Theorem 6.20, so it remains to consider the quantifier rules. For $(\Rightarrow\forall)$, suppose contrapositively that for some structure $\mathbf{M}$:

$$\sum_{C\in\Gamma}(\|C\|_{\mathbf{M}}-1)>(\|(\forall x)A(x)\|_{\mathbf{M}}-1)+\sum_{D\in\Delta}(\|D\|_{\mathbf{M}}-1)$$

Then this inequation holds also when $\|(\forall x)A(x)\|_{\mathbf{M}}$ is replaced with $\|A(x)\|_{\mathbf{M},m[x\to d]}$ for some $d\in M$. But now for some free variable a not occurring in Γ, Δ, or $A(x)$, we can extend $\mathbf{M}$ so that $\|A(x)\|_{\mathbf{M},m[x\to d]}=\|A(a)\|_{\mathbf{M}}$. Hence as required:

$$\sum_{C\in\Gamma}(\|C\|_{\mathbf{M}}-1)>(\|A(a)\|_{\mathbf{M}}-1)+\sum_{D\in\Delta}(\|D\|_{\mathbf{M}}-1)$$

For $(\forall\Rightarrow)$, clearly if $\sum_{C\in\Gamma}(\|C\|_{\mathbf{M}}-1)+(\|A(t)\|_{\mathbf{M}}-1)\leq\sum_{D\in\Delta}(\|D\|_{\mathbf{M}}-1)$ for some term t, then $\sum_{C\in\Gamma}(\|C\|_{\mathbf{M}}-1)+(\|(\forall x)A(x)\|_{\mathbf{M}}-1)\leq\sum_{D\in\Delta}(\|D\|_{\mathbf{M}}-1)$. The cases of the rules for $\exists$ are very similar. □

However, it does not follow from this that $\vdash_{\mathrm{GŁ}\forall}\Rightarrow A$ implies $\vdash_{\mathrm{HŁ}\forall} A$, since HŁ∀ is itself not complete for first-order Łukasiewicz Logic. This requires an interpretation, omitted here, of sequents in terms of abelian ℓ-groups and a correspondence with MV-algebras. The interested reader is referred to [24] for details.

On the other hand, it is easy to show that GŁ∀ extended with (CUT) is complete with respect to HŁ∀. We can derive all the axioms of HŁ∀ in GŁ∀, (MP) and (ADJ) are admissible as shown in Chapter 4, and (GEN) is admissible using $(\Rightarrow\forall)$.

Theorem 8.47. $\vdash_{\mathrm{HŁ}\forall}\mathrm{I}(\mathcal{G})$ *iff* $\vdash_{\mathrm{GŁ}\forall+(\mathrm{CUT})}\mathcal{G}$.

This is all very well. However, unfortunately, cut elimination fails for GŁ∀ + (CUT) and cancellation elimination fails for GŁ∀ + (CAN), so we do not have an analytic calculus even for the system HŁ∀. For example, $(\exists x)(\forall y)(p(x)\to p(y))$ has the following proof in GŁ∀ + (CAN):

$$\dfrac{\dfrac{\dfrac{\dfrac{\dfrac{\dfrac{\dfrac{\overline{p(a)\Rightarrow p(a)}\;(\mathrm{ID})}{(\forall z)p(z)\Rightarrow p(a)}\;(\forall\Rightarrow)\qquad\dfrac{\overline{p(a)\Rightarrow p(a)}\;(\mathrm{ID})\qquad\dfrac{\overline{p(b)\Rightarrow p(b)}\;(\mathrm{ID})}{(\forall z)p(z)\Rightarrow p(b)}\;(\forall\Rightarrow)}{(\forall z)p(z),p(a)\Rightarrow p(b),p(a)}\;(\mathrm{MIX})}{(\forall z)p(z)\Rightarrow p(a)\to p(b),p(a)}\;(\Rightarrow\to)_{Ł}}{(\forall z)p(z)\Rightarrow(\forall y)(p(a)\to p(y)),p(a)}\;(\Rightarrow\forall)}{(\forall z)p(z)\Rightarrow(\exists x)(\forall y)(p(x)\to p(y)),p(a)}\;(\Rightarrow\exists)}{(\forall z)p(z)\Rightarrow(\exists x)(\forall y)(p(x)\to p(y)),(\forall z)p(z)}\;(\Rightarrow\forall)}{\Rightarrow(\exists x)(\forall y)(p(x)\to p(y))}\;(\mathrm{CAN})$$

But no GŁ∀-derivation exists for this formula. A simple induction shows that in such a derivation there would be a branch where any hypersequent $\mathcal{G}$ satisfies the following property: $\mathcal{G}$ contains a sequent with an occurrence of a subformula of

$(\exists x)(\forall y)(p(x) \to p(y))$ on the right that does not occur on the left in any sequent in $\mathcal{G}$. But then the branch could not contain an initial sequent, a contradiction.

To obtain a calculus that is complete for the full first-order Łukasiewicz Logic, we need an *infinitary* rule. Let us reason as in the approximate completeness proof for the Hilbert system above. If $\models_{\text{Ł}} (\exists \bar{x})P(\bar{x})$, then for all $k \in \mathbb{N}^+$: $\vdash_{\text{GŁ}\forall} \bot \Rightarrow [B]^k$ where $B = \bigvee_{i=1}^{n} P^F(\bar{t}_i)$ for some $\bar{t}_1, \ldots, \bar{t}_n \in \mathcal{U}(P^F)$. But then using Lemma 8.34 k times, we can apply the quantifier and structural rules of GŁ$\forall$ to derive $(\bot \Rightarrow [A]^k)$ from $(\bot \Rightarrow [B]^k)$.

Theorem 8.48. *Let A be a prenex formula. Then $\models_{\text{Ł}} A$ iff A is derivable in* GŁ$\forall$ *extended with the rule:*

$$\frac{\bot \Rightarrow [A]^k \;\; \text{for all } k \in \mathbb{N}^+}{\Rightarrow A}$$

Example 8.49. Consider our earlier problematic formula $(\exists x)(\forall y)(p(x) \to p(y))$. To derive this in our calculus we would have to perform an infinite number of derivations in GŁ$\forall$. E.g in the case where $n = 2$, we have:

$$\dfrac{\dfrac{\dfrac{\dfrac{\dfrac{\dfrac{\dfrac{}{\bot \Rightarrow p(b) \to p(c)} \quad \dfrac{\dfrac{\dfrac{}{\bot \Rightarrow p(b)} \quad \dfrac{\dfrac{}{\bot \Rightarrow p(c)} \quad \dfrac{}{p(b) \Rightarrow p(b)}}{\bot, p(b) \Rightarrow p(b), p(c)}\,(\text{MIX})}{\bot \Rightarrow p(b), p(b) \to p(c)}\,(\Rightarrow\to)_{\text{Ł}}}{\bot, p(a) \Rightarrow p(b), p(b) \to p(c)}\,(\text{W})}{\bot \Rightarrow p(a) \to p(b), p(b) \to p(c)}\,(\Rightarrow\to)_{\text{Ł}}}{\bot \Rightarrow p(a) \to p(b), (\forall y)(p(b) \to p(y))}\,(\Rightarrow\forall)}{\bot \Rightarrow p(a) \to p(b), (\exists x)(\forall y)(p(x) \to p(y))}\,(\Rightarrow\exists)}{\bot \Rightarrow (\forall y)(p(a) \to p(y)), (\exists x)(\forall y)(p(x) \to p(y))}\,(\Rightarrow\forall)}{\bot \Rightarrow (\exists x)(\forall y)(p(x) \to p(y)), (\exists x)(\forall y)(p(x) \to p(y))}\,(\Rightarrow\exists)$$

We remark finally that it is possible to extend the above theorem to all hypersequents, proceeding similarly to Lemma 8.34. However, this is rather complicated and since all formulas admit a prenex form, we omit the details here.

8.6 Historical Remarks

Łukasiewicz Logic was the first fuzzy logic to be investigated at the first-order level. The disappointing but crucial result that the set of valid formulas is not recursively enumerable was established by Scarpellini in 1962 [192], and later sharpened by Ragaz in his 1981 thesis to Π_2-completeness [184]. Axiomatizations with an infinitary rule were obtained nevertheless, by Hay [117] and Belluce and Chang [32] in 1963 (see also [158, 31]). Pavelka's Łukasiewicz Logic with (rational) constants, developed in the late 1970s [181] was extended to the first-order level by Novák [167] and Hájek [105] in the 1990s. Interesting fragments of first-order Łukasiewicz

Logic have also been studied. The decidability result for the one-variable fragment given in this chapter was first obtained by Rutledge in 1959 [191]. Satisfiability for the monadic fragment (where all predicate symbols are nullary or unary) was shown to be Π_1-complete by Ragaz [184], but the complexity and even decidability for the valid formulas of this fragment remain important open problems. Finally, the axiomatization given in this chapter of the Σ_1-complete logic of (safe structures for) MV-chains is taken from Hájek's [105].

The first axiomatization and completeness proof for first-order Gödel Logic, as the logic based on linearly ordered Heyting algebras, was provided by Horn in 1969 [123]. An alternative axiomatization with the density rule was introduced independently by Takeuti and Titani in 1984 [205] as Intuitionistic Fuzzy Logic and used as the basis for a fuzzy set theory. Other axiomatizations closer to the presentation in this chapter were given by Takano [203] and Hájek [105]. First-order Gödel logics understood in a wider sense where the set of truth values forms a closed subset of $[0,1]$ including 0 and 1, and connectives are interpreted as in the standard algebra, have been studied intensively by Baaz, Preining, and Zach (see e.g. [26]). With this reading, infinite-valued propositional Gödel logics coincide but separate into infinitely many different logics at the first-order level. Finally, we mention that Corsi in 1992 [64] investigated the first-order logic characterized by formulas valid in linearly ordered Kripke models, axiomatized by removing the axiom $(\forall 3)$ from $\mathsf{HG}\forall$. This compares with the fact (shown by Gabbay in 1972 [85]) that the standard Gödel Logic studied in this chapter is the first-order logic of linearly ordered Kripke models with constant domains.

First-order Monoidal t-norm Logic was introduced by Esteva and Godo in 2001 [77]. Standard completeness for their axiomatization was established by Montagna and Ono a year later [155]. A general argument for the undecidability of the prenex fragments of first-order fuzzy logics, similar to the one presented here, was provided by Baaz et al. in [16]. Indeed it has been shown that of the fuzzy logics based on continuous t-norms, only Gödel Logic is recursively enumerable. Product Logic and Basic Logic, and uncountably many other such logics, do not even fit into the arithmetical hierarchy. The key results were proved by Montagna in 2001 [152] (see also [110] for a survey).

The axiomatic approach to first-order fuzzy logics based on restricting to safe structures described in this chapter is taken from Hajek's [105] and the 2006 paper of Cintula and Hájek [62] (see also the survey paper [61]). The completeness with respect to dense chains (sometimes called rational completeness) of such axiomatizations extended with the density rule is established in a more general setting in the 2008 paper of Ciabattoni and Metcalfe [55]. Our discussion of the formula $C = (\forall x)(p(x) \odot q) \to ((\forall x)p(x) \odot q)$ is based on two sources: first, a footnote in Hájek's [105], in which C is proved valid in all dense BL-chains, and second, a footnote in Esteva and Godo's [77], where a BL-chain (attributed to Bou) is given for which C is not valid.

There has been relatively little work on Gentzen systems for first-order fuzzy logics. The hypersequent calculus presented here for Gödel Logic was introduced by Baaz and Zach in 2000 [29] as an extension of the propositional system of

Avron [11]. The authors use the system to establish the mid-hypersequent theorem and hence the Herbrand theorem for the prenex fragment of this logic, and obtain a syntactic elimination of the density rule. An alternative proof of the Herbrand theorem is given by Baaz, Ciabattoni, and Fermüller in [14] together with general conditions (similar to those adopted here) for the prenex fragment to admit Skolemization. Many of these techniques and results (but not density elimination) were extended to first-order Monoidal t-norm Logic by Baaz, Ciabattoni, and Montagna in [17]. General conditions for single-conclusion hypersequent calculi to admit density elimination were provided in Ciabattoni and Metcalfe's [55].

Finally, the proof-theoretic results obtained here for Ł∀, in particular the approximate Herbrand theorem, Skolemization, and the infinitary hypersequent calculus, are taken from the papers of Baaz and Metcalfe [24, 25]. We note only that an alternative proof of the approximate Herbrand theorem was given by Novák in [165], and that related results have been obtained for so-called Continuous Logic (which allows any connective interpreted by a continuous function) (see e.g. [33]).

Chapter 9
Further Topics

This final chapter is devoted to awkward cases logics that do not quite fit the earlier presentations. There are two points of divergence here. On the one hand, we extend our standard languages for fuzzy logics with modalities and propositional quantifiers. On the other, we generalize the semantic basis of fuzzy logics to cover finite or non-linearly ordered sets of truth values, and non-commutative conjunctions. As a fitting end to both the chapter and the book, we then discuss open problems, in particular, the problematic case of Hájek's Basic Logic BL.

Due to the lack of homogeneity in this material, we proceed rather differently here to previous chapters. We offer the reader just a taste of each topic, and insert references and historical comments throughout the text as appropriate.

9.1 Modalities and Truth Stressers

Modalities – unary connectives that modify the meaning of sentences – play a key role in Logic, dealing with such diverse notions as necessity, obligation, provability, space, and time (see e.g. [44] for an authoritative survey). Fuzzy logics are no exception in this respect. There exist many different possible interpretations for modalities in these contexts, including the following special examples:

- *Truth stressers*. Natural language expressions like "very true" or "more or less true" that increase or decrease the truth value of a sentence are an important topic in Fuzzy Logic (see e.g. [106, 224]). Such "truth stresser" modalities can be modelled by reading $\Box A$ as e.g. "A is very true" and making use of axioms such as $\Box A \to A$ to capture properties like "A is very true" is less true than A".
- *Globalization*. One truth stresser widely studied in Fuzzy Logic (and other contexts) is "globalization" – also known as the (Baaz) Delta connective – where $\Box A$ is interpreted as "A is completely (classically) true" [13, 166, 206]. Globalization allows fuzzy logics to model not just vague sentences, but also sentences with classical or "crisp" truth values.

- *Exponentials*. Like the exponentials ! and ? of Linear Logic [10, 97] and modalities added to substructural logics [101, 185], $\Box$ can be used to specify properties of certain classes of formulas. For example, axioms such as $\Box A \to (\Box A \odot \Box A)$ permit the contraction of "boxed formulas". This extra flexibility can be used to embed other logics into the extended logic, just as Intuitionistic Logic can be embedded into Linear Logic (see e.g. [209]) or S4 (see e.g. [44]). Such modalities can also be used to encode useful semantic properties, an example being Montagna's "storage operator" for logics extending BL [153].

These examples have one thing in common, however. The intended interpretation of the $\Box$ modality is a unary function on the reals. Conversely, one of the characteristic features of modalities in Classical Logic is that modalities are *not* truth-functional. This is the case also in Fuzzy Logic for modalities such as necessity and probability (see e.g [105]). However here, since the above examples already provide plenty of motivation, we will restrict our attention to these truth-functional "truth stresser" interpretations.

9.1.1 Axioms and Algebras

The first step in investigating modal fuzzy logics is to expand our propositional languages with a unary connective $\Box$. E.g. restricting attention to the bounded language $\mathcal{L}_{\mathrm{B}}$, we let:

$$\mathcal{L}_{\Box} = \{\wedge, \vee, \odot, \to, \mathrm{f}, \mathrm{e}, \bot, \top, \Box\}$$

We mention also (but will not use) the dual connective $\Diamond A =_{\mathrm{def}} \neg\Box\neg A$, having a reciprocal meaning to the $\Box$ connective such as "quite true" or "possibly true".

The key rule for axiomatizing modal logics is "necessitation", which tells us that if a formula is true, then it is also necessarily true, very true, completely true, etc.

Definition 9.1 (Necessitation Rule).

$$\frac{A}{\Box A}\ (\textsc{nec})$$

A Hilbert System HK for the basic classical modal logic K is obtained by adding (NEC) and the following transitivity axiom schema to an axiomatization such as HCL of Classical Logic:

$$(\mathrm{K}_{\Box})\ \ \Box(A \to B) \to (\Box A \to \Box B)$$

However, axiomatizations for fuzzy modal logics (at least the truth-functional kind studied here), require also "shifting law of modalities" axioms to preserve the characteristic linearity of the truth value set:

$$(\vee_{\Box})\ \ \Box(A \vee B) \to (\Box A \vee \Box B)$$

Table 9.1 Modal axioms

Label	Axioms
$(K_\Box)$	$\Box(A \to B) \to (\Box A \to \Box B)$
$(\vee_\Box)$	$\Box(A \vee B) \to (\Box A \vee \Box B)$
$(T_\Box)$	$\Box A \to A$
$(4_\Box)$	$\Box A \to \Box\Box A$
$(C_\Box)$	$\Box A \to (\Box A \odot \Box A)$
$(W_\Box)$	$(\Box A \odot \Box B) \to \Box A$
$(S_\Box)$	$\Box A \vee \neg\Box A$

Table 9.2 Some fuzzy logics with modalities

Hilbert system	Axioms and rules
$\mathrm{HLK}^{\mathrm{r}}$	HL + $(K_\Box)$ + $(\vee_\Box)$ + (NEC)
$\mathrm{HLKT}^{\mathrm{r}}$	$\mathrm{HLK}^{\mathrm{r}}$ + $(T_\Box)$
$\mathrm{HLK4}^{\mathrm{r}}$	$\mathrm{HLK}^{\mathrm{r}}$ + $(4_\Box)$
$\mathrm{HLS4}^{\mathrm{r}}$	$\mathrm{HLK}^{\mathrm{r}}$ + $(T_\Box)$ + $(4_\Box)$
HL_Δ	$\mathrm{HLS4}^{\mathrm{r}}$ + $(S_\Box)$
HL!	$\mathrm{HLS4}^{\mathrm{r}}$ + $(W_\Box)$ + $(C_\Box)$

This can be compared with the "shifting law of quantifiers" axiom schema $(\forall 3)$ of the last chapter. Adding (NEC), $(K_\Box)$, and $(\vee_\Box)$ to a HUL-extension HL gives a Hilbert system $\mathrm{HLK}^{\mathrm{r}}$ that is sound and complete with respect to linearly ordered algebras (see below). The superscript "r" stands here for "representability", emphasizing that Hilbert systems defined without $(\vee_\Box)$ are not generally complete with respect to linearly-ordered algebras.

The modality $\Box$ can be characterized, as in the classical case, by adding further axioms such as those listed in Table 9.1 (referring to [44] for many other options). Adding $(T_\Box)$ and $(4_\Box)$ to HK gives an axiomatization for the modal logic S4; adding just $(T_\Box)$ gives KT and just $(4_\Box)$ gives K4. The other axioms only really make sense for substructural logics. For example, the weakening and contraction axioms for boxed formulas, $(C_\Box)$ and $(W_\Box)$, play an important role in Linear Logic, which can be axiomatized as HMALL extended with (NEC), $(K_\Box)$, $(T_\Box)$, $(4_\Box)$, $(W_\Box)$, and $(C_\Box)$ [10, 97]. Finally, the law of excluded middle for boxed formulas $(S_\Box)$ can be used to axiomatize logics with the globalization connective [13, 166, 206]. Table 9.2 lists some of these options for an arbitrary HUL-extension HL.[1]

Algebras for these logics are bpcrls extended with a unary operation $\Box$.

Definition 9.2. A $\mathrm{ULK}^{\mathrm{r}}$-*algebra* is an algebra $\langle L, \wedge, \vee, \odot, \to, \mathrm{e}, \mathrm{f}, \bot, \top, \Box\rangle$ where $\langle L, \wedge, \vee, \odot, \to, \mathrm{e}, \mathrm{f}, \bot, \top\rangle$ is a UL-algebra and $\Box$ is a unary operation satisfying:

[1] Note that it is more common in the literature to define modal fuzzy logics semantically via *Kripke models*. E.g. in [105] a logic K(L) is defined for a fuzzy logic L by defining Kripke models where the non-modal connectives are defined locally for each node of the model as in L. The question then is how to axiomatize such logics.

1. $\Box(x \to y) \leq \Box x \to \Box y$.
2. $\Box(x \vee y) = \Box x \vee \Box y$.
3. $\mathrm{e} \leq \Box \mathrm{e}$.

Other definitions and terminology carry over directly from Chapter 3. That is, valuations for UL^{r}-algebras are defined as for UL-algebras except that also $v(\Box A) = \Box v(A)$, and for any $\mathsf{HULK}^{\mathsf{r}}$-extension HL, an L-algebra $\mathbf{A}$ is any $\mathsf{ULK}^{\mathsf{r}}$-algebra such that the axioms of HL are $\mathbf{A}$-valid.

Establishing soundness and completeness for an $\mathsf{HULK}^{\mathsf{r}}$-extension HL with respect to L-algebras proceeds as in Chapter 3. Namely, we extend the definition of a Lindenbaum algebra $\mathsf{LIND}_T^{\mathrm{L}}$ with $\Box$, and show again that $T \vdash_{\mathrm{HL}} A$ iff A is $\mathsf{LIND}_T^{\mathrm{L}}$-valid. To obtain completeness for HL with respect to the class of L-chains $\mathsf{LIN}(\mathrm{L})$, the key step is to show that these logics retain the proof-by-cases property; i.e. if $T \cup \{A\} \vdash_{\mathrm{HL}} C$ and $T \cup \{B\} \vdash_{\mathrm{HL}} C$, then $T \cup \{A \vee B\} \vdash_{\mathrm{HL}} C$. As in Chapter 3, it is easy to prove this using a local deduction theorem, extending the notion of a confusion to that of a modal confusion.

Definition 9.3. A *modal confusion* of a theory T is defined inductively by:

(1) e, $\top$ (if in the language), and any element of T are confusions of T.
(2) If C_1 and C_2 are confusions of T, then so are $\Box C_1$, $C_1 \odot C_2$, and $C_1 \wedge C_2$.

Theorem 9.4. *Let* HL *be an* $\mathsf{HULK}^{\mathsf{r}}$*-extension:*

(a) $T \vdash_{\mathrm{HL}} A$ *for any modal confusion* A *of* T.
(b) $T \cup \{A\} \vdash_{\mathrm{HL}} B$ *iff* $T \vdash_{\mathrm{HL}} C \to B$ *for some modal confusion* C *of* $\{A\}$.
(c) $T \vdash_{\mathrm{HL}} B$ *iff* $\vdash_{\mathrm{HL}} C \to B$ *for some modal confusion* C *of* T.
(d) HL *has the proof-by-cases property.*

Proof. For (a), we proceed as in the proof of Lemma 3.42 by induction on $\mathrm{cp}(A)$. The only new case occurs when A is $\Box A'$ for some modal confusion A' of T. But then by the induction hypothesis, $T \vdash_{\mathrm{HL}} A'$ and so by (NEC), $T \vdash_{\mathrm{HL}} \Box A'$.

For (b) and (c), we show that $T_1 \cup T_2 \vdash_{\mathrm{HL}} B$ iff $T_1 \vdash_{\mathrm{HL}} C \to B$ for some modal confusion C of T_2. The right-to-left direction follows almost immediately using part (a). For the left-to-right direction, we proceed as in the proof of Theorem 3.43 by induction on the height of a derivation for $T_1 \cup T_2 \vdash_{\mathrm{HL}} B$. The only new case occurs when B is $\Box B'$ and the last step in the derivation is an application of (NEC), and $T_1 \cup T_2 \vdash_{\mathrm{HL}} B'$. Then by the induction hypothesis, $T_1 \vdash_{\mathrm{HL}} C' \to B'$ for some modal confusion C' of T_2. But then by (NEC), $T_1 \vdash_{\mathrm{HL}} \Box(C' \to B')$. Hence using ($\mathrm{K}_\Box$), $T_1 \vdash_{\mathrm{HL}} \Box C' \to \Box B'$ where $\Box C'$ is a modal confusion of T_2 as required.

Finally, for (d) we follow the same proof as in Lemma 3.54, replacing confusions with modal confusions. The only change comes in the proof of the claim: if A' is a modal confusion of $\{A\}$ and B' is a modal confusion of $\{B\}$, then $\vdash_{\mathrm{HL}} E \to (A' \vee B')$ for some modal confusion E of $\{A \vee B\}$, proved by induction on $\mathrm{cp}(A') + \mathrm{cp}(B')$. Suppose that A' is $\Box A_1$ and B' is $\Box B_1$ where A_1 and B_1 are modal confusions of $\{A\}$ and $\{B\}$, respectively. Then by the induction hypothesis $\vdash_{\mathrm{HL}} E' \to (A_1 \vee B_1)$ for some modal confusion E' of $\{A \vee B\}$. But then easily using ($\mathrm{K}_\Box$), $\vdash_{\mathrm{HL}}$

$\Box E' \to \Box(A_1 \vee B_1)$. So using $(\vee_\Box)$, $\vdash_{\text{HL}} \Box E' \to (\Box A_1 \vee \Box B_1)$ where $\Box E'$ is a modal confusion of $\{A \vee B\}$ as required. $\Box$

The same theorem holds with essentially the same proofs, if we replace HL with HL^{D}. The proofs of the following results then proceed exactly as in Chapter 3.

Theorem 9.5. *For any* $\mathsf{HULK}^{\mathsf{r}}$*-extension* HL*:*

$$T \vdash_{\text{HL}} A \;\; \textit{iff} \;\; T \models_{\mathsf{LIN}(\text{L})} A \qquad \textit{and} \qquad T \vdash_{\text{HL}^{\text{D}}} A \;\; \textit{iff} \;\; T \models_{\mathsf{DEN}(\text{L})} A.$$

9.1.2 Gentzen Systems

Hypersequent calculi for fuzzy logics with modalities (the representable kind) are easy enough to define. We simply add introduction rules for $\Box$ to our regular calculi plus extra structural rules characterizing its behaviour. For a finite multiset of formulas $\Gamma = [A_1, \ldots, A_n]$, let $\Box\Gamma$ stand for $[\Box A_1, \ldots, \Box A_n]$.

Definition 9.6 (Modal Logical Rules).

$$\frac{\mathcal{G} \mid \Gamma \Rightarrow A}{\mathcal{G} \mid \Box\Gamma \Rightarrow \Box A}\ (\Box) \qquad \frac{\mathcal{G} \mid \Gamma, A \Rightarrow \Delta}{\mathcal{G} \mid \Gamma, \Box A \Rightarrow \Delta}\ (\Box\Rightarrow) \qquad \frac{\mathcal{G} \mid \Box\Gamma \Rightarrow A}{\mathcal{G} \mid \Box\Gamma \Rightarrow \Box A}\ (\Rightarrow\Box)$$

Example 9.7. The $(\text{K}_\Box)$ axioms are derivable using the implication rules and $(\Box)$:

$$\dfrac{\dfrac{\dfrac{\dfrac{\dfrac{}{A \Rightarrow A}\,(\text{ID}) \quad \dfrac{}{B \Rightarrow B}\,(\text{ID})}{A \to B, A \Rightarrow B}\,(\to\Rightarrow)}{\Box(A \to B), \Box A \Rightarrow \Box B}\,(\Box)}{\Box(A \to B) \Rightarrow \Box A \to \Box B}\,(\Rightarrow\to)}{\Rightarrow \Box(A \to B) \to (\Box A \to \Box B)}\,(\Rightarrow\to)$$

It is also easy to see that $(\text{T}_\Box)$ and $(4_\Box)$ are derivable using the implication rules with $(\Box\Rightarrow)$ and $(\Rightarrow\Box)$.

The rules $(\Box)$, $(\Box\Rightarrow)$, and $(\Rightarrow\Box)$ are just hypersequent versions of rules familiar from sequent calculi for the modal logics K, KT, K4, and S4 (see e.g. [168, 174]). However, all is not quite what it seems here. The step up to hypersequents allows more modal formulas to be proved than in the classical case.

Example 9.8. The axiom $(\vee_\Box)$, which is not derivable even in the modal logic S4, is derivable here using $(\Box)$ together with the usual logical rules, (COM), (EW), and (EC), the top hypersequent being derived as in Example 4.35:

Table 9.3 Matching rules for modal axioms

Rule	Matching axioms
$(\Box\Rightarrow)$	$(\mathrm{T}_\Box)$ $\Box A \to A$
$(\Rightarrow\Box)$	$(4_\Box)$ $\Box A \to \Box\Box A$
$(\mathrm{W})_\Box$	$(\mathrm{W}_\Box)$ $(\Box A \odot \Box B) \to \Box A$
$(\mathrm{C})_\Box$	$(\mathrm{C}_\Box)$ $\Box A \to (\Box A \odot \Box A)$
$(\mathrm{SPLIT})_\Box$	$(\mathrm{S}_\Box)$ $\Box A \vee \neg\Box A$

$$\dfrac{\dfrac{\dfrac{\dfrac{A\vee B\Rightarrow A \mid A\vee B\Rightarrow B}{A\vee B\Rightarrow A \mid \Box(A\vee B)\Rightarrow \Box B}\,(\Box)}{\Box(A\vee B)\Rightarrow \Box A \mid \Box(A\vee B)\Rightarrow \Box B}\,(\Box)}{\Box(A\vee B)\Rightarrow \Box A\vee\Box B}\,(\Rightarrow\vee)}{\Rightarrow \Box(A\vee B)\to(\Box A\vee\Box B)}\,(\Rightarrow\to)$$

One way to think of this is that (COM) corresponds not to the prelinearity and distributivity axioms, but to the completeness of the logic with respect to chains.

The structural behaviour of $\Box$ is characterized further by adding "modal versions" of the usual structural rules. Let us just consider a selection of these:

Definition 9.9 (Modal Structural Rules).

$$\dfrac{\mathcal{G} \mid \Gamma\Rightarrow\Delta}{\mathcal{G} \mid \Gamma,\Box\Pi\Rightarrow\Delta}\,(\mathrm{W})_\Box \qquad \dfrac{\mathcal{G} \mid \Gamma,\Box\Pi,\Box\Pi\Rightarrow\Delta}{\mathcal{G} \mid \Gamma,\Box\Pi\Rightarrow\Delta}\,(\mathrm{C})_\Box \qquad \dfrac{\mathcal{G} \mid \Box\Gamma,\Pi\Rightarrow\Sigma}{\mathcal{G} \mid \Box\Gamma\Rightarrow \mid \Pi\Rightarrow\Sigma}\,(\mathrm{SPLIT})_\Box$$

The structural rules $(\mathrm{C})_\Box$ and $(\mathrm{W})_\Box$ are hypersequent versions of rules used for the exponential ! in Linear Logic [97].

Example 9.10. $(\mathrm{SPLIT})_\Box$ ensures that boxed formulas obey the law of excluded middle $(\mathrm{S}_\Box)$:

$$\dfrac{\dfrac{\dfrac{\dfrac{}{\Box A\Rightarrow\Box A}\,(\mathrm{ID})}{\Rightarrow\Box A \mid \Box A\Rightarrow}\,(\mathrm{SPLIT})_\Box}{\Rightarrow\Box A \mid \Rightarrow\neg\Box A}\,(\Rightarrow\neg)}{\Rightarrow\Box A\vee\neg\Box A}\,(\Rightarrow\vee)$$

However, note that for the interpretation where $\Box x$ is $\top$ for $x=\top$ and $\bot$ otherwise in logics without weakening, a slightly different rule is needed:

$$\dfrac{\mathcal{G} \mid \Box\Gamma_1,\Pi\Rightarrow\Sigma}{\mathcal{G} \mid \Box\Gamma_1,\Gamma_2\Rightarrow\Delta \mid \Pi\Rightarrow\Sigma}\,(\mathrm{SPLIT})_\Box^w$$

These rules are connected with matching axioms as listed in Table 9.3. Soundness and completeness results with respect to matching systems follow as in Chapter 4.

Theorem 9.11. *For an* HUL*-extension* HL *with matching Gentzen system* GL, *let* $\mathrm{GL}_{\Box}$ *be* GL *plus* ($\Box$) *and any selection of rules from Table 9.3 and let* $\mathrm{HL}_{\Box}$ *be* $\mathrm{HLK}^{\mathrm{r}}$ *plus the matching axioms for these rules. Then* $\vdash_{\mathrm{GL}_{\Box}} \mathcal{G}$ *iff* $\vdash_{\mathrm{HL}_{\Box}} \mathrm{I}(\mathcal{G})$.

Cut elimination for these systems proceeds essentially as in Chapter 5 (although note that the rules for $\Box$ require a little more care); we refer to [56] for details. This paper also includes a number of algebraic standard completeness proofs for modal logics based on HMTL-extensions. The density elimination method also works in this context, but we leave the details for the reader to investigate.

9.1.3 Embeddings

One motivation for extending a logic with modalities is the opportunity to embed other logics into this extended logic. Historically, the most famous examples are the various embeddings of Intuitionistic Logic into S4, also extended to intermediate logics and extensions of S4 (see e.g. [44]), and of Intuitionistic Logic again into Linear Logic (see e.g. [209]). In this section we show how the proof theory developed above allows us to embed certain fuzzy logics into other fuzzy logics with S4-like modalities.

Let us consider first two mappings found in the literature on modal and substructural logics (see e.g. [185, 209]), where a is any atom:

$$\begin{aligned} a^* &= a & a^{\Box} &= \Box a \\ (A \to B)^* &= \Box A^* \to B^* & (A \to B)^{\Box} &= \Box(A^{\Box} \to B^{\Box}) \\ (A \odot B)^* &= \Box A^* \odot \Box B^* & (A \odot B)^{\Box} &= \Box(A^{\Box} \odot B^{\Box}) \\ (A \wedge B)^* &= A^* \wedge B^* & (A \wedge B)^{\Box} &= A^{\Box} \wedge B^{\Box} \\ (A \vee B)^* &= A^* \vee B^* & (A \vee B)^{\Box} &= A^{\Box} \vee B^{\Box} \end{aligned}$$

These two mappings can be related as follows:

Lemma 9.12. *For any* $\mathrm{HULS4}^{\mathrm{r}}$*-extension* HL*:*

(a) $\vdash_{\mathrm{HL}} \Box A^* \leftrightarrow A^{\Box}$.

(b) $\vdash_{\mathrm{HL}} A^*$ *iff* $\vdash_{\mathrm{HL}} A^{\Box}$.

Proof. We prove (a) by induction on $\mathrm{cp}(A)$. If A is atomic, then the result follows by definition. If A is $B \to C$, then $\Box A^* = \Box(\Box B^* \to C^*)$. It is easy to show that $\vdash_{\mathrm{HL}} \Box(\Box B^* \to C^*) \leftrightarrow \Box(\Box B^* \to \Box C^*)$. Hence, using the induction hypothesis twice, $\vdash_{\mathrm{HL}} \Box(\Box B^* \to C^*) \leftrightarrow \Box(B^{\Box} \to C^{\Box})$ as required. If A is $B \odot C$, then $\Box A^* = \Box(\Box B^* \odot \Box C^*)$ and the result follows by the induction hypothesis twice. Suppose that A is $B \star C$ for $\star \in \{\wedge, \vee\}$. Then $\Box(A \star B)^* = \Box(B^* \star C^*)$. In each case, $\vdash_{\mathrm{HL}} \Box(B^* \star C^*) \leftrightarrow \Box B^* \star \Box C^*$. Hence, using the induction hypothesis twice, $\vdash_{\mathrm{HL}} \Box(B^* \star C^*) \leftrightarrow (B^{\Box} \star C^{\Box})$ as required. (b) follows from (a), since $\vdash_{\mathrm{HL}} A^*$ iff (using (NEC) and ($\mathrm{T}_{\Box}$)) $\vdash_{\mathrm{HL}} \Box A^*$ iff $\vdash_{\mathrm{HL}} A^{\Box}$. $\Box$

Suppose now that HL_1 is HUL extended with axiom schema taken from (SPLIT), (W), and (C). Let $\mathrm{HL}_2^\Box$ be $\mathsf{HIULS4^r}$ extended with the corresponding modal axiom schema from $(\mathrm{SPLIT})_\Box$, $(\mathrm{W})_\Box$, and $(\mathrm{C})_\Box$. Then we can embed HL_1 into $\mathrm{HL}_2^\Box$ using either of the above mappings. For concreteness, let us consider a direct analogue of the embedding of Intuitionistic Logic into Linear Logic. We will embed Gödel Logic into the "Fuzzy Linear Logic" IUL! (see Table 9.2).

Theorem 9.13. *Let $\mathcal{G}$ be a single-conclusion hypersequent in the language $\mathcal{L}_{\mathsf{G}} = \{\wedge, \vee, \to, \bot, \top\}$. Then $\vdash_{\mathsf{GG}} \mathcal{G}$ iff $\vdash_{\mathsf{GIUL!}} \mathcal{G}^*$ where:*

$$\Gamma^* = [A^* : A \in \Gamma] \quad \textit{and} \quad (\Gamma_1 \Rightarrow \Delta_1 \mid \ldots \mid \Gamma_n \Rightarrow \Delta_n)^* = \Box\Gamma_1^* \Rightarrow \Delta_1^* \mid \ldots \mid \Box\Gamma_n^* \Rightarrow \Delta_n^*$$

Proof. For the left-to-right direction, we proceed by induction on the height of a GG°-derivation of $\mathcal{G}$. The base cases are easy. E.g. if $\mathcal{G}$ is $\mathcal{H} \mid A \Rightarrow A$, then $\mathcal{G}^*$ is $\mathcal{H}^* \mid \Box A^* \Rightarrow A^*$ which is GIUL!-derivable using $(\Box\Rightarrow)$ and (ID). Notice also that for the inductive step, if the last application of a rule is (EW), (EC), or (COM), we can just apply the induction hypothesis and the same rule for GIUL!. Similarly, for (W) or (C), we can just apply the induction hypothesis and then the corresponding modal rule. For the logical rules, let us just consider the implication rules.

Suppose that the derivation ends with:

$$\frac{\mathcal{H} \mid \Pi, A \Rightarrow B}{\mathcal{H} \mid \Pi \Rightarrow A \to B}\ (\Rightarrow\to)$$

By the induction hypothesis, we obtain a GIUL!-derivation ending with:

$$\frac{\mathcal{H}^* \mid \Box\Pi^*, \Box A^* \Rightarrow B^*}{\mathcal{H}^* \mid \Box\Pi^* \Rightarrow \Box A^* \to B^*}\ (\Rightarrow\to)$$

Suppose that the derivation ends with:

$$\frac{\mathcal{H} \mid \Pi_1 \Rightarrow A \quad \mathcal{H} \mid \Pi_2, B \Rightarrow \Sigma}{\mathcal{H} \mid \Pi_1, \Pi_2, A \to B \Rightarrow \Sigma}\ (\to\Rightarrow)$$

Using the induction hypothesis twice, we obtain a GIUL!-derivation ending with:

$$\frac{\dfrac{\mathcal{H}^* \mid \Box\Pi_1^* \Rightarrow A^*}{\mathcal{H}^* \mid \Box\Pi_1^* \Rightarrow \Box A^*}\ (\Rightarrow\Box) \quad \mathcal{H}^* \mid \Box\Pi_2^*, \Box B^* \Rightarrow \Sigma^*}{\mathcal{H}^* \mid \Box\Pi_1^*, \Box\Pi_2^*, \Box A^* \to \Box B^* \Rightarrow \Sigma^*}\ (\to\Rightarrow)$$

But $\vdash_{\mathsf{GIUL!}} \mathcal{H}^* \mid \Box(\Box A^* \to B^*) \Rightarrow \Box A^* \to \Box B^*$, so by (CUT) we have as required:

$$\vdash_{\mathsf{GIUL!}} \mathcal{H}^* \mid \Box\Pi_1^*, \Box\Pi_2^*, \Box(\Box A^* \to B^*) \Rightarrow \Sigma^*$$

For the right-to-left direction, suppose that $\vdash_{\mathsf{GIUL!}} \mathcal{G}^*$. We claim that there exists a derivation of $\mathcal{G}^*$ in $\mathsf{GIUL!}^\circ$ where $(\Rightarrow\to)$ is restricted to single-conclusion sequents. Using cut elimination for GIUL! and the invertibility of the rules $(\Rightarrow\to)$, $(\Rightarrow\vee)$, and $(\Rightarrow\wedge)$, we obtain a cut-free derivation of $\mathcal{G}^*$ where $(\rightarrow\Rightarrow)$ is restricted to cases where all sequents in the conclusion contain only atoms and boxed formulas. It then follows by an easy induction that all sequents in such a derivation of $\mathcal{G}^*$ are either of this special form or single-conclusion. But this means that $(\Rightarrow\rightarrow)$ is restricted to single-conclusion sequents. Then finally we can show by induction on the height of a derivation that if $\mathcal{G}^*$ is derivable in GIUL! with the restricted use of $(\Rightarrow\rightarrow)$, then $\mathcal{G}$ is derivable in GG. □

Corollary 9.14. $\vdash_{\mathsf{G}} A$ *iff* $\vdash_{\mathsf{IUL!}} A^*$ *iff* $\vdash_{\mathsf{IUL!}} A^{\Box}$.

9.2 Propositional Quantifiers

We turn now to another interesting extension of languages. The addition of propositional quantifiers $\forall p$ and $\exists p$ – intuitively "for all propositions p" and "for some proposition p" – can be useful for expressing properties of truth value sets. For example, instead of expressing density using a rule, we might use "density axioms" of the form:

$$(\forall p)((A \rightarrow p) \vee (p \rightarrow B)) \rightarrow (A \rightarrow B)$$

More generally, this extra flexibility can be used to express topological properties of sets of real numbers such as the existence of limit points or successor elements.

In Classical Logic, propositional quantifiers are little more than a notational convenience: $(\forall p)A(p)$ and $(\exists p)A(p)$ are interpreted by infima and suprema over the truth value set $\{0,1\}$ and can therefore be defined as just $A(\bot) \wedge A(\top)$ and $A(\bot) \vee A(\top)$. However, in non-classical logics propositional quantifiers can increase expressive power quite considerably, an important example of this being Quantified Intuitionistic Logic [86]. For fuzzy logics, we will consider only the already tricky case of Quantified Gödel Logic, introduced by Baaz, Fermüller, and Veith in [21] (see also [15, 18, 28]).

Let us make use of the language $\mathcal{L}_{\mathrm{G}} = \{\wedge, \vee, \rightarrow, \bot, \top\}$ and add to the definition of a formula the condition that if A is a formula and p a variable, then $(\forall p)A$ and $(\exists p)A$ are formulas. As in the first-order case, we distinguish between bound variables, denoted p, q, and free variables, denoted a, b. We also write $A(\bar{a})$ to denote that the free variables of A are among those in $\bar{a}$, using $A(B)$ to denote distinguished occurrences of a formula B in A.

G-valuations are extended to such formulas as follows, where for each valuation v and $r \in [0,1]$, $v[r/p]$ is defined as $v[r/p](q) = r$ if $q = p$ and $v[r/p](q) = v(q)$ otherwise:

$$v((\exists p)A) = \sup\{v[r/p](A) : r \in [0,1]\}$$
$$v((\forall p)A) = \inf\{v[r/p](A) : r \in [0,1]\}$$

A formula A is QG-*valid*, written $\models_{\mathrm{QG}} A$, if $v(A) = 1$ for all G-valuations v.

A Hilbert system for QG is obtained as an extension of an axiomatization of Quantified Intuitionistic Logic defined by Gabbay in [86].

Definition 9.15. HQG consists of the axioms and rules of HG extended with:

$$
\begin{array}{ll}
(\mathrm{Q}\exists)\ A(B) \to (\exists q)A(q) & \\
(\forall \mathrm{Q})\ (\forall q)A(q) \to A(B) & \\
(\mathrm{Q}\vee)\ (\forall q)(A \vee B) \to (A \vee (\forall q)B) & q \text{ not occurring in } A \\
(\mathrm{QD})\ (\forall q)((A \to q) \vee (q \to B)) \to (A \to B) & q \text{ not occurring in } A \text{ or } B
\end{array}
$$

$$
\frac{A(a) \to B}{(\exists q)A(q) \to B}\ (\exists \mathrm{Q}) \qquad \frac{A \to B(a)}{A \to (\forall q)B(q)}\ (\mathrm{Q}\forall)
$$

where a does not occur in A or B.

It is straightforward to check that HQG is sound with respect to QG-validity. However, the crucial result for HQG is the following (we omit the rather involved proof):

Theorem 9.16 ([28]). HQG *admits* quantifier elimination*: for every formula A there is a quantifier-free formula B whose variables occur in A such that* $\vdash_{\mathsf{HQG}} A \leftrightarrow B$.

By Theorem 3.56, $\models_{\mathsf{QG}} B$ iff $\vdash_{\mathsf{HQG}} B$ for any quantifier-free formula B. Hence, combining this with quantifier elimination:

Corollary 9.17. $\vdash_{\mathsf{HQG}} A$ *iff* $\models_{\mathsf{QG}} A$.

In fact it follows from the proof of Theorem 9.16 in [28] that this last result holds for HQG even when the B in $(\forall\mathrm{Q})$ and $(\mathrm{Q}\exists)$ are restricted to quantifier-free formulas. Moreover, as shown in [27], quantifier elimination also provides an easy proof of the interpolation property for Gödel Logic.

Corollary 9.18. *Gödel Logic* G *admits* uniform interpolation*; i.e. if* $\vdash_{\mathsf{HG}} A \to B$*, then there exists a formula C with variables occurring in both A and B such that* $\vdash_{\mathsf{HG}} A \to C$ *and* $\vdash_{\mathsf{HG}} C \to B$.

Proof. Suppose that $\vdash_{\mathsf{HG}} A(\bar{a},\bar{b}) \to B(\bar{b},\bar{c})$ where $\bar{a}$ and $\bar{b}$ and $\bar{b}$ and $\bar{c}$ are the variables occurring in A and B, respectively. Then easily $\vdash_{\mathsf{HQG}} A(\bar{a},\bar{b}) \to (\exists \bar{p})A(\bar{p},\bar{b})$ and $\vdash_{\mathsf{HQG}} (\exists \bar{p})A(\bar{p},\bar{b}) \to B(\bar{b},\bar{c})$. But then by quantifier elimination, there exists a propositional formula $C(\bar{b})$ such that $\vdash_{\mathsf{HQG}} (\exists \bar{p})A(\bar{p},\bar{b}) \leftrightarrow C(\bar{b})$ as required. □

A hypersequent calculus for QG is obtained rather easily. We add quantifier rules to the standard calculus for G, and, crucially in this case, the hypersequent version of the density rule:

Definition 9.19. GQG consists of GG extended with:

$$
\frac{\mathcal{G} \mid \Gamma, A(B) \Rightarrow \Delta}{\mathcal{G} \mid \Gamma, (\forall p)A(p) \Rightarrow \Delta}\ (\forall\Rightarrow)_{\mathrm{Q}} \qquad \frac{\mathcal{G} \mid \Gamma \Rightarrow A(a)}{\mathcal{G} \mid \Gamma \Rightarrow (\forall p)A(p)}\ (\Rightarrow\forall)_{\mathrm{Q}}
$$

$$
\frac{\mathcal{G} \mid \Gamma, A(a) \Rightarrow \Delta}{\mathcal{G} \mid \Gamma, (\exists p)A(p) \Rightarrow \Delta}\ (\exists\Rightarrow)_{\mathrm{Q}} \qquad \frac{\mathcal{G} \mid \Gamma \Rightarrow A(B)}{\mathcal{G} \mid \Gamma \Rightarrow (\exists p)A(p)}\ (\Rightarrow\exists)_{\mathrm{Q}}
$$

$$
\frac{\mathcal{G} \mid \Gamma_1, a \Rightarrow \Delta \mid \Gamma_2 \Rightarrow a}{\mathcal{G} \mid \Gamma_1, \Gamma_2 \Rightarrow \Delta}\ (\text{DENSITY})
$$

where B is quantifier-free in $(\forall\Rightarrow)_{\mathrm{Q}}$ and $(\Rightarrow\exists)_{\mathrm{Q}}$, and a does not occur in the premises of $(\Rightarrow\forall)_{\mathrm{Q}}$ or $(\exists\Rightarrow)_{\mathrm{Q}}$ or the conclusion of (DENSITY).

Example 9.20. We illustrate this calculus with a derivation of the density axioms (for space reasons, omitting the easy derivations of $A, A \to a \Rightarrow A$ and $a, a \to B \Rightarrow B$):

$$
\dfrac{\dfrac{\dfrac{\dfrac{\dfrac{\dfrac{\dfrac{A \to a, A \Rightarrow a \qquad \dfrac{\dfrac{A, A \to a \Rightarrow a \quad a \to B, a \Rightarrow B}{a \to B, A \Rightarrow a \mid A \to a, a \Rightarrow B}(\textsc{com}) \qquad a \to B, a \Rightarrow B}{a \to B, A \Rightarrow a \mid (A \to a) \vee (a \to B), a \Rightarrow B}(\vee\Rightarrow)^*}{(A \to a) \vee (a \to B), A \Rightarrow a \mid (A \to a) \vee (a \to B), a \Rightarrow B}(\vee\Rightarrow)^*}{(A \to a) \vee (a \to B), A \Rightarrow a \mid (\forall q)((A \to q) \vee (q \to B)), a \Rightarrow B}(\forall\Rightarrow)_{\mathrm{Q}}}{(\forall q)((A \to q) \vee (q \to B)), A \Rightarrow a \mid (\forall q)((A \to q) \vee (q \to B)), a \Rightarrow B}(\forall\Rightarrow)_{\mathrm{Q}}}{(\forall q)((A \to q) \vee (q \to B)), (\forall q)((A \to q) \vee (q \to B)), A \Rightarrow B}(\textsc{density})}{(\forall q)((A \to q) \vee (q \to B)), A \Rightarrow B}(\textsc{c})}{(\forall q)((A \to q) \vee (q \to B)) \Rightarrow A \to B}(\Rightarrow\to)}{\Rightarrow (\forall q)((A \to q) \vee (q \to B)) \to (A \to B)}(\Rightarrow\to)
$$

Notice that the contraction rule (C) is essential for this derivation. To find axiomatizations of quantified fuzzy logics lacking contraction, alternative density axioms are required, or the presence of the density rule itself.

Soundness and completeness proofs for GQG follow the usual pattern. It is easy to see that the quantifier rules preserve validity, so an induction on the height of a derivation gives soundness, while completeness follows as before from the fact that the extra axioms of HQG are derivable, and the extra rules are admissible.

Theorem 9.21. $\vdash_{\mathsf{GQG}} \mathcal{G}$ *iff* $\vdash_{\mathsf{QG}} \mathrm{I}(\mathcal{G})$.

Cut elimination also follows the same pattern as described in Chapter 5, but here there are a couple of important extra points to consider.

Theorem 9.22. *Cut elimination holds for* GQG.

Proof. As in Theorem 5.9, the result is a consequence of the following:

Claim. For any hypersequent $\mathcal{G}$ and hypersequent $\mathcal{H}$ with marked formula A:

$$\text{If } d_{\mathcal{G}} \vdash_{\mathsf{GQG}^\circ} \mathcal{G} \text{ and } d_{\mathcal{H}} \vdash_{\mathsf{GQG}^\circ} \mathcal{H}, \text{ then } \vdash_{\mathsf{GQG}^\circ} \mathrm{CUT}(\mathcal{G}, \mathcal{H}).$$

However, this time we have to add an extra parameter to our induction hypothesis to cope with the fact that the complexity of the cut-formula treated can increase when stepping e.g. from $(\forall p)B(p)$ to $B(C)$. We let $\mathrm{q}(A)$ be the number of quantifier occurrences in A, and prove the claim by induction on the lexicographically ordered quadruple:

$$\langle \mathrm{q}(A), \mathrm{cp}(A), \mathrm{e}(d_{\mathcal{H}}), \mathrm{ht}(d_{\mathcal{G}}) \rangle$$

Notice first that we can assume, similarly to the first-order case, that new variables introduced by the density rule are completely new, i.e. do not occur elsewhere in the derivations of $\mathcal{G}$ and $\mathcal{H}$. Given this assumption, cases involving the density rule proceed in the same way as for other structural rules. Where we really need the extra parameter in the induction hypothesis is the case where both branches end with a universal (or existential) quantifier rule applied to an occurrence of $A = (\forall p)B(p)$ (or $A = (\exists p)B(p)$). E.g.

$$\dfrac{\dfrac{\vdots}{\mathcal{G}' \mid \Gamma, B(C), [(\forall p)B(p)]^{n-1} \Rightarrow \Delta}}{\mathcal{G}' \mid \Gamma, [(\forall p)B(p)]^{n} \Rightarrow \Delta}\ (\forall\Rightarrow)_{\mathsf{Q}} \qquad \dfrac{\dfrac{\vdots}{\mathcal{H}' \mid \Pi \Rightarrow B(a)}}{\mathcal{H}' \mid \Pi \Rightarrow (\forall p)B(p)}\ (\Rightarrow\forall)_{\mathsf{Q}}$$

Consider a member of $\mathrm{CUT}(\mathcal{G},\mathcal{H})$ of the form (the only tricky case):

$$\mathcal{G}'' \mid \Gamma, \Pi^{n} \Rightarrow \Delta$$

where $\mathcal{G}'' \in \mathrm{CUT}(\mathcal{G}',\mathcal{H})$. Then by the induction hypothesis:

$$\vdash_{\mathsf{GQG}^{\circ}} \mathcal{G}'' \mid \Gamma, B(C), \Pi^{n-1} \Rightarrow \Delta$$

We can substitute C for a in the derivation of $\mathcal{H}' \mid \Pi \Rightarrow B(a)$ (an easy induction) to obtain a derivation of $\mathcal{H}' \mid \Pi \Rightarrow B(C)$. But $\mathrm{q}(B(C)) < \mathrm{q}((\forall p)B(p))$. So by the induction hypothesis again and (EC), $\vdash_{\mathsf{GQG}^{\circ}} \mathcal{G}'' \mid \Gamma, \Pi^{n} \Rightarrow \Delta$ as required. □

As already remarked, density elimination does not hold for this system. However, GQG without (DENSITY) does enjoy cut elimination. Indeed, it is conjectured in [15] that this system corresponds to the intersection of all the finite-valued propositional quantified Gödel logics.

As at the first-order level, infinite-valued Gödel logics with truth values different to $[0,1]$ (such as, e.g. $\{1-1/n : n \in \mathbb{N}^{+}\}$) that coincide with G at the propositional level, can diverge when extended with propositional quantifiers. In fact, there are uncountably many different quantified Gödel logics. For an account of this and many other interesting facts, we refer to [18].

Finally, note that for other fuzzy logics such as MTL, it is not hard to establish soundness, completeness, and cut elimination for corresponding hypersequent calculi extended as above with respect to a suitably extended Hilbert system. The tricky part is to show that such calculi are sound and complete in turn with respect to the semantics for propositional quantification. The only other fuzzy logic investigated at this level so far is propositional quantified Łukasiewicz Logic, studied from a purely semantic perspective in [5].

9.3 Non-Commutative Logics

One of the most natural assumptions made in this book is the commutativity of conjunction: that "*A* and *B*" and "*B* and *A*" always have the same truth value. However, dropping this assumption has played an important role in the development of substructural logics. The Lambek calculus – a sequent calculus with sequents as ordered pairs of sequences that lacks exchange rules – was introduced by Lambek in the 1950s to model the assignment of types to linguistic expressions in natural language [130]. Related systems have also been widely studied, including the extension (with additive connectives) to the Full Lambek Calculus FL (see e.g. [176]) and non-commutative versions of Linear Logic (see e.g. [1, 190]). In Algebra, many important classes such as lattice-ordered groups and residuated lattices are non-commutative [66, 127, 210]. The non-commutative case has also received attention in Fuzzy Logic, spawning generalizations of MV-algebras [91, 94] and related classes of algebras [108, 109, 126]. Here we consider briefly the possible implications and interest of non-commutativity for the proof theory of fuzzy logics.

9.3.1 Residuated Lattices

Residuated lattices, originating in the work of Ward and Dilworth in the 1930s [217], provide a suitable algebraic semantics for a wide range of substructural logics, encompassing also other important classes of algebras such as lattice-ordered groups. A thorough investigation of these algebras and their scope, unifying previous approaches, has been undertaken by Blount, Jipsen, and Tsinakis in [127, 210]. Pointed residuated lattices (often called FL-algebras after the Full Lambek Calculus), their subvarieties and corresponding logics, are also the subject of the 2007 monograph of Galatos, Jipsen, Kowalski, and Ono [90].

Definition 9.23. A *pointed residuated lattice (prl for short)* is an algebra:

$$\mathbf{A} = \langle L_{\mathbf{A}}, \wedge, \vee, \odot, \backslash, /, \mathrm{e}, \mathrm{f}\rangle$$

with universe $L_{\mathbf{A}}$, binary operations $\wedge$, $\vee$, $\odot$, $\backslash$, $/$, and constants e, f such that:

1. $\langle L_{\mathbf{A}}, \wedge, \vee\rangle$ is a lattice.
2. $\langle L_{\mathbf{A}}, \odot, \mathrm{e}\rangle$ is a monoid.
3. $x \odot y \leq z$ iff $x \leq z/y$ iff $y \leq x\backslash z$ for all $x, y, z \in L_{\mathbf{A}}$.

Bounded pointed residuated lattices (bprls) are algebras $\langle L_{\mathbf{A}}, \wedge, \vee, \odot, \backslash, /, \mathrm{e}, \mathrm{f}, \bot, \top\rangle$ such that $\langle L_{\mathbf{A}}, \wedge, \vee, \odot, \backslash, /, \mathrm{e}, \mathrm{f}\rangle$ is a prl with top and bottom elements $\top$ and $\bot$.

As in the commutative case, the classes of residuated lattices $\mathcal{RL}$ and bounded residuated lattices $\mathcal{BRL}$ can be identified respectively with the classes of prls $\mathcal{RL}^+$ and bprls $\mathcal{BRL}^+$ satisfying $\mathrm{f} = \mathrm{e}$. Moreover, as shown in [210], all of these classes are varieties, axiomatized by the equations for (bounded) lattices and monoids of Chapter 2 together with:

$$\begin{array}{ll}
x = x \wedge (((x \odot y) \vee z)/y) & x \odot (y \vee z) = (x \odot y) \vee (x \odot z) \\
y = y \wedge (x\backslash((x \odot y) \vee z)) & (y \vee z) \odot x = (y \odot x) \vee (z \odot x) \\
x = x \vee ((x/y) \odot y) & x = x \vee (y \odot (y\backslash x))
\end{array}$$

$$
\begin{array}{rlrl}
(\mathrm{id}_l) & A\backslash A & (\odot\wedge) & ((A\wedge \mathrm{e})\odot(B\wedge \mathrm{e}))\backslash(A\wedge B) \\
(\mathrm{id}_l) & (A\backslash B)\backslash((C\backslash A)\backslash(C\backslash B)) & (\wedge\backslash) & (A\wedge B)\backslash A \\
(\mathrm{as}_{ll}) & A\backslash((B/A)\backslash A) & (\wedge\backslash) & (A\wedge B)\backslash B \\
(\mathrm{a}) & ((B\backslash C)/A)\backslash(B\backslash(C/A)) & (\vee\backslash) & A\backslash(A\vee B) \\
(\odot\backslash/) & ((B\odot(B\backslash A))/B)\backslash(A/B) & (\vee\backslash) & B\backslash(A\vee B) \\
(\backslash\odot) & B\backslash(A\backslash(A\odot B)) & (\backslash\wedge) & ((A\backslash B)\wedge(A\backslash C))\backslash(A\backslash(B\wedge C)) \\
(\odot\backslash) & (B\backslash(A\backslash C))\backslash((A\odot B)\backslash C) & (\vee\backslash) & ((A\backslash C)\wedge(B\backslash C))\backslash((A\vee B)\backslash C) \\
(\mathrm{e}) & \mathrm{e} & (\mathrm{e}\backslash) & \mathrm{e}\backslash(A\backslash A) \\
(\backslash\mathrm{e}) & A\backslash(\mathrm{e}\backslash A) & (\backslash\top) & A\backslash\top \\
(\bot\backslash) & \bot\backslash A & &
\end{array}
$$

$$
\frac{A \quad A\backslash B}{B}\ (\mathrm{mp}_l) \qquad \frac{A \quad B}{A\wedge B}\ (\mathrm{ADJ}) \qquad \frac{A}{B\backslash(A\odot B)}\ (\mathrm{pn}_l) \qquad \frac{A}{(B\odot A)/B}\ (\mathrm{pn}_r)
$$

Fig. 9.1 The Hilbert system HBFL

For prls satisfying the commutativity equation $x\odot y = y\odot x$, the residuals $\backslash$ and $/$ collapse to one operation and are hence (unsurprisingly) term equivalent to pcrls.

Important non-commutative varieties of prls include $\mathcal{LG}$, the class of residuated lattices satisfying $x\odot(x\backslash\mathrm{e}) = \mathrm{e}$, term equivalent to lattice-ordered groups, and, particularly interesting from the fuzzy logic perspective, the variety generated by *bounded pointed residuated chains* $\mathcal{BRL}^{+\mathcal{C}}$.

Theorem 9.24 ([210]). *$\mathcal{BRL}^{+\mathcal{C}}$ is the class of bprls satisfying:*

$$\mathrm{e} = \lambda_u((x\vee y)\backslash x)\vee\rho_v((x\vee y)\backslash y)$$

where $\lambda_u(x) = (u\backslash(x\odot u))\wedge\mathrm{e}$ and $\rho_u(x) = ((u\odot x)/u)\wedge\mathrm{e}$, called the left and right conjugates of x with respect to u, *respectively.*

As in the commutative case, we can speak about "standard" bprls where $L_{\mathbf{A}}$ is the real unit interval $[0,1]$ with the usual order. In this case the operation $\odot$ could be called a *residuated pseudo uninorm*: an increasing associative binary function on $[0,1]$ with unit element e and residuals $\backslash$ and $/$. In particular, when $\mathrm{e}=1$, $\odot$ is a *pseudo t-norm*. Such functions and their algebras have been investigated in several papers; see e.g. [82, 108, 109].

9.3.2 Hilbert Systems

Hilbert systems are developed for classes of residuated lattices much as in the commutative case. For simplicity of exposition, we will just consider bprls, making use of the language $\mathcal{L}_{\mathrm{N}} = \{\wedge,\vee,\odot,\to,\backslash,/,\mathrm{e},\mathrm{f},\bot,\top\}$. Our starting point is then the Hilbert system HBFL taken from [90] and displayed in Fig. 9.1.

The correspondence between classes of bprls and HBFL-extensions (axiomatic extensions of HBFL) is developed exactly as in Chapter 3. Let $\mathbf{A}$ be a bprl. Then as before, an $\mathbf{A}$-*valuation* is a function $v : \mathrm{Fm}_{\mathcal{L}_{\mathrm{N}}} \to L_{\mathbf{A}}$ satisfying $v(\star(A_1,\ldots,A_n)) =$

$\star(v(A_1),\ldots,v(A_n))$ for each n-ary $\star \in \mathcal{L}_N$ and a formula A is $\mathbf{A}$-*valid* if $v(A) \geq \mathrm{e}$ for all $\mathbf{A}$-valuations v. Also an $\mathbf{A}$-valuation v is an $\mathbf{A}$-*model* of an $\mathcal{L}_N$-theory T if $v(A) \geq \mathrm{e}$ for all $A \in T$, and we write $T \models_{\mathbf{A}} A$ if every $\mathbf{A}$-model of T is an $\mathbf{A}$-model of $\{A\}$. For a class of algebras $\mathcal{K}$, we write $T \models_{\mathcal{K}} A$ if $T \models_{\mathbf{A}} A$ for all $\mathbf{A} \in \mathcal{K}$. Then for any HFL-extension HL, a bprl $\mathbf{A}$ is an L-*algebra* iff all the axioms of HL are $\mathbf{A}$-valid, and GEN(L) is the set of all L-algebras.

Following the same steps as in Chapter 3:

Theorem 9.25. *$T \vdash_{\mathrm{HL}} A$ iff $T \models_{\mathsf{GEN}(\mathrm{L})} A$ for any* HBFL*-extension* HL.

A Hilbert system for the variety $\mathcal{BRL}^{+\mathcal{C}}$ of bounded pointed residuated chains is obtained by adding axioms corresponding to the condition in Theorem 9.24.

Definition 9.26. HpsUL consists of HBFL extended with:

$$\text{(psPRL)} \quad (C\backslash(((A \vee B)\backslash A) \odot C)) \vee ((C \odot ((A \vee B)\backslash B))/C)$$

It is easy to see that a bprl is a psUL-algebra iff it satisfies the condition in Theorem 9.24. So letting LIN(L) be the set of all L-chains and either using this theorem or proceeding as in Chapter 3 using the proof-by-cases property:

Theorem 9.27. *$T \vdash_{\mathrm{HL}} A$ iff $T \models_{\mathsf{LIN}(\mathrm{L})} A$ for any* HpsUL*-extension* HL.

9.3.3 Gentzen Systems

We will illustrate the extension of proof theory to non-commutative logics with a base case: a Gentzen system for the logic of bprls. First, a new definition is needed for sequents to reflect the fact that we now care about the ordering of formulas. For convenience and familiarity (and just for this section), let Γ,Δ,Π,Σ represent *sequences* of formulas, and let Γ,A and Γ,Π denote the concatenation of Γ and (A), and Γ and Π, respectively. We also write $\Gamma[A]$ and $\Gamma[\Pi]$ for sequences Γ of the form Γ_1,A,Γ_2 and Γ_1,Π,Γ_2.

Definition 9.28. A *non-commutative sequent* is an ordered pair of finite sequences of formulas, written $\Gamma \Rightarrow \Delta$, and a *non-commutative hypersequent* is a finite multiset of non-commutative sequents.

The single-conclusion (non-commutative) hypersequent calculus GpsUL is displayed in Fig. 9.2. Its sequent version is essentially the Bounded Full Lambek Calculus BFL, a calculus for the variety of bprls studied in particular by Ono and co-authors in [90, 176]. Moreover, a calculus for UL is obtained by adding the exchange rule:

$$\frac{\mathcal{G} \mid \Gamma_1,\Pi_2,\Pi_1,\Gamma_2 \Rightarrow \Delta}{\mathcal{G} \mid \Gamma_1,\Pi_1,\Pi_2,\Gamma_2 \Rightarrow \Delta}\ (\text{EX})$$

Initial Sequents

$$\frac{}{\Gamma[\bot] \Rightarrow \Delta}\ (\bot) \qquad \frac{}{\Gamma \Rightarrow \top}\ (\Rightarrow\top) \qquad \frac{}{A \Rightarrow A}\ (\text{ID})$$

Structural Rules

$$\frac{\mathcal{G}}{\mathcal{G} \mid \mathcal{H}}\ (\text{EW}) \qquad \frac{\mathcal{G} \mid \mathcal{H} \mid \mathcal{H}}{\mathcal{G} \mid \mathcal{H}}\ (\text{EC}) \qquad \frac{\mathcal{G} \mid \Gamma_1[\Pi_2] \Rightarrow \Delta \quad \mathcal{G} \mid \Gamma_2[\Pi_1] \Rightarrow \Sigma}{\mathcal{G} \mid \Gamma_1[\Pi_1] \Rightarrow \Delta \mid \Gamma_2[\Pi_2] \Rightarrow \Sigma}\ (\text{COM})$$

Logical Rules

$$\frac{\mathcal{G} \mid \Gamma \Rightarrow \Delta}{\mathcal{G} \mid \Gamma[\mathrm{e}] \Rightarrow \Delta}\ (\mathrm{e}\Rightarrow) \qquad \frac{}{\mathcal{G} \mid \Rightarrow \mathrm{e}}\ (\Rightarrow\mathrm{e})$$

$$\frac{}{\mathcal{G} \mid \mathrm{f} \Rightarrow}\ (\mathrm{f}\Rightarrow) \qquad \frac{\mathcal{G} \mid \Gamma \Rightarrow}{\mathcal{G} \mid \Gamma \Rightarrow \mathrm{f}}\ (\Rightarrow\mathrm{f})$$

$$\frac{\mathcal{G} \mid \Pi \Rightarrow A \quad \mathcal{G} \mid \Gamma[B] \Rightarrow \Delta}{\mathcal{G} \mid \Gamma[\Pi, A\backslash B] \Rightarrow \Delta}\ (\backslash\Rightarrow) \qquad \frac{\mathcal{G} \mid A, \Gamma \Rightarrow B}{\mathcal{G} \mid \Gamma \Rightarrow A\backslash B}\ (\Rightarrow\backslash)$$

$$\frac{\mathcal{G} \mid \Pi \Rightarrow A \quad \mathcal{G} \mid \Gamma[B] \Rightarrow \Delta}{\mathcal{G} \mid \Gamma[B/A, \Pi] \Rightarrow \Delta}\ (/\Rightarrow) \qquad \frac{\mathcal{G} \mid \Gamma, A \Rightarrow B}{\mathcal{G} \mid \Gamma \Rightarrow B/A}\ (\Rightarrow/)$$

$$\frac{\mathcal{G} \mid \Gamma[A, B] \Rightarrow \Delta}{\mathcal{G} \mid \Gamma[A \odot B] \Rightarrow \Delta}\ (\odot\Rightarrow) \qquad \frac{\mathcal{G} \mid \Gamma_1 \Rightarrow A \quad \mathcal{G} \mid \Gamma_2 \Rightarrow B}{\mathcal{G} \mid \Gamma_1, \Gamma_2 \Rightarrow A \odot B}\ (\Rightarrow\odot)$$

$$\frac{\mathcal{G} \mid \Gamma[A] \Rightarrow \Delta}{\mathcal{G} \mid \Gamma[A \wedge B] \Rightarrow \Delta}\ (\wedge\Rightarrow)_1 \qquad \frac{\mathcal{G} \mid \Gamma[B] \Rightarrow \Delta}{\mathcal{G} \mid \Gamma[A \wedge B] \Rightarrow \Delta}\ (\wedge\Rightarrow)_2$$

$$\frac{\mathcal{G} \mid \Gamma[A] \Rightarrow \Delta \quad \mathcal{G} \mid \Gamma[B] \Rightarrow \Delta}{\mathcal{G} \mid \Gamma[A \vee B] \Rightarrow \Delta}\ (\vee\Rightarrow) \qquad \frac{\mathcal{G} \mid \Gamma \Rightarrow A \quad \mathcal{G} \mid \Gamma \Rightarrow B}{\mathcal{G} \mid \Gamma \Rightarrow A \wedge B}\ (\Rightarrow\wedge)$$

$$\frac{\mathcal{G} \mid \Gamma \Rightarrow A}{\mathcal{G} \mid \Gamma \Rightarrow A \vee B}\ (\Rightarrow\vee)_1 \qquad \frac{\mathcal{G} \mid \Gamma \Rightarrow B}{\mathcal{G} \mid \Gamma \Rightarrow A \vee B}\ (\Rightarrow\vee)_2$$

Cut Rule

$$\frac{\mathcal{G} \mid \Gamma[A] \Rightarrow \Delta \quad \mathcal{G} \mid \Pi \Rightarrow A}{\mathcal{G} \mid \Gamma[\Pi] \Rightarrow \Delta}\ (\text{CUT})$$

Fig. 9.2 The hypersequent calculus GpsUL

Example 9.29. The key prelinearity axioms of HpsUL are derivable in GpsUL as follows, where the top hypersequent is derived exactly as in Example 4.35 (recall that $(\vee\Rightarrow)$ is derivable using $(\vee\Rightarrow)_1$, $(\vee\Rightarrow)_2$, and (EC), and $(\Rightarrow\odot)^*$ denotes the combination of $(\Rightarrow\odot)$ with applications of (EC) and (EW)):

$$\dfrac{\dfrac{}{C \Rightarrow C}\,(\text{ID}) \quad \dfrac{\dfrac{}{C \Rightarrow C}\,(\text{ID}) \quad \dfrac{\dfrac{A \vee B \Rightarrow A \mid A \vee B \Rightarrow B}{A \vee B \Rightarrow A \mid \Rightarrow (A \vee B)\backslash B}\,(\Rightarrow/)}{\Rightarrow (A \vee B)\backslash A \mid \Rightarrow (A \vee B)\backslash B}\,(\Rightarrow/)}{\Rightarrow (A \vee B)\backslash A \mid C \Rightarrow C \odot ((A \vee B)\backslash B)}\,(\Rightarrow\odot)^*}{\dfrac{C \Rightarrow ((A \vee B)\backslash A) \odot C \mid C \Rightarrow C \odot ((A \vee B)\backslash B)}{\dfrac{C \Rightarrow ((A \vee B)\backslash A) \odot C \mid \Rightarrow (C \odot ((A \vee B)\backslash B))/C}{\dfrac{\Rightarrow C\backslash(((A \vee B)\backslash A) \odot C) \mid \Rightarrow (C \odot ((A \vee B)\backslash B))/C}{\Rightarrow (C\backslash(((A \vee B)\backslash A) \odot C)) \vee ((C \odot ((A \vee B)\backslash B))/C)}\,(\Rightarrow\vee)}\,(\Rightarrow\backslash)}\,(\Rightarrow/)}\,(\Rightarrow\odot)^*$$

Soundness and completeness proofs with respect to HpsUL follow the same procedure as in Chapter 4. We keep the same definition of a confusion of a set of formulas, but redefine the standard interpretation of a sequent as $\mathrm{I}(\Gamma \Rightarrow \Delta) = \odot\Gamma \setminus \oplus\Delta$. We show that for each rule $\mathcal{G}_1 \ldots \mathcal{G}_n \ / \ \mathcal{G}$ of GpsUL that $\vdash_{\mathsf{HpsUL}} C\backslash\mathrm{I}(\mathcal{G})$ for some confusion C of $\{\mathrm{I}(\mathcal{G}_1), \ldots, \mathrm{I}(\mathcal{G}_n)\}$. The soundness of GpsUL then follows as usual by an induction on the height of a derivation. For completeness, we check that each axiom of HpsUL is GpsUL-derivable and that the rules are GpsUL-admissible. The rest of the proof follows exactly as before.

Theorem 9.30. $\vdash_{\mathsf{GpsUL}} \mathcal{G}$ *iff* $\vdash_{\mathsf{HpsUL}} \mathrm{I}(\mathcal{G})$.

The general characterization of cut elimination in terms of substitutivity and reductivity can be extended to cover non-commutative single-conclusion calculi (see e.g. [207] for extensions of FL with sequent structural rules). Here for clarity, we just sketch the particular case of GpsUL.

Theorem 9.31. *Cut elimination holds for* GpsUL.

Proof. First, we generalize the definition of $\mathrm{CUT}(\mathcal{G},\mathcal{H})$ for a (possibly marked) hypersequent $\mathcal{G}$ and marked hypersequent $\mathcal{H}$ in the obvious way to deal with sequences rather than multisets. Then as before it is sufficient to establish the following:

Claim. For any hypersequent $\mathcal{G}$ and hypersequent $\mathcal{H}$ with marked formula A:

$$\text{If } d_{\mathcal{G}} \vdash_{\mathsf{GpsUL}^\circ} \mathcal{G} \text{ and } d_{\mathcal{H}} \vdash_{\mathsf{GpsUL}^\circ} \mathcal{H}, \text{ then} \vdash_{\mathsf{GpsUL}^\circ} \mathrm{CUT}(\mathcal{G},\mathcal{H}).$$

We proceed by induction on the lexicographically ordered triple $\langle \mathrm{cp}(A), \mathrm{e}(d_{\mathcal{H}}), \mathrm{ht}(d_{\mathcal{G}})\rangle$, recalling that $\mathrm{e}(d)$ is 0 if d ends with a logical rule applied to a marked formula, and 1 otherwise. If $d_{\mathcal{G}}$ does not end with a rule application having A as principal formula, then we can use the induction hypothesis and the substitutivity (suitably revised) of the rule. For example, suppose that $\mathcal{H} = \mathcal{H}' \mid \Pi \Rightarrow \underline{A}$ and $d_{\mathcal{G}}$ ends with:

$$\dfrac{\mathcal{G}_1 \mid \Gamma_1(A), \Sigma_2(A), \Gamma_3(A) \Rightarrow \Delta_1 \quad \mathcal{G}_1 \mid \Sigma_1(A), \Gamma_2(A), \Sigma_3(A) \Rightarrow \Delta_2}{\mathcal{G}_1 \mid \Gamma_1(A), \Gamma_2(A), \Gamma_3(A) \Rightarrow \Delta_1 \mid \Sigma_1(A), \Sigma_2(A), \Sigma_3(A) \Rightarrow \Delta_2}\,(\text{COM})$$

where $\Gamma(A)$ indicates some but perhaps not all occurrences of A in Γ. Then the required derivation for a member of $\mathrm{CUT}(\mathcal{G},\mathcal{H})$ is of the form:

$$\frac{\mathcal{G}'_1 \mid \Gamma_1(\Pi),\Sigma_2(\Pi),\Gamma_3(\Pi) \Rightarrow \Delta_1 \quad \mathcal{G}'_1 \mid \Sigma_1(\Pi),\Gamma_2(\Pi),\Sigma_3(\Pi) \Rightarrow \Delta_2}{\mathcal{G}'_1 \mid \Gamma_1(\Pi),\Gamma_2(\Pi),\Gamma_3(\Pi) \Rightarrow \Delta_1 \mid \Sigma_1(\Pi),\Sigma_2(\Pi),\Sigma_3(\Pi) \Rightarrow \Delta_2} \;(\textsc{com})$$

where $\mathcal{G}'_1 \in \mathrm{CUT}(\mathcal{G}_1,\mathcal{H})$ and the premises are derivable by the induction hypothesis.

Suppose now that both $d_{\mathcal{G}}$ and $d_{\mathcal{H}}$ end with applications of rules where the marked formula A in $\mathcal{H}$ is principal. Then we can use the reductivity (again, suitable revised) of the logical rules. For example, suppose that $A = B\backslash C$ and $d_{\mathcal{G}}$ and $d_{\mathcal{H}}$ end with, respectively:

$$\frac{\mathcal{G}_1 \mid \Gamma_2(A) \Rightarrow B \quad \mathcal{G}_1 \mid \Gamma_1(A),C,\Gamma_3(A) \Rightarrow \Delta}{\mathcal{G}_1 \mid \Gamma_1(A),\Gamma_2(A),B\backslash C,\Gamma_3(A) \Rightarrow \Delta} \qquad \frac{\mathcal{H}' \mid B,\Pi \Rightarrow C}{\mathcal{H}' \mid \Pi \Rightarrow B\backslash C}$$

Then using the induction hypothesis and (EW), the following are GpsUL°-derivable:

$$\mathcal{G}'_1 \mid \Gamma_2(\Pi) \Rightarrow B \qquad \mathcal{G}'_1 \mid \Gamma_1(\Pi),C,\Gamma_3(\Pi) \Rightarrow \Delta \qquad \mathcal{G}'_1 \mid B,\Pi \Rightarrow C$$

where $\mathcal{G}'_1 \in \mathrm{CUT}(\mathcal{G}_1,\mathcal{H})$. But now again by the induction hypothesis, we can cut on the smaller complexity subformulas, first on B and then on C, to get that $\mathcal{G}'_1 \mid \Gamma_1(\Pi),\Gamma_2(\Pi),\Pi,\Gamma_3(\Pi) \Rightarrow \Delta$ is GpsUL°-derivable as required. □

Adding structural rules such as weakening and contraction to GpsUL does not spoil cut elimination. However, the multiple-conclusion case is problematic already at the sequent level and requires further investigation.

The obvious question here is whether the systems HpsUL and GpsUL are standard complete. A semantic proof of this fact has been given in [126] for HpsMTL, the extension of HpsUL with weakening axioms of the form $(A\backslash \mathrm{e}) \wedge (\mathrm{f}\backslash A)$. However, there seems to be no difficulty also to extend the density elimination approach to the non-commutative case. As a starting point we should adapt the Hilbert system version of the density rule to:

$$\frac{(A\backslash p) \vee (p\backslash B) \vee C}{(A\backslash B) \vee C}$$

In other respects, the proofs should be fairly similar to the commutative case.

There are in fact a number of further interesting questions here. For example, can we find Gentzen systems for the important (non-commutative) classes of lattice-ordered groups or representable lattice-ordered groups? For such classes, the standard set of logical rules will not be sufficient. We will require a set of rules closer to those given for Abelian Logic in Chapter 5.

9.4 Finite-Valued Logics

In previous chapters, we assumed a continuum of truth values: typically, the real unit interval $[0,1]$. However, sometimes a finite set of values is enough. In this case we could try restricting a t-norm or uninorm based fuzzy logic to:

$$\left\{0, \frac{1}{n-1}, \ldots, \frac{n-2}{n-1}, 1\right\} \qquad \text{for } n = 2, 3, \ldots$$

For Ł and G, this gives us the well-known families of n-valued Łukasiewicz and Gödel logics. For P, this only works if $n = 2$ (when we get Classical Logic) since for $n \geq 3$, the set is not closed under multiplication.

More generally, an n-valued logic is fixed by any algebra with n elements, where some elements are distinguished as "true". Surprisingly perhaps, there exists a procedure to find a Gentzen-style system for *any* such logic, as well as more elegant calculi for particular cases.

9.4.1 Logical Matrices

Finite-valued logics are usually defined via *logical matrices*. The method is close to the algebraic style presentation, but typically just one algebra is specified and the "true" truth values are determined arbitrarily (i.e. not just as those greater than a particular element). More precisely:

Definition 9.32. A *logical matrix* $\mathcal{M} = [\mathcal{N}, \mathcal{D}, \mathcal{C}]$ for a language $\mathcal{L}$ consists of:

1. A non-empty set of *truth-values* $\mathcal{N}$.
2. A set $\mathcal{D} \subseteq \mathcal{N}$ of *designated truth values*.
3. A set of *truth-functions* $\mathcal{C} = \{\star^i : \mathcal{N}^m \to \mathcal{N} : \star \in \mathcal{L} \text{ with arity } m\}$.

$\mathrm{L} = (\mathcal{L}, \mathcal{M})$ is called a *(matrix-defined) propositional many-valued logic*. If $\mathcal{N}$ contains exactly n elements, L is called an *n-valued logic* and in this case, we also assume without loss of generality that $\mathcal{N}$ is a *sequence* $(w_0, \ldots, w_{n-1})$.

L-*valuations* are functions $v : \mathrm{Fm}_{\mathcal{L}} \to \mathcal{N}$ satisfying:

$$v(\star(A_1, \ldots, A_m)) = \star^i(v(A_1), \ldots, v(A_m))$$

and $A \in \mathrm{Fm}_{\mathcal{L}}$ is L-*valid*, written $\models_{\mathrm{L}} A$, iff $v(A) \in \mathcal{D}$ for all L-valuations v.

Example 9.33. Three-valued Łukasiewicz Logic Ł3 based on the language $\{\neg, \to\}$ is defined by the matrix:

$$\left[\left(0, \frac{1}{2}, 1\right), \{1\}, \{\neg^i, \to^i\}\right]$$

where the functions $\neg^i$ and $\to^i$ are defined using the truth tables:

$\neg^i$	
1	0
$\frac{1}{2}$	$\frac{1}{2}$
0	1

$\to^i$	1	$\frac{1}{2}$	0
1	1	$\frac{1}{2}$	0
$\frac{1}{2}$	1	1	$\frac{1}{2}$
0	1	1	1

More generally, n-valued Łukasiewicz logics $\mathsf{Ł_n}$ for $n = 2,3,\ldots$ based on the language $\mathcal{L}_{\mathrm{T}} = \{\odot, \to, \bot\}$ are defined by the matrices:

$$\left[\left(0, \frac{1}{n-1}, \ldots, \frac{n-2}{n-1}, 1\right), \{1\}, \{\to^i, \odot^i, \bot^i\}\right]$$

where $\odot^i$ and $\to^i$ are the Łukasiewicz t-norm and its residuum respectively, and $\bot^i = 0$. Similarly, n-valued Gödel logics are defined by the same matrix where $\odot^i$ and $\to^i$ are the Gödel t-norm and its residuum. Of course, letting $\mathcal{N}$ be $[0,1]$ in these cases gives us matrix presentations of the fuzzy logics Ł and G.

Truth values do not have to be numbers (although if the logic is finite-valued, we can always number them). The logic FDE mentioned way back in Chapter 2 has the usual truth values 0 and 1 plus b for "both true and false" and n for "neither true nor false". For the language $\{\neg, \wedge, \vee\}$, the FDE matrix is:

$$[\{0, \mathrm{n}, \mathrm{b}, 1\}, \{\mathrm{b}, 1\}, \{\neg^i, \wedge^i, \vee^i\}]$$

with truth-functions:

$\neg^i$	
1	0
b	b
n	n
0	1

$\wedge^i$	1	b	n	0
1	1	b	n	0
b	b	b	0	0
n	n	0	n	0
0	0	0	0	0

$\vee^i$	1	b	n	0
1	1	1	1	1
b	1	b	1	b
n	1	1	n	n
0	1	b	n	0

Interestingly, FDE has no valid formulas: for any formula A, the valuation sending all variables to n, also sends A to n.

9.4.2 n-sequents

There is a vast literature on proof methods for finite-valued logics. As well as tailored systems for particular cases, there are several general approaches which provide a calculus for *any* finite-valued logic.[2] These appear in a number of frameworks – (labelled) tableaux [104], (labelled) sequents [102], n-sequents [20, 222], etc. – but the underlying ideas (first proposed by Schröter [194] in the 1950s and Rousseau [188, 189] in the 1960s) are very similar in each case.

Here we will briefly describe the n-sequent approach, following the elegant presentation of Zach in [222]. While Classical Logic requires two "slots" (think of the two sides of a sequent, or a label T or F in a tableaux system), an n-valued logic requires n slots: putting a formula into the ith slot means that it has the ith truth value. This motivates the following definition:

2 One such approach will even rather cynically output a LATEX paper presenting the calculus; see http://www.logic.at/multlog.

Definition 9.34. An *n-sequent* S for a language $\mathcal{L}$ is an n-tuple of finite sets of $\mathcal{L}$-formulas, written:

$$T_0 \mid \ldots \mid T_{n-1}$$

Definition 9.35. Let $\mathrm{L} = (\mathcal{L}, [\mathcal{N}, \mathcal{D}, \mathcal{C}])$ be an n-valued logic with $\mathcal{N} = (w_0, \ldots, w_{n-1})$. Then S is *satisfied* by an L-valuation v iff $v(A) = w_i$ for some $A \in T_i$ and $i \in \{0 \ldots n-1\}$. S is L-*valid*, written $\models_{\mathrm{L}} S$, iff S is satisfied by every L-valuation v.

Example 9.36. For a logic with truth values $(0, \frac{1}{2}, 1)$, let S be the 3-sequent:

$$p \rightarrow q \mid q \mid p, q$$

If $v(p) = 1$ or $v(q) \geq \frac{1}{2}$, then S is satisfied, no matter how the connectives are interpreted. If $v(p) = \frac{1}{2}$ and $v(q) = 0$, then for Ł3, $v(p \rightarrow q) = \frac{1}{2}$ and S is not satisfied. For G3 on the other hand, $v(p \rightarrow q) = 0$ so S is satisfied by v.

Defining rules for connectives for logics is not particularly hard in this context: we just follow the truth values. That is, let $\star$ be a connective of arity k. Then for every ith slot of the n-sequent, we derive an introduction rule for $\star$ where the premises are determined by the values of $v(A_1), \ldots, v(A_k)$ such that $v(\star(A_1, \ldots, A_k))$ takes the ith truth value. More precisely:

Definition 9.37. Let $\mathrm{L} = (\mathcal{L}, [\mathcal{N}, \mathcal{D}, \mathcal{C}])$ with $\mathcal{N} = (w_0, \ldots, w_{n-1})$. An *i-th-cf (conjunctive form)* for a k-ary $\star \in \mathcal{C}$ is a set $K_i^\star$ of n-sequents containing $p_1, \ldots, p_k$ such that for all L-valuations v: $v(\star(p_1, \ldots, p_k)) = w_i$ iff each $S \in K_i$ is satisfied by v.

Note that such forms can be found automatically for the connectives of *any* finite-valued logic. For example, a related disjunctive form for $\star$ and truth value w_i is obtained by considering all valuations v such that $v(\star(p_1, \ldots, p_k)) = w_i$ and dealing with the values of $v(p_1), \ldots, v(p_k)$. A conjunctive form is then defined (rather inefficiently) by distributing conjunctions over disjunctions. We refer to [222] for a more detailed explanation.

Definition 9.38. Let $\mathrm{L} = (\mathcal{L}, [\mathcal{N}, \mathcal{D}, \mathcal{C}])$ with $\mathcal{N} = (w_0, \ldots, w_{n-1})$ and an i-th-cf $K_i^\star$ for each $\star \in \mathcal{C}$ and $i = 0 \ldots n-1$. Then the *ith introduction rule* for $\star \in \mathcal{L}$ is:

$$\frac{\{(T_0, T_0' \mid \ldots \mid T_i, T_i' \mid \ldots \mid T_{n-1}, T_{n-1}') : (T_0' \mid \ldots \mid T_{n-1}') \in K_i^\star(A_1, \ldots, A_k)\}}{T_0 \mid \ldots \mid T_{i-1} \mid T_i, \star(A_1, \ldots, A_k) \mid T_{i+1} \mid \ldots \mid T_{n-1}} \; (\star_i)$$

where $K_i^\star(A_1, \ldots, A_k)$ is $K_i^\star$ with each p_j replaced by A_j for $j = 1 \ldots k$.

Example 9.39. For any G3-valuation v, we have that $v(p \rightarrow q) = 0$ iff $v(q) = 0$ and $v(p) \geq \frac{1}{2}$, which hold iff $(q \mid \emptyset \mid \emptyset)$ and $(\emptyset \mid p \mid p)$ are satisfied by v. Hence the 0th introduction rule for $\rightarrow$ is:

$$\frac{T_0, B \mid T_1 \mid T_2 \quad T_0 \mid T_1, A \mid T_2, A}{T_0, A \rightarrow B \mid T_1 \mid T_2} \; (\rightarrow_0)$$

A calculus for a logic then consists of the introduction rules for all its connectives and a set of axioms capturing the fact that any formula must take some value in $\mathcal{N}$.

Definition 9.40. Let $\mathrm{L} = (\mathcal{L}, [\mathcal{N}, \mathcal{D}, \mathcal{C}])$ with $\mathcal{N} = (w_0, \ldots, w_{n-1})$ and an i-th-cf $K_i^\star$ for each $\star \in \mathcal{C}$ and $i = 0 \ldots n-1$. Then GL^{F} is the n-sequent calculus consisting of the introduction rules for each $\star \in \mathcal{L}$ and axioms:

$$\frac{}{T_0, A \mid \ldots \mid T_{n-1}, A}\ (\mathrm{ID})_n$$

Example 9.41. We obtain the following calculus for $Ł_3$:

$$\frac{T_0 \mid T_1 \mid T_2, A \quad T_0, B \mid T_1 \mid T_2}{T_0, A \to B \mid T_1 \mid T_2}\ (\to 0) \qquad \frac{T_0 \mid T_1 \mid T_2, A}{T_0, \neg A \mid T_1 \mid T_2}\ (\neg 0)$$

$$\frac{T_0 \mid T_1, A \mid T_2, A \quad T_0 \mid T_1, A, B \mid T_2 \quad T_0, B \mid T_1 \mid T_2, A}{T_0 \mid T_1, A \to B \mid T_2}\ (\to 1) \qquad \frac{T_0 \mid T_1, A \mid T_2}{T_0 \mid T_1, \neg A \mid T_2}\ (\neg 1)$$

$$\frac{T_0, A \mid T_1, A \mid T_2, B \quad T_0, A \mid T_1, B \mid T_2, B}{T_0 \mid T_1 \mid T_2, A \to B}\ (\to 2) \qquad \frac{T_0, A \mid T_1 \mid T_2}{T_0 \mid T_1 \mid T_2, \neg A}\ (\neg 2)$$

The beauty of this definition is that it always gives a sound and complete calculus. Just notice that every rule is sound and invertible for L (by definition) and that the multiset complexity of each premise is strictly lower than the multiset complexity of the conclusion. Hence applying these rules upwards exhaustively terminates with n-sequents containing only variables. But then it is easy to see that such n-sequents are valid iff some variable occurs in every slot (otherwise just define a valuation where variables take the value of a slot in which they do not occur).

Theorem 9.42. *Let* $\mathrm{L} = (\mathcal{L}, [\mathcal{N}, \mathcal{D}, \mathcal{C}])$ *with* $\mathcal{N} = (w_0, \ldots, w_{n-1})$ *and an i-th-cf* $K_i^\star$ *for each* $\star \in \mathcal{C}$ *and* $i = 0 \ldots n-1$. *Then:*

$$\models_{\mathrm{L}} T_0 \mid \ldots \mid T_{n-1} \quad \textit{iff} \quad \vdash_{\mathrm{GL}^{\mathrm{F}}} T_0 \mid \ldots \mid T_{n-1}$$

Hence for any formula A, letting $T_i^A = \{A\}$ *if* $m_i \in \mathcal{D}$ *and* $T_i^A = \emptyset$ *otherwise:*

$$\models_{\mathrm{L}} A \quad \textit{iff} \quad \vdash_{\mathrm{GL}^{\mathrm{F}}} T_0^A \mid \ldots \mid T_{n-1}^A$$

Example 9.43. As a simple example, which also shows how tedious proofs can be in this framework, consider the following derivation of the involution axioms in the calculus for $Ł_3$ given in Example 9.41:

$$\dfrac{\dfrac{\dfrac{\dfrac{\dfrac{\dfrac{}{A \mid A \mid A}\ (\mathrm{ID})_3}{A \mid \neg A \mid A}\ (\neg 1)}{A \mid \neg\neg A \mid A}\ (\neg 1)}{\emptyset \mid \neg\neg A \mid A, \neg A}\ (\neg 2)}{\neg\neg A \mid \neg\neg A \mid A}\ (\neg 0) \qquad \dfrac{\dfrac{\dfrac{}{A \mid A \mid A}\ (\mathrm{ID})_3}{\emptyset \mid A \mid A, \neg A}\ (\neg 2)}{\neg\neg A \mid A \mid A}\ (\neg 0)}{\emptyset \mid \emptyset \mid \neg\neg A \to A}\ (\to 2)$$

We have used sets of formulas here in our definitions of n-sequents. However, we could just as well have used multisets (or sequences) and given axioms of the form $A \mid \ldots \mid A$ with weakening and contraction rules:

$$\frac{T_0 \mid \ldots \mid T_i \mid \ldots \mid T_{n-1}}{T_0 \mid \ldots \mid T_i, A \mid \ldots \mid T_{n-1}}\ (\mathrm{w}_i) \qquad \frac{T_0 \mid \ldots \mid T_i, A, A \mid \ldots \mid T_{n-1}}{T_0 \mid \ldots \mid T_i, A \mid \ldots \mid T_{n-1}}\ (\mathrm{c}_i)$$

We could also have added cut rules of the form (for each $i \neq j$):

$$\frac{T_0 \mid \ldots \mid T_i, A \mid \ldots \mid T_{n-1} \quad T'_0 \mid \ldots \mid T'_j, A \mid \ldots \mid T'_{n-1}}{T_0, T'_0 \mid \ldots \mid T_{n-1}, T'_{n-1}}\ (\mathrm{CUT})_{ij}$$

The elimination of such rules for first-order finite-valued logics was investigated by Baaz, Fermüller, and Zach in [22].

Evidently, there is a great deal of redundancy in the general approach described here. For particular families of logics, the number of rules can be reduced dramatically by interpreting n-sequents differently, e.g. by defining satisfaction for $T_0 \mid \ldots \mid T_{n-1}$ by a valuation v as $v(A) \geq m_i$ for some $A \in T_i$ (a useful simplification for linearly ordered sets of truth values). More generally, formulas can be labelled with sets of truth values and efficient strategies developed for manipulating these sets. This approach is developed extensively by Hähnle in the monograph [104]. Other methods may be found in the Handbook of Automated Reasoning chapter [20].

Sometimes, calculi for finite-valued logics can even be used to reason in infinite-valued logics. All we need is a way of calculating for any given formula A, a finite-valued logic such that A is a tautology of this logic iff it is a tautology of the given infinite-valued logic. This methodology was applied by Aguzzoli and Ciabattoni in [4] to obtain a calculus for Łukasiewicz Logic by calculating an upper bound for the finite-valued Łukasiewcz logic falsifying formulas of a given complexity. This was extended to other continuous t-norm based logics in [3]. However, while the connections with finite-valued logics are interesting, this approach does not really provide a proof-theoretic perspective on fuzzy logics or constitute a viable framework for automated reasoning.

9.4.3 Hypersequents

In some special cases, the proof theory of finite-valued logics can be tackled by adding rules to existing hypersequent calculi. This gives a genuine algorithmic interpretation of the logic in a framework with other non-classical (not just finite-valued) logics, and allows us to make use of the techniques developed in previous chapters.

As a first example, observe that a calculus for the n-valued Gödel Logic $\mathsf{G_n}$ can be obtained by extending GG (or just a hypersequent calculus for Intuitionistic Logic) with one extra rule (see [15, 51]):

Definition 9.44. $\mathsf{GG_n}$ is GG extended with:

$$\frac{\mathcal{G} \mid \Gamma_1,\Gamma_2 \Rightarrow \Delta_1 \quad \dots \quad \mathcal{G} \mid \Gamma_{n-1},\Gamma_n \Rightarrow \Delta_{n-1}}{\mathcal{G} \mid \Gamma_1 \Rightarrow \Delta_1 \mid \dots \mid \Gamma_n \Rightarrow \Delta_n}\ (\mathrm{G_n})$$

Example 9.45. The extra rule of $\mathsf{GG_n}$ corresponds to the characteristic axiom schema $A_1 \vee (A_1 \to A_2) \vee \dots \vee (A_{n-1} \to A_n)$ axiomatizing $\mathsf{G_n}$ as an extension of HG. E.g.

$$\dfrac{\dfrac{\dfrac{\dfrac{\dfrac{\dfrac{}{A \Rightarrow A}\,(\mathrm{ID}) \quad \dfrac{}{B \Rightarrow B}\,(\mathrm{ID})}{\Rightarrow A \mid A \Rightarrow B \mid B \Rightarrow C}\,(\mathrm{G_3})}{\Rightarrow A \mid A \Rightarrow B \mid\Rightarrow B \to C}\,(\Rightarrow\to)}{\Rightarrow A \mid\Rightarrow A \to B \mid\Rightarrow B \to C}\,(\Rightarrow\to)}{\Rightarrow A \mid\Rightarrow (A \to B) \vee (B \to C)}\,(\Rightarrow\vee)}{\Rightarrow A \vee (A \to B) \vee (B \to C)}\,(\Rightarrow\vee)$$

Unfortunately, such a simple extension has not been found for the family of n-valued Łukasiewicz logics. However, elegant hypersequent calculi have been defined for the three-valued case [11, 52]:

Definition 9.46. $\mathsf{GŁ_3}$ is GIMTL extended with:

$$\frac{\mathcal{G} \mid \Gamma_1,\Pi \Rightarrow \Sigma,\Delta_1 \quad \mathcal{G} \mid \Gamma_2,\Pi \Rightarrow \Sigma,\Delta_2}{\mathcal{G} \mid \Gamma_1,\Gamma_2 \Rightarrow \Delta_1,\Delta_2 \mid \Pi \Rightarrow \Sigma}\ (\mathrm{W_3})$$

Example 9.47. $(\mathrm{W_3})$ can be viewed as a generalization of the communication rule (COM) (which is derivable using $(\mathrm{W_3})$ and the weakening rules, and so is redundant in this calculus). It can be used to derive the characteristic axioms $((A \to \neg A) \to A) \to A$ of Ł_3 as follows:

$$\dfrac{\dfrac{\dfrac{\dfrac{\dfrac{}{\Rightarrow A \mid A \Rightarrow A}\,(\mathrm{ID}) \quad \dfrac{\dfrac{\dfrac{\dfrac{}{A \Rightarrow A}\,(\mathrm{ID}) \quad \dfrac{}{A \Rightarrow A}\,(\mathrm{ID})}{\Rightarrow A \mid A,A \Rightarrow}\,(\mathrm{W_3})}{\Rightarrow A \mid A \Rightarrow \neg A}\,(\Rightarrow\neg)}{\Rightarrow A \mid\Rightarrow A \to \neg A}\,(\Rightarrow\to)}{\Rightarrow A \mid (A \to \neg A) \to A \Rightarrow A}\,(\to\Rightarrow)}{(A \to \neg A) \to A \Rightarrow A \mid (A \to \neg A) \to A \Rightarrow A}\,(\mathrm{W})}{(A \to \neg A) \to A \Rightarrow A}\,(\mathrm{EC})}{\Rightarrow ((A \to \neg A) \to A) \to A}\,(\Rightarrow\to)$$

Soundness and completeness proofs for these systems follow the same pattern as Chapter 4. Namely, we show that the extra axioms are derivable in the systems, and that the rules of the systems are sound. Note also that unlike $\mathsf{GŁ}$ for infinite-valued Łukasiewicz Logic, $\mathsf{GŁ_3}$ is complete with respect to the standard interpretation of hypersequents. Cut elimination also proceeds as in Chapter 5, just checking the extra cases for the additional structural rules (see [11, 15, 52] for details).

9.5 Comparative Logics

Some logics, not fuzzy in the sense of being characterized by chains, nevertheless have certain "fuzzy features". In particular, logics for comparative reasoning introduced by Casari in the 1980s provide an alternative truth degree semantics for modelling vagueness [42, 43]. Algebras for these logics, called lattice-ordered pregroups (see [43, 179]), although not in general prelinear, have degrees of both truth and falsity related by an involutive negation, and (possibly) intermediate degrees between. In the language of residuated lattices, they are involutive pcrls satisfying $\mathrm{f} = \mathrm{f} \odot \mathrm{f}$ and $x \to x = \mathrm{e}$. The corresponding Hilbert System for the "basic comparative logic" in the language $\mathcal{L}_{\mathrm{F}} = \{\wedge, \vee, \odot, \to, \mathrm{f}, \mathrm{e}\}$ is defined as follows:

Definition 9.48. HC is HMALL^- extended with:

$$\text{(C1)}\ \mathrm{f} \leftrightarrow (\mathrm{f} \odot \mathrm{f})$$
$$\text{(C2)}\ (A \to A) \to \mathrm{e}$$

Casari also considered logics axiomatized by HC extended with (EM), (PRL) and (DIS), or (f), the latter giving an axiomatization for our old friend Abelian Logic A. By Theorem 3.51, all these logics are sound and complete with respect to corresponding classes of pcrls. By Theorem 3.56, the logic with (PRL) and (DIS), and of course A, are also complete with respect to the corresponding class of chains.

We will show here that interesting Gentzen systems can be defined for these logics. The starting point is a crucial connection between HC and HA:

Proposition 9.49. $\vdash_{\mathsf{HC}} A \oplus \mathrm{e}$ *iff* $\vdash_{\mathsf{HA}} A$.

Proof. Note first that the following are HC-derivable:

(i) $\mathrm{f} \to \mathrm{e}$
(ii) $A \to (A \oplus \mathrm{e})$
(iii) $((A \to B) \oplus \mathrm{e}) \to ((A \oplus \mathrm{e}) \to (B \oplus \mathrm{e}))$

(i) Using (C1) and ($\odot$1), $\vdash_{\mathsf{HC}} \mathrm{f} \to (\mathrm{f} \to \mathrm{f})$. So, since by (C2), $\vdash_{\mathsf{HC}} (\mathrm{f} \to \mathrm{f}) \to \mathrm{e}$, using Lemma 3.23 (iii), $\vdash_{\mathsf{HC}} \mathrm{f} \to \mathrm{e}$. (ii) By (e2), $\vdash_{\mathsf{HC}} (A \to A) \to (\mathrm{e} \to (A \to A))$ and by (I), $\vdash_{\mathsf{HC}} A \to A$. Hence by (MP), $\vdash_{\mathsf{HC}} \mathrm{e} \to (A \to A)$. Also $(\mathrm{e} \to \mathrm{f}) \to \mathrm{f}$ is an instance of (INV). So using (i) and Lemma 3.23 (iii) twice, $\vdash_{\mathsf{HC}} (\mathrm{e} \to \mathrm{f}) \to (A \to A)$. But then using Lemma 3.23 (vi), $\vdash_{\mathsf{HC}} A \to ((\mathrm{e} \to \mathrm{f}) \to A)$; i.e. $\vdash_{\mathsf{HC}} A \to (A \oplus \mathrm{e})$ as required. (iii) First note that $\vdash_{\mathsf{HC}} \mathrm{f} \to ((\mathrm{f} \to A) \to ((A \to (\mathrm{f} \to B)) \to (\mathrm{f} \to B)))$ using (B) and Lemma 3.23 (iii). Hence using Lemma 3.23 (vi), $\vdash_{\mathsf{HC}} (\mathrm{f} \to (A \to B)) \to ((\mathrm{f} \to A) \to (\mathrm{f} \to (\mathrm{f} \to B)))$. But $(\mathrm{f} \to (\mathrm{f} \to B)) \to (\mathrm{f} \to B)$ using (C1) and ($\odot$2). So by Lemma 3.23 (iii), $\vdash_{\mathsf{HC}} (\mathrm{f} \to (A \to B)) \to ((\mathrm{f} \to A) \to (\mathrm{f} \to B))$. Finally, using the fact that $\vdash_{\mathsf{HC}} (\mathrm{e} \to \mathrm{f}) \leftrightarrow \mathrm{f}$, we obtain $\vdash_{\mathsf{HC}} ((A \to B) \oplus \mathrm{e}) \to ((A \oplus \mathrm{e}) \to (B \oplus \mathrm{e}))$ as required.

The left-to-right direction of the proposition is straightforward. If $\vdash_{\mathsf{HC}} A \oplus \mathrm{e}$, then $\vdash_{\mathsf{HA}} A \oplus \mathrm{e}$, and since $\vdash_{\mathsf{HA}} (A \oplus \mathrm{e}) \to A$, we get $\vdash_{\mathsf{HA}} A$. For the right-to-left direction, we use the fact (easily checked) that $\vdash_{\mathsf{HA}} A$ iff $\vdash_{\mathsf{HA}^*} A$ where HA^* is HC extended with $\mathrm{f} \to \mathrm{e}$, and proceed by induction on the height of an HA^*-derivation of A. For the base case, if A is an axiom of HC then, since by (ii), $\vdash_{\mathsf{HC}} A \to (A \oplus \mathrm{e})$, also $\vdash_{\mathsf{HC}} A \oplus \mathrm{e}$. If A is $\mathrm{e} \to \mathrm{f}$, then $(\mathrm{e} \to \mathrm{f}) \oplus \mathrm{e}$ is $\neg\neg\mathrm{e} \to \mathrm{e}$, an instance of (INV). For the inductive step, A follows by either (ADJ) or (MP). For the former, $A = B \wedge C$ and B and C are derivable in HA^*. By the induction hypothesis twice we get $\vdash_{\mathsf{HC}} B \oplus \mathrm{e}$ and $\vdash_{\mathsf{HC}} C \oplus \mathrm{e}$. Hence by (ADJ) we have $\vdash_{\mathsf{HC}} (B \oplus \mathrm{e}) \wedge (C \oplus \mathrm{e})$ and by ($\wedge 3$), $\vdash_{\mathsf{HC}} (B \wedge C) \oplus \mathrm{e}$. For the case of (MP) we have $\vdash_{\mathsf{HA}^*} B$ and $\vdash_{\mathsf{HA}^*} B \to A$ so by the induction hypothesis twice we get $\vdash_{\mathsf{HC}} B \oplus \mathrm{e}$ and $\vdash_{\mathsf{HC}} (B \to A) \oplus \mathrm{e}$. So using (iii) and (MP) twice, $\vdash_{\mathsf{HC}} A \oplus \mathrm{e}$. □

We can use this result to define a Gentzen system for C (and other comparative logics) by "combining" the sequent calculus GMALL^- with the cut-free hypersequent calculus GA° for Abelian Logic. The crucial element here is a special mix rule that allows one of its premises to be derived in GA° (denoted by $\vdash_{\mathsf{GA}^\circ}$):

Definition 9.50. GC is GMALL^- plus the following combination rule:

$$\dfrac{\Gamma_1 \Rightarrow \Delta_1 \quad \vdash_{\mathsf{GA}^\circ} \Gamma_2 \Rightarrow \Delta_2}{\Gamma_1, \Gamma_2 \Rightarrow \Delta_1, \Delta_2}\ (\text{MIX})_{\mathsf{A}}$$

To view GC as a calculus in the sense defined in Chapter 3, we can read $\vdash_{\mathsf{GA}^\circ} \Gamma_2 \Rightarrow \Delta_2$ as denoting any sequent $\Gamma_2 \Rightarrow \Delta_2$ derivable in GA°. In practice of course, we would simply switch Gentzen systems mid-derivation.

Example 9.51. Consider the following derivation of the axiom (C2):

$$\dfrac{\dfrac{\dfrac{}{\Rightarrow \mathrm{e}}\ (\Rightarrow\mathrm{e}) \qquad \dfrac{\dfrac{}{\vdash_{\mathsf{GA}^\circ} A \Rightarrow A}\ (\text{ID})}{\vdash_{\mathsf{GA}^\circ} A \to A \Rightarrow}\ (\to\Rightarrow)_{\mathsf{A}}}{A \to A \Rightarrow \mathrm{e}}\ (\text{MIX})_{\mathsf{A}}}{\Rightarrow (A \to A) \to \mathrm{e}}\ (\Rightarrow\to)$$

Notice that in the left branch of the proof, the initial sequent ($\Rightarrow$e) is from GMALL^-, while in the right branch, the initial sequent (ID) and the rule $(\to\Rightarrow)_{\mathsf{A}}$ are from GA°. Similarly, in the proof below for (C1), the rule $(\text{MIX})_{\mathsf{A}}$ allows the more extensive rules of GA° to be used in one of the branches:

$$\dfrac{\dfrac{\dfrac{\dfrac{}{\mathrm{f} \Rightarrow}\ (\mathrm{f}\Rightarrow) \qquad \dfrac{\dfrac{\dfrac{\dfrac{}{\vdash_{\mathsf{GA}^\circ} \Rightarrow}\ (\text{EMP})}{\vdash_{\mathsf{GA}^\circ} \Rightarrow \mathrm{f}}\ (\Rightarrow\mathrm{f})}{\vdash_{\mathsf{GA}^\circ} \Rightarrow \mathrm{f}, \mathrm{f}}\ (\Rightarrow\mathrm{f})}{\vdash_{\mathsf{GA}^\circ} \Rightarrow \mathrm{f} \odot \mathrm{f}}\ (\Rightarrow\odot)_{\mathsf{A}}}{\mathrm{f} \Rightarrow \mathrm{f} \odot \mathrm{f}}\ (\text{MIX})_{\mathsf{A}}}{\Rightarrow \mathrm{f} \to (\mathrm{f} \odot \mathrm{f})}\ (\Rightarrow\to) \qquad \dfrac{\dfrac{\dfrac{\dfrac{}{\mathrm{f} \Rightarrow}\ (\mathrm{f}\Rightarrow) \qquad \dfrac{}{\vdash_{\mathsf{GA}^\circ} \mathrm{f} \Rightarrow \mathrm{f}}\ (\text{ID})}{\mathrm{f}, \mathrm{f} \Rightarrow \mathrm{f}}\ (\text{MIX})_{\mathsf{A}}}{\mathrm{f} \odot \mathrm{f} \Rightarrow \mathrm{f}}\ (\odot\Rightarrow)}{\Rightarrow (\mathrm{f} \odot \mathrm{f}) \to \mathrm{f}}\ (\Rightarrow\to)}{\Rightarrow (\mathrm{f} \to (\mathrm{f} \odot \mathrm{f})) \wedge ((\mathrm{f} \odot \mathrm{f}) \to \mathrm{f})}\ (\Rightarrow\wedge)$$

Completeness for GC follows as usual from the fact that we can derive all the axioms of HC, and the rules (MP) and (ADJ) are admissible. For soundness, the key step is to show that $(\text{MIX})_{\mathsf{A}}$ preserves derivability in HC. So suppose that $\vdash_{\mathsf{HC}} \mathrm{I}(\Gamma_1 \Rightarrow \Delta_1)$ and $\vdash_{\mathsf{HA}} \mathrm{I}(\Gamma_2 \Rightarrow \Delta_2)$. Then, using the link between HA and HC established by Proposition 9.49, $\vdash_{\mathsf{HC}} \mathrm{I}(\Gamma_2 \Rightarrow \Delta_2) \oplus \mathrm{e}$. Hence $\vdash_{\mathsf{HC}} \mathrm{I}(\Gamma_1 \Rightarrow \Delta_1) \oplus \mathrm{I}(\Gamma_2 \Rightarrow \Delta_2)$, and it follows easily that $\vdash_{\mathsf{HC}} \mathrm{I}(\Gamma_1, \Gamma_2 \Rightarrow \Delta_1, \Delta_2)$.

Theorem 9.52. $\vdash_{\mathsf{GC}} \Gamma \Rightarrow \Delta$ *iff* $\vdash_{\mathsf{HC}} \mathrm{I}(\Gamma \Rightarrow \Delta)$.

Of course, the crucial result here is cut elimination.

Theorem 9.53. *Cut elimination holds for* GC.

Proof. We adapt the general approach presented in Chapter 5 and give a constructive proof that (CUT) is admissible for GC°. That is we prove:

Claim. If $d_1 \vdash_{\mathsf{GC}^\circ} \Gamma, A \Rightarrow \Delta$ and $d_2 \vdash_{\mathsf{GC}^\circ} \Pi \Rightarrow A, \Sigma$, then $\vdash_{\mathsf{GC}^\circ} \Gamma, \Pi \Rightarrow \Sigma, \Delta$.

We prove the claim by a double induction on the lexicographically ordered pair $\langle \mathrm{cp}(A), \mathrm{ht}(d_1) + \mathrm{ht}(d_2) \rangle$. The tricky case here is $(\text{MIX})_{\mathsf{A}}$ and we leave other steps to the reader. Let us suppose that d_1 (the situation for d_2 being very similar) ends with:

$$\frac{\Gamma_1 \Rightarrow \Delta_1 \quad \vdash_{\mathsf{GA}^\circ} \Gamma_2, A \Rightarrow \Delta_2}{\Gamma, A \Rightarrow \Delta}\ (\text{MIX})_{\mathsf{A}} \quad \text{or} \quad \frac{\Gamma_1, A \Rightarrow \Delta_1 \quad \vdash_{\mathsf{GA}^\circ} \Gamma_2 \Rightarrow \Delta_2}{\Gamma, A \Rightarrow \Delta}\ (\text{MIX})_{\mathsf{A}}$$

where $\Gamma = \Gamma_1 \uplus \Gamma_2$ and $\Delta = \Delta_1 \uplus \Delta_2$.

For the first case, we use the fact that GA° is a stronger system than GMALL$^-$ to get that $\vdash_{\mathsf{GA}^\circ} \Pi \Rightarrow A, \Sigma$. Hence, since (CUT) is GA°-admissible, $\vdash_{\mathsf{GA}^\circ} \Gamma_2, \Pi \Rightarrow \Sigma, \Delta_2$, and we obtain the GC°-derivation:

$$\frac{\Gamma_1 \Rightarrow \Delta_1 \quad \vdash_{\mathsf{GA}^\circ} \Gamma_2, \Pi \Rightarrow \Sigma, \Delta_2}{\Gamma, \Pi \Rightarrow \Sigma, \Delta}\ (\text{MIX})_{\mathsf{A}}$$

For the second case, by the induction hypothesis, $\vdash_{\mathsf{GC}^\circ} \Gamma_1, \Pi \Rightarrow \Sigma, \Delta_1$. So we obtain the GC°-derivation:

$$\frac{\Gamma_1, \Pi \Rightarrow \Sigma, \Delta_1 \quad \vdash_{\mathsf{GA}^\circ} \Gamma_2 \Rightarrow \Delta_2}{\Gamma, \Pi \Rightarrow \Sigma, \Delta}\ (\text{MIX})_{\mathsf{A}}$$

□

One nice result of cut elimination for GC is an algorithm for checking whether or not a formula A is a theorem of HC. We just search for a proof of the sequent $\Rightarrow A$ in GC° by applying rules (upwards) using e.g. depth-first search, and restricting applications of $(\text{MIX})_{\mathsf{A}}$ so that $\Gamma_2 \uplus \Delta_2 \neq []$ (otherwise the left premise is equal to the conclusion). But then, reading the right premise of $(\text{MIX})_{\mathsf{A}}$ as an initial sequent, there are only a finite number of possible proofs of $\Rightarrow A$ in GC°. Since each rule has a finite number of premises with strictly smaller complexity than the conclusion, this process terminates with a set of sequents to be proved in GA°. But the validity problem for A is decidable, and hence also:

Theorem 9.54. *The validity problem for* C *is decidable.*

This method of combining calculi with GA° allows us also to define Gentzen systems admitting cut elimination for other comparative logics. In the presence of (EM) we add the rule (SC_2) to GC, and for (PRL) we take the hypersequent version of GC extended with (EW), (EC), and (COM). More details may be found in [143].

9.6 Basic Logic and Other Open Problems

The main aim of this book has been to convince the reader that Gentzen systems admitting cut elimination – in particular, hypersequent calculi – provide an appropriate and useful framework for investigating fuzzy logics. Unfortunately such systems are not always easy to find. In some cases, like Łukasiewicz Logic and Product Logic, a fair amount of ingenuity is required to determine the right interpretation for hypersequents and then to establish cut-free completeness for a calculus. In other cases, a suitable Gentzen system has yet to be found, and may (in some restricted sense) not even exist.

Most notably, no Gentzen system has been presented here for Hájek's Basic Logic BL, the logic of continuous t-norms [105]. Basic Logic is one of the most important and widely studied fuzzy logics, and from an algebraic perspective, is quite natural. BL-algebras are prelinear integral pcrls (MTL-algebras) that satisfy the divisibility condition $x \wedge y = x \odot (x \to y)$. The variety of BL-algebras is generated not only by all standard (continuous t-norm based) BL-algebras, as shown by Cignoli et al. in [59], but even by just one such algebra.

Consider the binary function defined on the lexicographically ordered set $(\mathbb{N} \times [0,1)) \cup \{(\infty,\infty)\}$ by:

$$(n,x) \odot_{\mathsf{BL}} (m,y) = \begin{cases} (n, \max(0, x+y-1)) & \text{if } n = m \\ \min((n,x),(m,y)) & \text{otherwise} \end{cases}$$

It is easy to see that $\odot_{\mathsf{BL}}$ is order-isomorphic on this set to a continuous t-norm $*_{\mathsf{BL}}$: just take the mapping $\phi(n,x) = 1 - e^{-(n+x)}$. Moreover, it has been shown by Aglianò and Montagna in [2] that the algebra $\mathbf{A}(*_{\mathsf{BL}}, 0) = \langle [0,1], \min, \max, *_{\mathsf{BL}}, \to_{\mathsf{BL}}, 0, 1 \rangle$ generates the variety of BL-algebras. Moreover, this algebra provides one route to defining a calculus of sorts. Information about the positioning of the valuations of the formula can be encoded by structural features of the calculus and ultimately the question of the validity of a formula in BL can be reduced (as in Łukasiewicz Logic) to solving linear programming problems. This approach, developed in [37, 156], does not provide an elegant calculus for BL but does at least give a reasonable algorithm for deciding questions of validity in this logic.

The main difficulty faced in defining Gentzen systems for Basic Logic is to discover structural rules corresponding to divisibility or continuity. Already for Łukasiewicz Logic (in some sense a simple limit case of BL) our solution relies on a non-standard interpretation of hypersequents. A more promising approach is perhaps to introduce a little more structure, without simply encoding all the algebraic aspects of the standard algebra. In particular, recall the uniform "relational hypersequent" rules for Ł, G, and P of Fig. 7.1, which make use of two kinds of sequents. We can use this extra structure to find a reasonable calculus for the implicational fragment of BL (adding connectives $\wedge$, $\vee$, $\bot$, $\top$ is straightforward – the problem is $\odot$). First, we adapt the BL-algebra described above.

Consider a second binary function on the lexicographically ordered set $(\mathbb{N} \times [0,1)]) \cup \{(\infty, \infty)\}$:

$$(n,x) \odot_{\mathsf{PBL}} (m,y) = \begin{cases} (n, x \cdot y) & \text{if } n = m \\ \min((n,x),(m,y)) & \text{otherwise} \end{cases}$$

This function is again order-isomorphic to a continuous t-norm $*_{\mathsf{PBL}}$. Moreover, while the algebra $\mathbf{A}(*_{\mathsf{PBL}})$ does not generate the variety of BL-algebras, an *implicational* formula is valid in this algebra iff it is valid in all BL-algebras.

Now let us define:

$$\star^{v}_{\mathsf{BL}}([A_1, \ldots, A_n]) = v(A_1) *_{\mathsf{PBL}} \cdots *_{\mathsf{PBL}} v(A_n)$$

We can prove that the uniform implication rules are sound and invertible with respect to this interpretation. Then as for Ł, G, and P, this means that the question of the validity of an implicational formula in BL is reduced to the question of the validity of strictly atomic r-hypersequents. Structural rules and initial r-hypersequents for deciding this latter question – our (EMP), (ID), (EW), (EC), $(\text{SPLIT})_{\leq}$, (MIX), plus a rather complicated splitting rule – have been defined by Vetterlein in [213]. A calculus is also defined in this paper for the whole logic but at the cost of introducing a further modal connective into the language, requiring more complicated rules. We will not go into details here. It is reasonable to claim that r-hypersequents provide the right level of generality for defining calculi for BL and other continuous t-norm based logics, and not too optimistic to hope that more elegant systems can be obtained in the future.

Let us finish then with a list of other open problems in and around the area of the proof theory of fuzzy logics, some already the object of active research, others merely speculative:

- *Calculi for other (fuzzy) logics*. Basic Logic is the most important fuzzy logic lacking a Gentzen system, but other interesting cases have also resisted analysis. These include other logics such as Strict Basic Logic SBL that involve some form of divisibility, where success should be tied to progress on BL, and also logics such as Product Monoidal t-norm Logic PMTL [121] involving some notion of cancellativity. In this latter case, the problem and interest extends beyond fuzzy logics. Also, no Gentzen system is known for systems such as $\mathsf{HMAILL}^{-} + (\text{CAN})$, the logic of cancellative crls. Since the decidability of these logics (varieties) is also open (see [30] for further details), the development of Gentzen systems, perhaps using the techniques described in this book, could be very helpful.

- *Decidability and complexity issues.* We have used our calculi to answer decidability and complexity questions for a number of fuzzy logics, but there remain many open problems, in particular, the decidability of Uninorm Logic UL (and related logics), and the complexity of Monoidal t-norm Logic MTL (and related logics). As remarked previously, hypersequent calculi for these logics do not seem to help: the presence of the external contraction rule (EC) being the most obvious problem. Nevertheless, just as loop-checking mechanisms can be successful in taming internal contraction, it is possible that a similar approach could be useful in dealing with external contraction.
- *Density elimination and standard completeness.* The use of density elimination for hypersequent calculi to establish standard completeness results for fuzzy logics is one of the most important applications of the proof theory developed here. It also promises a characterization of those logics that are "fuzzy" (in the sense of being standard complete) and those that are not. In the case of single-conclusion calculi with weakening rules for example, density elimination is guaranteed by the substitutivity of the structural rules. It would be nice to obtain similar characterizations for multiple-conclusion calculi and calculi without weakening rules, noting that the particular case of proving density elimination for GIUL is still open. Perhaps even more challenging is to find conditions that are both necessary and sufficient (within some framework) for calculi to admit density elimination.
- *First-order logics and their fragments.* We have developed hypersequent calculi for a wide class of first-order fuzzy logics, including MTL and UL. However, first-order logics based on continuous t-norms are more problematic. First-order Łukasiewicz Logic is not recursively enumerable; our calculus for this logic uses an infinitary rule. First-order Product Logic and Basic Logic are not even arithmetical. More generally, since all first-order fuzzy logics are undecidable, it is important to investigate and identify fragments that are decidable or at least recursively enumerable. Of particular interest here are fragments suitable for fuzzy logic programming [214] and fuzzy description logics [111, 201].

References

1. V. M. Abrusci and P. Ruet. Non-commutative logic I: The multiplicative fragment. *Annals of Pure and Applied Logic*, 101:29–64, 2000.
2. P. Aglianò and F. Montagna. Varieties of BL-algebras I: General properties. *Journal of Pure and Applied Algebra*, 181:105–129, 2003.
3. S. Aguzzoli. Uniform description of calculi for all t-norm logics. In *Proceedings of ISMVL 2004*, pp. 38–43. IEEE Computer Society Press, California, 2004.
4. S. Aguzzoli and A. Ciabattoni. Finiteness in infinite-valued logic. *Journal of Logic, Language and Information*, 9(1):5–29, 2000.
5. S. Aguzzoli and D. Mundici. Weierstrass approximation theorem and Łukasiewicz formulas with one quantified variable. In M. Fitting and E. Orlowska, editors, *Beyond Two: Theory and Applications of Multiple Valued Logic*, pp. 315–335. Springer, New York, 2003.
6. A. R. Anderson and N. D. Belnap. *Entailment*, volume 1. Princeton University Press, Princeton, 1975.
7. A. R. Anderson, N. D. Belnap, and J. M. Dunn. *Entailment*, volume 2. Princeton University Press, Princeton, 1992.
8. A. Avellone, M. Ferrari, and P. Miglioli. Duplication-free tableau calculi and related cut-free sequent calculi for the interpolable propositional intermediate logics. *Logic Journal of the IGPL*, 7(4):447–480, 1999.
9. A. Avron. A constructive analysis of RM. *Journal of Symbolic Logic*, 52(4):939–951, 1987.
10. A. Avron. The semantics and proof theory of linear logic. *Theoretical Computer Science*, 57(2–3):161–184, 1988.
11. A. Avron. Hypersequents, logical consequence and intermediate logics for concurrency. *Annals of Mathematics and Artificial Intelligence*, 4(3–4):225–248, 1991.
12. A. Avron and B. Konikowska. Decomposition Proof Systems for Gödel-Dummett Logics. *Studia Logica*, 69(2):197–219, 2001.
13. M. Baaz. Infinite-valued Gödel logics with 0-1-projections and relativizations. In P. Hájek, editor, *Proceedings of GÖDEL 1996: Logical Foundations of Mathematics, Computer Science and Physics*, volume 6 of *Lecture Notes in Logic*, pp. 23–33. Springer, Berlin, 1996.
14. M. Baaz, A. Ciabattoni, and C. G. Fermüller. Herbrand's theorem for Prenex Gödel logic and its consequences for theorem proving. In *Proceedings of LPAR 2001*, volume 2250, pp. 201–216. Springer, Heidelberg, 2001.
15. M. Baaz, A. Ciabattoni, and C. G. Fermüller. Hypersequent calculi for Gödel logics: A survey. *Journal of Logic and Computation*, 13:1–27, 2003.
16. M. Baaz, A. Ciabattoni, C. G. Fermüller, and H. Veith. On the undecidability of some subclassical first-order logics. In *Proceedings of FST & TCS 1999*, volume 1738 of *LNCS*, pp. 258–268. Springer, Heidelberg, 1999.
17. M. Baaz, A. Ciabattoni, and F. Montagna. Analytic calculi for monoidal t-norm based logic. *Fundamenta Informaticae*, 59(4):315–332, 2004.

18. M. Baaz, A. Ciabattoni, and R. Zach. Quantified propositional Gödel logics. In *Proceedings of LPAR 2000*, volume 1955 of *LNAI*, pp. 240–257. Springer, Berlin, 2000.
19. M. Baaz and C. G. Fermüller. Analytic calculi for projective logics. In *Proceedings of TABLEAUX 1999*, volume 1617 of *LNAI*, pp. 36–50. Springer, Berlin, 1999.
20. M. Baaz, C. G. Fermüller, and G. Salzer. Automated deduction for many-valued logics. In A. Robinson and A. Voronkov, editors, *Handbook of Automated Reasoning*, volume II, Chapter 20, pp. 1355–1402. Elsevier Science B.V., Amsterdam, 2001.
21. M. Baaz, C. G. Fermüller, and H. Veith. An analytic calculus for quantified propositional Gödel logic. In *Proceedings of TABLEAUX 2000*, volume 1847 of *LNCS*, pp. 112–126. Springer, Heidelberg, 2000.
22. M. Baaz, C. G. Fermüller, and R. Zach. Elimination of cuts in first-order finite-valued logics. *Journal of Information Processing, Cybernet. EIK*, 29(6):333–355, 1994.
23. M. Baaz, P. Hájek, F. Montagna, and H. Veith. Complexity of t-tautologies. *Annals of Pure and Applied Logic*, 113(1):3–11, 2001.
24. M. Baaz and G. Metcalfe. Proof theory for first order Łukasiewicz logic. In N. Olivetti, editor, *Proceedings of TABLEAUX 2007*, volume 4548 of *LNAI*, pp. 24–28. Springer, Berlin, 2007.
25. M. Baaz and G. Metcalfe. Herbrand theorems and skolemization for prenex fuzzy logics. In *Proceedings of CiE 2008*, volume 5028 of *LNCS*, pp. 22–31. Springer, Heidelberg, 2008.
26. M. Baaz, N. Preining, and R. Zach. First-order Gödel logics. *Annals of Pure and Applied Logic*, 147:23–47, 2007.
27. M. Baaz and H. Veith. Interpolation in fuzzy logic. *Archive for Mathematical Logic*, 38:461–489, 1999.
28. M. Baaz and H. Veith. An axiomatization of quantified propositional Gödel logic using the Takeuti-Titani rule. In *Logic Colloquium 1999*, volume 13 of *Lecture Notes in Logic*, pp. 91–104. Springer, Berlin, 2000.
29. M. Baaz and R. Zach. Hypersequents and the proof theory of intuitionistic fuzzy logic. In *Proceedings of CSL 2000*, volume 1862 of *LNCS*, pp. 187–201. Springer, Heidelberg, 2000.
30. P. Bahls, J. Cole, N. Galatos, P. Jipsen, and C. Tsinakis. Cancellative residuated lattices. *Algebra Universalis*, 50(1):83–106, 2003.
31. L. P. Belluce. Further results on infinite valued predicate logic. *Journal of Symbolic Logic*, 29:69–78, 1964.
32. L. P. Belluce and C. C. Chang. A weak completeness theorem for infinite valued first order logic. *Journal of Symbolic Logic*, 28:43–50, 1963.
33. I. Ben Yaacov, A. Berenstein, C. Ward Henson, and A. Usvyatsov. Model theory for metric structures. In Z. Chatzidakis, D. Macpherson, A. Pillay, and A.Wilkie, editors, *Model Theory with Applications to Algebra and Analysis, Vol. II*, volume 350 of *Lecture Notes of the London Mathematical Society*, pp. 315–427. Cambridge University Press, Cambridge, 2008.
34. N. D. Belnap, Jr. Display logic. *Journal of Philosophical Logic*, 11(4):375–417, 1982.
35. W. J. Blok and C. J. Van Alten. The finite embeddability property for residuated lattices, pocrims and BCK-algebras. *Algebra Universalis*, 48(3):253–271, 2002.
36. W. J. Blok and D. Pigozzi. *Algebraizable Logics*. Number 396 in Memoirs of the American Mathematical Society, volume 77. American Mathematical Society, Providence, 1989.
37. S. Bova and F. Montagna. Proof search in Hájek's basic logic. *ACM Transactions on Computational Logic*, 9(3):1–26, 2008.
38. R. T. Brady. Gentzenization of relevant logics without distribution, I. *Journal of Symbolic Logic*, 61(2):353–378, 1996.
39. R. T. Brady. Gentzenization of relevant logics without distribution. II. *Journal of Symbolic Logic*, 61(2):379–401, 1996.
40. S. Burris and H. P. Sankappanavar. *A Course in Universal Algebra*. Springer, New York, 1981.
41. S. Buss, editor. *Handbook of Proof Theory*. Elsevier, Amsterdam 1998.
42. E. Casari. Comparative logics. *Synthese*, 73:421–449, 1987.
43. E. Casari. Comparative logics and abelian ℓ-groups. In C. Bonotto, R. Ferro, S. Valentini, and A. Zanardo, editors, *Logic Colloquium '88*, pp. 161–190. Elsevier, Amsterdam, 1989.

44. A. Chagrov and M. Zakharyaschev. *Modal Logic*. Oxford University Press, Oxford, 1996.
45. C. C. Chang. Algebraic analysis of many-valued logics. *Transactions of the American Mathematical Society*, 88:467–490, 1958.
46. C. C. Chang. Proof of an axiom of Łukasiewicz. *Transactions of the American Mathematical Society*, 87:55–56, 1958.
47. A. Church. A note on the Entscheidungsproblem. *Journal of Symbolic Logic*, 1:40–41, 1936.
48. A. Ciabattoni. Automated generation of analytic calculi for logics with linearity. In *Proceedings of CSL 2004*, volume 3210 of LNCS, pp. 503–517. Springer, Berlin, 2004.
49. A. Ciabattoni, F. Esteva, and L. Godo. T-norm based logics with n-contraction. *Neural Network World*, 12(5):441–453, 2002.
50. A. Ciabattoni, C. G. Fermüller, and G. Metcalfe. Uniform Rules and Dialogue Games for Fuzzy Logics. In *Proceedings of LPAR 2004*, volume 3452 of *LNAI*, pp. 496–510. Springer, Berlin, 2005.
51. A. Ciabattoni and M. Ferrari. Hypersequent calculi for some intermediate logics with bounded Kripke models. *Journal of Logic and Computation*, 11, 2001.
52. A. Ciabattoni, D. Gabbay, and N. Olivetti. Cut-free proof systems for logics of weak excluded middle. *Soft Computing*, 2(4):147–156, 1999.
53. A. Ciabattoni, N. Galatos, and K. Terui. From axioms to analytic rules in nonclassical logics. In *Proceedings of LICS 2008*, pp. 229–240, IEEE Computer Society 2008.
54. A. Ciabattoni and G. Metcalfe. Bounded Łukasiewicz logics. In M. Cialdea Mayer and F. Pirri, editors, *Proceedings of TABLEAUX 2003*, volume 2796 of *LNCS* pp. 32–48. Springer, Berlin, 2003.
55. A. Ciabattoni and G. Metcalfe. Density elimination. *Theoretical Computer Science*, 403:328–346, 2008.
56. A. Ciabattoni, G. Metcalfe, and F. Montagna. Adding modalities to MTL and its extensions. To appear in *Proceedings of the 26th Linz Symposium*.
57. A. Ciabattoni and K. Terui. Towards a semantic characterization of cut-elimination. *Studia Logica*, 82(1):95–119, 2006.
58. R. Cignoli, I. M. L. D'Ottaviano, and D. Mundici. *Algebraic Foundations of Many-Valued Reasoning*, volume 7 of *Trends in Logic*. Kluwer Academic Publishers, Dordrecht, 1999.
59. R. Cignoli, F. Esteva, L. Godo, and A. Torrens. Basic fuzzy logic is the logic of continuous t-norms and their residua. *Soft Computing*, 4(2):106–112, 2000.
60. P. Cintula. Weakly implicative (fuzzy) logics I: Basic properties. *Archive for Mathematical Logic*, 45:673–704, 2006.
61. P. Cintula and P. Hájek. Triangular norm based predicate logics. To appear in *Proceedings of the Linz Symposium 2005*.
62. P. Cintula and P. Hájek. On theories and models in fuzzy predicate logics. *Journal of Symbolic Logic*, 71(3):832–863, 2006.
63. S. A. Cook. The complexity of theorem-proving procedures. In *Proceedings of the 3rd Annual ACM Symposium on Theory of Computing*, pp. 151–158, ACM press, New York, 1971.
64. G. Corsi. Completeness theorem for Dummett's LC quantified. *Studia Logica*, 51:317–335, 1992.
65. E. Czogala and J. Drewniak. Associative monotonic operations in fuzzy set theory. *Fuzzy Sets and Systems*, 12(3):249–269, 1984.
66. M. Darnel. *Theory of Lattice-Ordered Groups*. Dekker, New York, 1995.
67. B. A. Davey and H. A. Priestley. *Introduction to Lattices and Order*. Cambridge University Press, Cambridge, second Edition, 2002.
68. B. De Baets. Idempotent uninorms. *European Journal of Operational Research*, 118:631–642, 1999.
69. B. De Baets and J. Fodor. Residual operators of uninorms. *Soft Computing*, 3:89–100, 1999.
70. N. Dershowitz and Z. Manna. Proving termination with multiset orderings. *Communications of the Association for Computing Machinery*, 22:465–476, 1979.
71. J. Dombi. A general class of fuzzy operators, the De Morgan class of fuzzy operators and fuzziness measures induced by fuzzy operators. *Fuzzy Sets and Systems*, 8:149–163, 1982.

72. D. Dubois and H. Prade, editors. *The Handbook of Fuzzy Sets (7 volumes)*. Elsevier, Amsterdam, 2000.
73. M. Dummett. A propositional calculus with denumerable matrix. *Journal of Symbolic Logic*, 24:97–106, 1959.
74. J. M. Dunn. A 'Gentzen' system for positive relevant implication. *Journal of Symbolic Logic*, 38:356–357, 1974. (Abstract).
75. R. Dyckhoff. A deterministic terminating sequent calculus for Gödel-Dummett logic. *Logic Journal of the IGPL*, 7(3):319–326, 1999.
76. F. Esteva, J. Gispert, L. Godo, and F. Montagna. On the standard and rational completeness of some axiomatic extensions of the monoidal t-norm logic. *Studia Logica*, 71(2):199–226, 2002.
77. F. Esteva and L. Godo. Monoidal t-norm based logic: Towards a logic for left-continuous t-norms. *Fuzzy Sets and Systems*, 124:271–288, 2001.
78. F. Esteva, L. Godo, P. Hájek, and F. Montagna. Hoops and fuzzy logic. *Journal of Logic and Computation*, 13(4):532–555, 2003.
79. F. Esteva, L. Godo, P. Hájek, and M. Navara. Residuated fuzzy logics with an involutive negation. *Archive for Mathematical Logic*, 39(2):103–124, 2000.
80. F. Esteva, L. Godo, and F. Montagna. The ŁΠ and ŁΠ$\frac{1}{2}$ logics: Two complete fuzzy systems joining Łukasiewicz and Product logics. *Archive for Mathematical Logic*, 40:39–67, 1999.
81. C. G. Fermüller and A. Ciabattoni. From intuitionistic logic to Gödel-Dummett logic via parallel dialogue games. In *Proceedings of the 33rd IEEE International Symposium on Multiple-Valued Logic*, pp. 188–193, Tokyo, May 2003.
82. P. Flondor, G. Georgescu, and A. Iorgulescu. Pseudo-t-norms and pseudo-BL-algebras. *Soft Computing*, 5:355–371, 2001.
83. J. Fodor, R. R. Yager, and A. Rybalov. Structure of uni-norms. *International Journal of Uncertainty, Fuzziness, and Knowledge-Based Systems*, 5:411–427, 1997.
84. G. Frege. Begriffsschrift, eine der arithmetischen nachgebildete formelsprache des reinen denkens (gekuerzter nachdruck). In K. Berka and L. Kreiser, editors, *Logik-Texte: Kommentierte Auswahl zur Geschichte der Modernen Logik (vierte Auflage)*, pp. 82–107. Akademie-Verlag, Berlin, 1986.
85. D. Gabbay. Decidability of some intuitionistic predicate theories. *Journal of Symbolic Logic*, 37:579–587, 1972.
86. D. Gabbay. *Semantical Investigations in Heyting's Intuitionistic Logic*, volume 148 of *Synthese Library*. D. Reidel Publishing Company, Dordrecht, 1981.
87. D. Gabbay and G. Metcalfe. Fuzzy logics based on $[0,1)$-continuous uninorms. *Archive for Mathematical Logic*, 46(6):425–469, 2007.
88. D. Gabbay and N. Olivetti. *Goal-Directed Proof Theory*. Kluwer Academic Publishers, Dordrecht, 2000.
89. D. Gabbay and U. Reyle. N-Prolog: An extension of Prolog with hypothetical implication. *Journal of Logic Programming*, 4:319–355, 1984.
90. N. Galatos, P. Jipsen, T. Kowalski, and H. Ono. *Residuated Lattices: An Algebraic Glimpse at Substructural Logics*. Elsevier, Amsterdam, 2007.
91. N. Galatos and C. Tsinakis. Generalized MV-algebras. *Journal of Algebra*, 283(1):254–291, 2005.
92. M. R. Garey and D. S. Johnson. *Computers and Intractability*. W.H. Freeman and Co., San Francisco, 1979.
93. G. Gentzen. Untersuchungen über das Logische Schliessen. *Math. Zeitschrift*, 39:176–210,405–431, 1935.
94. G. Georgescu and A. Iorgulescu. Pseudo-MV-algebras. *Multiple Valued Logic*, 6:65–135, 2001.
95. R. Giles. A non-classical logic for physics. *Studia Logica*, 4(33):399–417, 1974.
96. R. Giles. Łukasiewicz logic and fuzzy set theory. *International Journal of Man-Machine Studies*, 8(3):313–327, 1976.
97. J. Girard. Linear logic. *Theoretical Computer Science*, 50:1–101, 1987.

98. K. Gödel. Zum intuitionistischen Aussagenkalkül. *Anzeiger Akademie der Wissenschaften Wien, mathematisch-naturwiss. Klasse*, 32:65–66, 1932. Reprinted and translated in [99].
99. K. Gödel. *Collected Works: Publications 1929–1936*, volume 1. Oxford University Press, Oxford, 1986. Edited by S. Feferman, J. Dawson, and S. Kleene.
100. J. A. Goguen, Jr. The logic of inexact concepts. *Synthése*, 19:325–373, 1969.
101. R. Goré. Substructural logics on display. *Logic Journal of the IGPL*, 6(3):451–504, 1998.
102. S. Gottwald. *A Treatise on Many-Valued Logics*, volume 9 of *Studies in Logic and Computation*. Research Studies Press, Baldock, 2000.
103. V. N. Grishin. Predicate and set-theoretic calculi based on logics without contraction. *Mathematical URSSS Izvestija*, 18(1):41–59, 1982.
104. R. Hähnle. *Automated Deduction in Multiple-Valued Logics*. Oxford University Press, Oxford, 1993.
105. P. Hájek. *Metamathematics of Fuzzy Logic*. Kluwer Academic Publishers, Dordrecht, 1998.
106. P. Hájek. On very true. *Fuzzy Sets and Systems*, 124:329–334, 2001.
107. P. Hájek. Observations on the monoidal t-norm logic. *Fuzzy Sets and Systems*, 132(1):107–112, 2002.
108. P. Hájek. Fuzzy logics with noncommutative conjunctions. *Journal of Logic and Computation*, 13(4):469–479, 2003.
109. P. Hájek. Observations on non-commutative fuzzy logic. *Soft Computing*, 8(1):38–43, 2003.
110. P. Hájek. Arithmetical complexity of fuzzy logic – A survey. *Soft Computing*, 9:935–941, 2005.
111. P. Hájek. Making fuzzy description logic more general. *Fuzzy Sets and Systems*, 154(1):1–15, 2005.
112. P. Hájek, L. Godo, and F. Esteva. A complete many-valued logic with product-conjunction. *Archive for Mathematical Logic*, 35:191–208, 1996.
113. P. Hájek and J. Valdés. An analysis of MYCIN-like expert systems. *Mathware & Soft Computing*, 1:45–68, 1994.
114. Z. Hanikova. A note on the complexity of tautologies of individual t-algebras. *Neural Network World*, 12(5):453–460, 2002.
115. J. Harland and D. Pym. The uniform proof-theoretic foundation of linear logic programming. In *Proceedings of the 1991 International Logic Programming Symposium*, pp. 304–318. MIT Press, Cambridge, 1991.
116. J. Hart, L. Rafter, and C. Tsinakis. The structure of commutative residuated lattices. *International Journal of Algebra and Computation*, 12(4):509–524, 2002.
117. L. S. Hay. Axiomatization of the infinite-valued predicate calculus. *Journal of Symbolic Logic*, 28(1):77–86, 1963.
118. D. Hilbert and W. Ackermann. *Grundzüge der theoretischen Logik*. Springer, Berlin, 1928.
119. U. Höhle. Commutative, residuated ℓ-monoids. In U. Höhle and E. P. Klement, editors, *Non-Classical Logics and their Applications to Fuzzy Subsets*, pp. 53–106. Kluwer Academic Publishers, Dordrecht, 1995.
120. R. Horčík. Decidability of cancellative extension of monoidal t-norm based logic. *Logic Journal of the IGPL*, 14(6):827–843, 2006.
121. R. Horčík. Standard completeness theorem for PiMTL. *Archive for Mathematical Logic*, 44:413–424, 2005.
122. R. Horčík, C. Noguera, and M. Petrik. On n-contractive fuzzy logics. *Mathematical Logic Quarterly*, 53(3):268–288, 2007.
123. A. Horn. Logic with truth values in a linearly ordered Heyting algebra. *Journal of Symbolic Logic*, 34(3):395–409, 1969.
124. P. G. Jeavons, D. A. Cohen, and M. Gyssens. Closure properties of constraints. *The Journal of the ACM*, 44:527–548, 1997.
125. S. Jenei and F. Montagna. A proof of standard completeness for Esteva and Godo's MTL logic. *Studia Logica*, 70(2):183–192, 2002.
126. S. Jenei and F. Montagna. A proof of standard completeness for non-commutative monoidal t-norm logic. *Neural Network World*, 13(5):481–490, 2003.

127. P. Jipsen and C. Tsinakis. A survey of residuated lattices. In J. Martinez, editor, *Ordered Algebraic Structures*, pp. 19–56. Kluwer Academic Publishers, Dordrecht, 2002.
128. L. G. Khachiyan. A polynomial algorithm in linear programming. *Soviet Mathematics Doklady*, 20:191–194, 1979.
129. E. P. Klement, R. Mesiar, and E. Pap. *Triangular Norms*, volume 8 of *Trends in Logic*. Kluwer Academic Publishers, Dordrecht, 2000.
130. J. Lambek. The mathematics of sentence structure. *Amer. Math. Monthly*, 65:154–170, 1958.
131. D. Larchey-Wendling. Counter-model search in Gödel-Dummett logics. In *Proceedings of IJCAR 2004*, pp. 274–288. Springer, Heidelberg, 2004.
132. P. Lorenzen. Logik und Agon. In *Atti Congr. Internaz. di Filosofia*, pp. 187–194. Sansoni, Firenze, 1960.
133. J. Łukasiewicz. O logice tròjwartościowej. *Ruch Filozoficzny*, 5:169–171, 1920. Reprinted and translated in [134].
134. J. Łukasiewicz. *Jan Łukasiewicz, Selected Writings*. North-Holland Publishing Company, Amsterdam, 1970. Edited by L. Borowski.
135. J. Łukasiewicz and A. Tarski. Untersuchungen über den Aussagenkalkül. *Comptes Rendus des Séances de la Societé des Sciences et des Lettres de Varsovie, Classe III*, 23, 1930. Reprinted and translated in [134].
136. H. M. MacNeille. Partially ordered sets. *Transactions of the American Mathematical Society*, 42:416–460, 1937.
137. G. Malinowski. *Many-Valued Logics*, volume 25 of *Oxford Logic Guides*. Oxford University Press, Oxford, 1993.
138. E. Marchioni and F. Montagna. On triangular norms and uninorms definable in $Ł\Pi\frac{1}{2}$. *International Journal of Approximate Reasoning*, 47(2):179–201, 2008.
139. R. McNaughton. A theorem about infinite-valued sentential logic. *Journal of Symbolic Logic*, 16(1):1–13, 1951.
140. K. Menger. Statistical metrics. *Proceedings of the National Academy of Sciences U.S.A*, 28:535–537, 1942.
141. C. A. Meredith. The dependence of an axiom of Łukasiewicz. *Transactions of the American Mathematical Society*, 87:54, 1958.
142. G. Metcalfe. *Proof Theory for Propositional Fuzzy Logics*. PhD thesis, King's College, London, 2003.
143. G. Metcalfe. Proof calculi for Casari's comparative logics. *Journal of Logic and Computation*, 16(4):405–422, 2006.
144. G. Metcalfe and F. Montagna. Substructural fuzzy logics. *Journal of Symbolic Logic*, 72(3):834–864, 2007.
145. G. Metcalfe and N. Olivetti. Goal-directed methods for fuzzy logics. In S. Artemov, H. Barringer, L. Lamb, and J. Woods, editors, *We Will Show Them: Festschrift in honour of the 60th birthday of Dov Gabbay*. King's College Publications, London, 2005.
146. G. Metcalfe, N. Olivetti, and D. Gabbay. Analytic proof calculi for product logics. *Archive for Mathematical Logic*, 43(7):859–889, 2004.
147. G. Metcalfe, N. Olivetti, and D. Gabbay. Łukasiewicz logic: From proof systems to logic programming. *Logic Journal of the IGPL*, 12:561–585, 2005.
148. G. Metcalfe, N. Olivetti, and D. Gabbay. Sequent and hypersequent calculi for abelian and Łukasiewicz logics. *ACM Transactions on Computational Logic*, 6(3):578–613, 2005.
149. R. K. Meyer and J. K. Slaney. Abelian logic from A to Z. In G. Priest et al., editor, *Paraconsistent Logic: Essays on the Inconsistent*, pp. 245–288. Philosophia Verlag, Munich, 1989.
150. D. Miller, G. Nadathur, F. Pfenning, and A. Scedrov. Uniform proofs as a foundation for logic programming. *Annals of Pure and Applied Logic*, 51:125–157, 1991.
151. G. Minc. Cut-elimination theorem in relevant logics. *The Journal of Soviet Mathematics*, 6:422–428, 1976.
152. F. Montagna. Three complexity problems in quantified fuzzy logic. *Studia Logica*, 68:143–152, 2001.
153. F. Montagna. Storage operators and multiplicative quantifiers in many-valued logics. *Journal of Logic and Computation*, 14(2):299–322, 2004.

154. F. Montagna, C. Noguera, and R. Horčík. On weakly cancellative fuzzy logics. *Journal of Logic and Computation*, 16(4):423–450, 2006.
155. F. Montagna and H. Ono. Kripke semantics, undecidability and standard completeness for Esteva and Godo's logic MTL∀. *Studia Logica*, 71(2):227–245, 2002.
156. F. Montagna, G. M. Pinna and E. B. P. Tiezzi. A tableau calculus for Hájek's logic BL. *Journal of Logic and Computation*, 13(2):241–259, 2003.
157. P. S. Mostert and A. L. Shields. On the structure of semigroups on a compact manifold with boundary. *Annals of Mathematics*, 65:117–143, 1957.
158. A. Mostowski. Axiomatizability of some many valued predicate calculi. *Fundamentica Mathematica*, 50:165–190, 1961.
159. D. Mundici. Satisfiability in many-valued sentential logic is NP-complete. *Theoretical Computer Science*, 52(1–2):145–153, 1987.
160. D. Mundici. The logic of Ulam's game with lies. In C. Bicchieri and M.L. Dalla Chiara, editors, *Knowledge, belief and strategic interaction*, pp. 275–284. Cambridge University Press, Cambridge, 1992.
161. D. Mundici. Ulam's game, Łukasiewicz logic and C*-algebras. *Fundamenta Informaticae*, 18:151–161, 1993.
162. D. Mundici and N. Olivetti. Resolution and model building in the infinite-valued calculus of Łukasiewicz. *Theoretical Computer Science*, 200(1–2):335–366, 1998.
163. S. Negri and J. von Plato. *Structural Proof Theory*. Cambridge University Press, Cambridge, 2001.
164. T. Nipkow. An inductive proof of the wellfoundedness of the multiset order. Technical report, http://www4.informatik.tu-muenchen.de/nipkow/misc/index.html, 1998.
165. V. Novák. On the Hilbert-Ackermann theorem in fuzzy logic. *Acta Mathematica et Informatica Universitatis Ostraviensis*, 4:57–74, 1996.
166. V. Novák. On fuzzy type theory. *Fuzzy Sets and Systems*, 149(2):235–273, 2005.
167. V. Novák, I. Perfilieva, and J. Mockor. *Mathematical Principles of Fuzzy Logic*. Springer, Heidelberg, 1999.
168. M. Ohnishi and K. Matsumoto. Gentzen methods in modal calculi. *Osaka Mathematical Journal*, 9:113–130, 1957.
169. M. Okada. Phase semantics for higher order completeness, cut-elimination and normalization proofs. *Electronic Notes in Theoretical Computer Science*, 3, 1996.
170. M. Okada. A uniform semantic proof for cut-elimination and completeness of various first and higher order logics. *Theoretical Computer Science*, 281(1–2):471–498, 2002.
171. N. Olivetti. Tableaux for Łukasiewicz infinite-valued logic. *Studia Logica*, 73(1):81–111, 2003.
172. H. Ono. The finite embeddability property for MTL. Unpublished note.
173. H. Ono. Semantics for substructural logics. In P. Schroeder-Heister and K. Došen, editors, *Substructural Logics*, pp. 259–291. Oxford University Press, Oxford, 1993.
174. H. Ono. Proof-theoretic methods in nonclassical logic – an introduction. In M. Takahashi et al., editor, *Theories of Types and Proofs*, pp. 207–254. Mathematical Society of Japan, 1998.
175. H. Ono. Closure operators and complete embeddings of residuated lattices. *Studia Logica*, 74(3):427–440, 2003.
176. H. Ono. Substructural logics and residuated lattices – an introduction. In V. F. Hendricks and J. Malinowski, editors, *Trends in Logic: 50 years of Studia Logics*, volume 20, pp. 177–212. Kluwer Academic Publishers, Dordrecht, 2003.
177. H. Ono and Y. Komori. Logics without the contraction rule. *Journal of Symbolic Logic*, 50:169–201, 1985.
178. G. Panti. A geometric proof of the completeness of the Łukasiewicz calculus. *Journal of Symbolic Logic*, 60(2):563–578, 1995.
179. F. Paoli. A common abstraction of MV-algebras and abelian ℓ-groups. *Studia Logica*, 65:355–366, 2000.
180. F. Paoli. *Substructural Logics: A Primer*. Kluwer Academic Publishers, Dordrecht, 2002.

181. J. Pavelka. On fuzzy logic I, II, III. *Zeitschrift für mathematische Logik und Grundlagen der Mathematik*, 25:45–52, 119–134, 447–464, 1979.
182. G. Pottinger. Uniform, cut-free formulations of T, S4 and S5. *Journal of Symbolic Logic*, 48(3):900, 1983. (Abstract)
183. A. Prijatelj. Bounded contraction and Gentzen-style formulation of Łukasiewicz logics. *Studia Logica*, 57(2–3):437–456, 1996.
184. M. E. Ragaz. *Arithmetische Klassifikation von Formelmengen der unendlichwertigen Logik*. PhD thesis, ETH Zürich, 1981.
185. G. Restall. Modalities in substructural logics. *Logique et Analyse*, 35:303–321, 1992.
186. G. Restall. *An Introduction to Substructural Logics*. Routledge, London, 1999.
187. A. Rose and J. B. Rosser. Fragments of many-valued statement calculi. *Transactions of the American Mathematical Society*, 87:1–53, 1958.
188. G. Rousseau. Sequents in many valued logic I. *Fundamenta Mathematicæ*, LX:23–33, 1967.
189. G. Rousseau. Sequents in many valued logic II. *Fundamenta Mathematicæ*, LXVII:125–131, 1970.
190. P. Ruet. Non-commutative logic II: Sequent calculus and phase semantics. *Mathematical Structures in Computer Science*, 10(2):277–312, 2000.
191. J. D. Rutledge. *A preliminary investigation of the infinitely many-valued predicate calculus*. PhD thesis, Cornell University, Ithaca, 1959.
192. B. Scarpellini. Die Nichtaxiomatisierbarkeit des unendlichwertigen Prädikatenkalküls von Łukasiewicz. *Journal of Symbolic Logic*, 27(2):159–170, 1962.
193. A. Schrijver. *Theory of Linear and Integer Programming*. John Wiley and Sons, New York, 1987.
194. K. Schröter. Methoden zur axiomatisierung beliebiger aussagen und prädikatenkalküle. *Zeitschrift für math. Logik und Grundlagen der Mathematik*, pp. 241–251, 1955.
195. K. Schütte. *Beweistheorie*, volume 103 of *Die Grundlehren der mathematischen Wissenschaften in Einzeldarstellungen mit besonderer Berücksichtigung der Anwendungsgebiete*. Springer, Berlin, 1960.
196. B. Schweizer and A. Sklar. *Probabilistic Metric Spaces*. North Holland Publishing Company, Amsterdam, 1983.
197. D. Scott. Completeness and Axiomatizability in Many-Valued Logic. In L. Henkin et al., editors, *Proceedings of Symposia in Pure Mathematics Volume XXV: Proceedings of the Tarski Symposium*, pp. 411–435. American Mathematical Society, Providence, 1974.
198. E. H. Shortliffe. *Computer-Based Medical Consultation: MYCIN*. American Elsevier, Amsterdam, 1976.
199. W. Silvert. Symmetric summation: A class of operations on fuzzy sets. *IEEE Transactions on Systems, Man Cybernetics*, 9:657–659, 1979.
200. O. Sonobe. A Gentzen-type formulation of some intermediate propositional logics. *Journal of Tsuda College*, 7:7–14, 1975.
201. U. Straccia. Reasoning within fuzzy description logics. *Journal of Artificial Intelligence Research*, 14:137–166, 2001.
202. W. W. Tait. Normal derivability in classical logic. In J. Barwise, editor, *The Syntax and Semantics of Infinitary Languages*, Lecture Notes in Mathematics, pp. 204–236. Springer, Berlin, 1968.
203. M. Takano. Another proof of the strong completeness of the intuitionistic fuzzy logic. *Tsukuba Journal of Mathematics*, 11:851–866, 1984.
204. G. Takeuti. *Proof theory*, volume 81 of *Studies in Logic and the Foundations of Mathematics*. North Holland Publishing Company, Amsterdam, second edition, 1987.
205. G. Takeuti and T. Titani. Intuitionistic fuzzy logic and intuitionistic fuzzy set theory. *Journal of Symbolic Logic*, 49(3):851–866, 1984.
206. G. Takeuti and T. Titani. Global intuitionistic fuzzy set theory. In *The Mathematics of Fuzzy Systems*, pp. 291–301. TUV-Verlag, Germany, 1986.
207. K. Terui. Which structural rules admit cut elimination? —an algebraic criterion. *Journal of Symbolic Logic*, 72(3):738–754, 2007.

208. A. S. Troelstra. *Lectures on Linear Logic*. CSLI Lecture Notes 29, Center for the Study of Language and Information Stanford, California, 1992.
209. A. S. Troelstra and H. Schwichtenberg. *Basic Proof Theory*. Cambridge University Press, Cambridge, second edition, 2000.
210. C. Tsinakis and K. Blount. The structure of residuated lattices. *International Journal of Algebra and Computation*, 13(4):437–461, 2003.
211. E. Turunen. *Mathematics Behind Fuzzy Logic*. Advances in Soft Computing. Springer, Heidelberg, 1999.
212. A. Urquhart. Many-valued logics. In D. Gabbay and F. Günthner, editors, *Handbook of Philosophical Logic*, volume 3, pp. 71–116. D. Reidel Publishing Company, Dordrecht, 1986.
213. T. Vetterlein. Analytic calculi for logics of ordinal multiples of standard t-norms. *Journal of Logic and Computation*, 18:35–57, 2008.
214. P. Vojtáš. Fuzzy logic programming. *Fuzzy Sets and Systems*, 124:361–370, 2001.
215. H. Wagner. A new resolution calculus for the infinite-valued propositional logic of Łukasiewicz. In *Proceedings of FTP 98*, Technical Report E1852-GS-981, pp. 234–243. TU Wien, Austria, 1998.
216. H. Wansing. Translation of hypersequents into display sequents. *Logic Journal of the IGPL*, 6(5):719–733, 1998.
217. M. Ward and R. P. Dilworth. Residuated lattices. *Transactions of the American Mathematical Society*, 38:335–354, 1939
218. E. C. Weinberg. Free abelian lattice-ordered groups. *Mathematische Annalen*, 151:187–199, 1963.
219. V. Weispfenning. The complexity of the word problem for Abelian l-groups. *Theoretical Computer Science*, 48(1):127–132, 1986.
220. S. Willard. *General Topology*. Dover, New York, 2004.
221. R. R. Yager and A. Rybalov. Uninorm aggregation operators. *Fuzzy Sets and Systems*, 80:111–120, 1996.
222. R. Zach. Proof theory of finite-valued logics. Master's thesis, TU Wien, Austria, 1993.
223. L. A. Zadeh. Fuzzy sets. *Information and Control*, 8:338–353, 1965.
224. L. A. Zadeh. Fuzzy logics and approximate reasoning. *Synthese*, 30:407–428, 1975.

Index

Printed by Printforce, the Netherlands